AF595076

DNA Damage, Oxidative Stress and Related Metabolic By-Products in Cancer and Environmental Studies

DNA Damage, Oxidative Stress and Related Metabolic By-Products in Cancer and Environmental Studies

Editors

Marco E. M. Peluso
Andrea Galli
Tommaso Mello

MDPI • Basel • Beijing • Wuhan • Barcelona • Belgrade • Manchester • Tokyo • Cluj • Tianjin

Editors
Marco E. M. Peluso
Prevention and Oncology
Network Institute (ISPRO)
Italy

Andrea Galli
University of Florence
Italy

Tommaso Mello
University of Florence
Italy

Editorial Office
MDPI
St. Alban-Anlage 66
4052 Basel, Switzerland

This is a reprint of articles from the Special Issue published online in the open access journal *International Journal of Molecular Sciences* (ISSN 1422-0067) (available at: http://www.mdpi.com).

For citation purposes, cite each article independently as indicated on the article page online and as indicated below:

LastName, A.A.; LastName, B.B.; LastName, C.C. Article Title. *Journal Name* **Year**, *Volume Number*, Page Range.

ISBN 978-3-0365-1270-9 (Hbk)
ISBN 978-3-0365-1271-6 (PDF)

Contents

About the Editors

Marco Peluso, MsC, PhD holds a PhD in Biochemistry. He received his training in Molecular Epidemiology by Prof. Bartch at the International Agency for Research on Cancer, World Health Organization, Lyon, France. Later, he spent more than ten years at the National Research Institute of Genoa, Ita-ly. Currently, he is the Referent of the Research and Development Branch at the ISPRO—Study, Prevention and Oncology Network Institute, Florence, Italy. Main research interests are on pros-tate, breast, lung, colorectal cancer screening, and environmental carcinogenesis. He has a large experience in setting and management of epidemiological research projects, as well as on execu-tion and critical evaluation of omics analysis from observational studies.

Andrea Galli, M.D. PhD (Full professor, Dept. of Clinical and Experimental Biomedical Sciences "Mario Serio", University of Florence). Andrea Galli is a Professor of Gastroenterology at the University of Florence and a physician at the local University Hospital. His research interests range from the pathophysiology of fibrosis in the gastroenteric system to the microbioma, as a therapeutic target in and causative agent of colon cancer.

Tommaso Mello, Ph.D. (Senior Research Associate, Department of Clinical and Experimental Biomedical Sciences "Mario Serio, University of Florence): Dr. Mello obtained his Master Degree in Biology and Ph.D. in Physiology and Nutritional Sciences at the University of Florence in 1999 and 2004, respectively. He conducted part of his Ph.D. research in the lab of W.F. Bosron, Indiana University School of Medicine, in 2003 and 2004. Between 2004 he was appointed a research fellowship at the Fiorgen Farmacogenomic Foundation (Florence, Italy) after which he returned at the at the University of Florence in 2007. His research interests span over several aspects of liver patophysiology: from cellular and molecular mechanisms in liver and pancreatic fibrosis to nuclear receptor biology in hepatic and pancreatic cancer. Dr. Mello current research focuses on the molecular mechanisms of metabolic rewiring in liver cancer.

International Journal of
Molecular Sciences

Editorial

Oxidative Stress and DNA Damage in Chronic Disease and Environmental Studies

Marco Peluso [1,*], Valentina Russo [1], Tommaso Mello [2] and Andrea Galli [2]

1 Research Branch, Regional Cancer Prevention Laboratory, ISPRO-Study, Prevention and Oncology Network Institute, 50139 Florence, Italy; valrusso.bio@gmail.com

2 Department of Experimental and Clinical Biomedical Sciences, University of Florence, 50139 Florence, Italy; tommaso.mello@unifi.it (T.M.); a.galli@dfc.unifi.it (A.G.)

* Correspondence: m.peluso@ispro.toscana.it

Received: 11 September 2020; Accepted: 18 September 2020; Published: 21 September 2020

Humans are continually exposed to a large number of environmental carcinogens [1], some of them share a specific toxicity model which acts via the enhancement of cellular levels of reactive oxygen species (ROS), such as superoxide anion and hydroxyl radicals [2]. ROS are chemicals capable of inducing oxidative stress, a condition where free radical generation overwhelms antioxidant protection in the body's cells, causing oxidative damage of proteins, lipids and nucleic acids [2]. One biological consequence is the decline in physiological mechanisms designed to maintain cell repair and metabolic homeostasis, which can lead to tissue injury and cell transformation [3]. Oversensitive reactions to endogenous and extrinsic agents are associated with phenotypes characterized by elevated levels of genetic alterations [4,5], a process that can initiate carcinogenesis [6]. From a general perspective, the study of these factors can be useful to develop personalized medicine and to fill the knowledge gap of the molecular mechanisms behind the pathogenesis of chronic diseases. In this context, the current Special Issue's challenge is to provide an overview on the topic of ROS, oxidative DNA damage and related-DNA repair factors in chronic diseases and environmental studies.

The study by Vodicka et al. [7] analyzed the role of DNA damage and DNA repair mechanisms in colorectal carcinogenesis. In colorectal cancer (CRC), ROS overproduction can induce oxidative stress and oxidative DNA damage, that, if unrepaired, can cause mutations, including G>T transversion, leading, ultimately, to cancer. Evidence of the relationship between oxidative stress and CRC is coming from *MUTYH*-associated polyposis and *NTHL1*-associated tumor syndrome. The latter are two hereditary syndromes whose germline mutations cause the loss-of-function in glycosylases, a DNA repair defect which induces oxidative DNA damage accumulation and the transition from early adenoma to malignant cells. Other support for the oxidative stress hypothesis is provided from genetic studies showing an association between polymorphisms in base excision repair (BER) genes, e.g., 326Ser/Cys *OGG1*, 324Gln/His and 324His/His *MUTYH* genotypes and CRC risk as well as a link between the rates of BER-individual DNA excision repair capacity in non-malignant adjacent mucosa and the rates of overall and relapse-free survival in CRC. Furthermore, a diet rich in antioxidants and bioactive compounds appears to modify CRC risk, e.g., by increasing repair capacity, by decreasing DNA damage or by modifying the metabolic profile of gut microbiome. Specifically, the human microbiome can affect CRC risk through the production of ROS and genotoxic compounds or through the activation of pro-inflammatory cascades and cellular transformation. Thus, CRC treatment could be improved in various ways, e.g., by targeting ROS production and antioxidant defense or by exploiting acquired/inherited defects in DNA repair together with the use of conventional chemotherapeutics.

The review by Bhardwaj et al. [8] addressed the study of abnormal energy metabolism in cancer cells that show a dependence on glycolysis, a cytoplasmic pathway that generates ATP, rather than on mitochondrial oxidative phosphorylation in the presence of oxygen for their energy requirement. In tumors, energy metabolic alterations are associated with the high production of ROS levels, which need to be kept under certain limits by increasing the action of the glycolysis pathway, the pentose phosphate pathway and the tricarboxylic acid cycle in order to maintain redox homeostasis and to prevent ROS-mediated cancer cell death. In tumors, the interplay between alterations of energy metabolism and ROS production plays a major role in regulating the tumor's response to chemotherapeutic drugs. In particular, tumor resistance can be induced by various mechanisms, e.g., by high activity of p-glycoprotein, an important ATP-binding cassette transporter involved in pumping chemotherapeutic drugs out of cells, by hexokinase 2 overexpression which confers drug resistance enhancing ROS-induced autophagy, by stimulating the pentose phosphate pathway activity or by inducing high levels of glucose 6-phosphate dehydrogenase. Therefore, targeting metabolic deregulations in tumors with personalized cancer treatments could enhance the tumor's response to chemotherapeutic drugs by using inhibitors capable of blocking the MEK/ERK pathway enhanced by the hexokinase 2 activity or by the inhibition of glucose 6-phosphate dehydrogenase with chemical inhibitors.

In the review by Lee et al. [9], large interest was on the study of the repair of oxidative DNA damage via the nucleotide excision repair (NER), rather than through the classical BER pathways. In simple terms, ROS-induced oxidative DNA damage can be grouped into two categories (a) non-bulky single-base lesions, if the damage is associated to non-helix distorting modifications, and (b) bulky lesions, if the DNA damage causes helix distorting lesions. On one hand, the BER pathway is charged to repair non-bulky single-base lesions in the form of small chemical adducts, such as oxidation, alkylation and deamination damage; on the other hand, NER fixes bulky oxidative lesions, such as purine 5′,8-cyclonucleosides, interstrand cross-links and DNA–protein cross-links. NER factors can also participate in the BER process of lesion recognition and interact with BER proteins to stimulate enzyme activity and improve DNA repair rates. Hence, newly therapeutic treatments could be developed by targeting DNA repair defects or by using small molecule inhibitors or modulators of NER/BER factors.

To investigate whether the accumulation of ROS-induced DNA damage and defective DNA damage response (DDR) rates are associated with chronic myeloid leukemia (CML), a myeloproliferative neoplasm, the study by Popp et al. [10] examined the levels of double strand breaks (DSBs) and DDR in CML patients by immunofluorescence microscopy and Western blotting. High DSB production, detected by the γH2AX foci analysis, was found in chronic, accelerated and blast phases of CML patients as well as in those with loss of major molecular response in respect to controls or CML patients with deep molecular response. Increased frequencies of erroneous non-homologous end joining and microhomology-mediated end joining repair mechanisms, measured by the co-localization of γH2AX/53BP1 foci, were observed across the spectrum from the chronic towards the blast phases of CML. DDR decline was also detected by the defective expression of (p-)ATM and (p-)CHK2. Lastly, the development of genetic instability in the CML appears to be due to DNA damage accumulation in the course from the chronic phase of CML towards the blast phase of CML and, more importantly, DNA damage together with the progressive activation of error-prone DSB repair mechanisms and the DDR decline could play a role in the blastic transformation and the disease progression of CML.

The study by Souliotis [11] addressed the analysis of the interplay between DNA damage response and repair (DDR/R) with the innate immune response to endogenous and exogenous agents, in order to understand how DDR/R deregulation can act together with immune activation in the pathogenesis of systemic autoimmunity. In this field, evidence of the interaction between DNA metabolism and innate immune response was initially provided from the Aicardi–Goutières syndrome, a disease characterized by mutations in RNaseH2, a protein that removes ribonucleotides from DNA to maintain DNA integrity. Further evidence was obtained from the telengiectasia, an autosomal recessive disorder characterized by mutations in the *ATM* gene, a defect that can cause chronic accumulation of DNA damage capable to activate the immune system. The occurrence of an interplay between DDR/R defects and the innate immune response was also demonstrated in systemic lupus erythematosus, an autoimmune disease where patients carry mutations in DNA repair enzymes, DNA damage accumulation, high rates of apoptosis and increased levels of antibodies against DNA repair proteins. Moreover, this interplay was also found in the pathogenesis of rheumatoid arthritis. Therefore, new therapies against autoimmune diseases could be developed targeting this interaction, e.g., by utilizing DDR/R inhibitors, such as HDACi, givinostat, or vorinostat, or by applying a combination of p53 activators and CHK1/2 inhibitors.

In the effort to explain the higher incidence of thyroid cancer in populations who live near volcanos, the review by Malandrino et al. [12] examined the toxic effects caused by exposure to heavy metals, chemical compounds capable of inducing ROS overproduction, in volcanic areas. Interestingly, the concentrations of several heavy metals were not increased in water and lichen, whereas elevated amounts of boron, tungsten, molybdenum and palladium were detected in the subjects who were living in residential areas near the Etna volcano, in Sicily, Italy. To explain how small increases in environmental metals can be associated to an enhanced risk of thyroid cancer in volcanic areas, authors suggested that this relationship was more mediated by hormesis effects resulting in inverted U-shaped dose-response curves rather than a linear dose-response relationship. Additionally, life-long and selective accumulation of one or more heavy metals could be behind the carcinogenic effects that are predominantly found in the thyroid gland.

The study by Kucharova et al. [13] evaluated the genotoxic effects of anesthetic substances used for general anesthesia or neuraxial anesthesia treatments in a cohort of patients undergoing orthopedic traumatological surgery. The frequency of single-strand DNA breaks, oxidized pyrimidine bases and oxidized purine bases was measured by comet assay before and after anesthesia procedures. High levels of DNA damage were detected in patients undergoing general anesthesia, indicating that halogenated gases isoflurane and sevoflurane can induce oxidative DNA damage, whereas the treatment with epidural and subarachnoid anesthesia was not toxic.

Cellai et al. [14] examined the genotoxic mechanisms underlying the development of nasal and sinonasal cancer (SNC) in occupational settings characterized by carcinogenic exposure to wood dust, a recognized cause of SNC. In that study, the association between the generation of 3-(2-deoxy -α-D-erythro-pentafuranosyl)pyrimido[1,2-β]purin-10(3H)-one deoxyguanosine, a major-peroxidation -derived DNA adduct, and the wood dusts was analyzed in woodworkers in respect to controls living in Tuscany, Italy, by chromatographic and mass spectrometry techniques. High levels of oxidative DNA damage were found in woodworkers exposed to average levels of 1.48 mg/m^3 wood dust, especially when concomitant carcinogen co-exposure occurred.

In the exciting exploration of the anti-genotoxic effects of dietary flavonoids, the study of Jee et al. [15] analyzed the beneficial properties of silymarin, a natural flavonoid known for its anti-oxidant and anti-inflammatory properties, in relation to the exposure to airborne carcinogens. Specifically, the toxicity of benzo[a]pyrene (B[a]P), a carcinogen contained in tobacco smoke and air pollution and capable of causing both DNA damage and ROS production, was analyzed by ELISA, HPLC, immunofluorescence and Western blot in experimental cells. The co-treatment of B[a]P with silymarin significantly reduced the levels of DNA damage, indicating that this natural flavonoid has antigenotoxic properties. Interestingly, the protective property of silymarin was mainly related to its capacity of up-regulating the expression of *Nrf2* and *PxR* rather than to antioxidant effects.

In conclusion, the collection of this Special Issue covers topics representative of several key research areas in the field of cancer, autoimmune diseases and environmental studies. Further effort is indeed needed for developing innovative models of therapy—based on personalized medicine—that identify the most efficient therapeutic intervention for patients with low response to chemotherapeutics. For example, chemotherapeutic response could be improved by targeting the overproduction of ROS in malignant cells to reduce multidrug resistance or by using modulators of NER/BER proteins acting against acquired or inherited defects in DNA repair pathways. Therapies for autoimmune diseases could also benefit from utilizing DNA repair inhibitors and/or p53 activators. Next, the collection shows that DNA damage accumulation and/or DNA repair can be behind the blastic transformation and the disease progression of CML; that general anesthesia as well as the environmental exposure to wood dust can induce DNA damage; and that dietary flavonoids can act against the production of DNA damage. Lastly, inverted U-shaped dose-response curves and/or life-long and selective accumulation of heavy metals could be associated with the risk of thyroid cancer in populations living in volcanic areas. The data produced in those and future studies could be translated in the real environment to develop new therapies, to fill knowledge gaps and to be the source for further methodological and application advances.

Funding: This research project was funded by Tuscany Region.

Conflicts of Interest: The authors declare no conflict of interest.

References

1. Drakvik, E.; Altenburger, R.; Aoki, Y.; Backhaus, T.; Bahadori, T.; Barouki, R.; Brack, W.; Cronin, M.T.D.; Demeneix, B.; Hougaard Bennekou, S.; et al. Statement on advancing the assessment of chemical mixtures and their risks for human health and the environment. *Environ. Int.* **2020**, *134*, 105267. [CrossRef] [PubMed]
2. Marnett, L.J. Oxy radicals, lipid peroxidation and DNA damage. *Toxicology* **2002**, *181*, 219–222. [CrossRef]
3. Peluso, M.E.; Munnia, A.; Srivatanakul, P.; Jedpiyawongse, A.; Sangrajrang, S.; Ceppi, M.; Godschalk, R.W.; van Schooten, F.J.; Boffetta, P. DNA adducts and combinations of multiple lung cancer at-risk alleles in environmentally exposed and smoking subjects. *Environ. Mol. Mutagenes.* **2013**, *54*, 375–383. [CrossRef] [PubMed]
4. Brancato, B.; Munnia, A.; Cellai, F.; Ceni, E.; Mello, T.; Bianchi, S.; Catarzi, S.; Risso, G.G.; Galli, A.; Peluso, M.E. 8-Oxo-7, 8-dihydro-2-deoxyguanosine and other lesions along the coding strand of the exon 5 of the tumour suppressor gene P53 in a breast cancer case-control study. *DNA Res. Int. J. Rapid Publ. Rep. Genes Genomes* **2016**, *23*, 395–402. [CrossRef] [PubMed]
5. Peluso, M.; Munnia, A.; Risso, G.G.; Catarzi, S.; Piro, S.; Ceppi, M.; Giese, R.W.; Brancato, B. Breast fine-needle aspiration malondialdehyde deoxyguanosine adduct in breast cancer. *Free Radic Res.* **2011**, *45*, 477–482. [CrossRef] [PubMed]
6. Munnia, A.; Giese, R.W.; Polvani, S.; Galli, A.; Cellai, F.; Peluso, M.E. Bulky DNA Adducts, Tobacco Smoking, Genetic Susceptibility, and Lung Cancer Risk. *Adv. Clin. Chem.* **2017**, *81*, 231–277. [PubMed]
7. Vodicka, P.; Urbanova, M.; Makovicky, P.; Tomasova, K.; Kroupa, M.; Stetina, R.; Opattova, A.; Kostovcikova, K.; Siskova, A.; Schneiderova, M.; et al. Oxidative Damage in Sporadic Colorectal Cancer: Molecular Mapping of Base Excision Repair Glycosylases in Colorectal Cancer Patients. *Int. J. Mol. Sci.* **2020**, *21*, 2473. [CrossRef] [PubMed]
8. Bhardwaj, V.; He, J. Reactive Oxygen Species, Metabolic Plasticity, and Drug Resistance in Cancer. *Int. J. Mol. Sci.* **2020**, *21*, 3412. [CrossRef] [PubMed]
9. Lee, T.-H.; Kang, T.-H. DNA Oxidation and Excision Repair Pathways. *Int. J. Mol. Sci.* **2020**, *20*, 6092. [CrossRef] [PubMed]
10. Popp, H.D.; Kohl, V.; Naumann, N.; Flach, J.; Brendel, S.; Kleiner, H.; Weiss, C.; Seifarth, W.; Saussele, S.; Hofmann, W.-K.; et al. DNA Damage and DNA Damage Response in Chronic Myeloid Leukemia. *Int. J. Mol. Sci.* **2020**, *21*, 1177. [CrossRef] [PubMed]
11. Souliotis, V.L.; Vlachogiannis, N.I.; Pappa, M.; Argyriou, A.; Ntouros, P.A.; Sfikakis, P.P. DNA Damage Response and Oxidative Stress in Systemic Autoimmunity. *Int. J. Mol. Sci.* **2020**, *21*, 55. [CrossRef] [PubMed]
12. Malandrino, P.; Russo, M.; Gianì, F.; Pellegriti, G.; Vigneri, P.; Belfiore, A.; Rizzarelli, E.; Vigneri, R. Increased Thyroid Cancer Incidence in Volcanic Areas: A Role of Increased Heavy Metals in the Environment? *Int. J. Mol. Sci.* **2020**, *21*, 3425. [CrossRef] [PubMed]
13. Kucharova, M.; Astapenko, D.; Zubanova, V.; Koscakova, M.; Stetina, R.; Zadak, Z.; Hronek, M. Does Neuraxial Anesthesia as General Anesthesia Damage DNA? A Pilot Study in Patients Undergoing Orthopedic Traumatological Surgery. *Int. J. Mol. Sci.* **2020**, *21*, 84. [CrossRef] [PubMed]

14. Cellai, F.; Capacci, F.; Sgarrella, C.; Poli, C.; Arena, L.; Tofani, L.; Giese, R.W.; Peluso, M. A Cross-Sectional Study on 3-(2-Deoxy-β-D-Erythro-Pentafuranosyl)Pyrimido[1,2-α]Purin-10(3H)-One Deoxyguanosine Adducts among Woodworkers in Tuscany, Italy. *Int. J. Mol. Sci.* **2019**, *20*, 2763. [CrossRef] [PubMed]
15. Jee, S.-C.; Kim, M.; Sung, J.-S. Modulatory Effects of Silymarin on Benzo[a]pyrene-Induced Hepatotoxicity. *Int. J. Mol. Sci.* **2020**, *21*, 2369. [CrossRef] [PubMed]

International Journal of
Molecular Sciences

Article

Oxidative DNA Damage, Inflammatory Signature, and Altered Erythrocytes Properties in Diamond-Blackfan Anemia

Katarina Kapralova [1,†], Ondrej Jahoda [1,†], Pavla Koralkova [1], Jan Gursky [1], Lucie Lanikova [2], Dagmar Pospisilova [3], Vladimir Divoky [1] and Monika Horvathova [1,*]

1 Department of Biology, Faculty of Medicine and Dentistry Palacky University, 775 15 Olomouc, Czech Republic; katka.kapralova@gmail.com (K.K.); ondrej.jahoda@upol.cz (O.J.); pavla.koralkova@gmail.com (P.K.); jan.gursky@upol.cz (J.G.); vladimir.divoky@upol.cz (V.D.)

2 Institute of Molecular Genetics of the Czech Academy of Sciences, 142 20 Prague, Czech Republic; lucie.lanikova@img.cas.cz

3 Department of Pediatrics, Palacky University and University Hospital Olomouc, 779 00 Olomouc, Czech Republic; dagmar.pospisilova@fnol.cz

* Correspondence: monika.horvathova@upol.cz; Tel.: +420-(585)-63-2342

† These authors contributed equally to this work.

Received: 3 December 2020; Accepted: 15 December 2020; Published: 17 December 2020

Abstract: Molecular pathophysiology of Diamond-Blackfan anemia (DBA) involves disrupted erythroid-lineage proliferation, differentiation and apoptosis; with the activation of p53 considered as a key component. Recently, oxidative stress was proposed to play an important role in DBA pathophysiology as well. CRISPR/Cas9-created Rpl5- and Rps19-deficient murine erythroleukemia (MEL) cells and DBA patients' samples were used to evaluate proinflammatory cytokines, oxidative stress, DNA damage and DNA damage response. We demonstrated that the antioxidant defense capacity of Rp-mutant cells is insufficient to meet the greater reactive oxygen species (ROS) production which leads to oxidative DNA damage, cellular senescence and activation of DNA damage response signaling in the developing erythroblasts and altered characteristics of mature erythrocytes. We also showed that the disturbed balance between ROS formation and antioxidant defense is accompanied by the upregulation of proinflammatory cytokines. Finally, the alterations detected in the membrane of DBA erythrocytes may cause their enhanced recognition and destruction by reticuloendothelial macrophages, especially during infections. We propose that the extent of oxidative stress and the ability to activate antioxidant defense systems may contribute to high heterogeneity of clinical symptoms and response to therapy observed in DBA patients.

Keywords: Diamond-Blackfan anemia; reactive oxygen species; 8-oxoguanine; DNA damage response; inflammatory cytokines; erythrocyte lifespan

1. Introduction

Diamond-Blackfan anemia (DBA) is a rare congenital bone marrow failure syndrome characterized by severe anemia, erythroblastopenia, congenital malformations and predisposition to cancer. DBA is mostly related to pathogenic variants in ribosomal protein (RP) genes, which cause their haploinsufficiency and a consequent defect in ribosome biogenesis [1]. The most frequent ones are mutations in *RPS19* (25%) and *RPL5* (7%) [1]. The molecular pathophysiology of DBA is attributed to both p53-dependent and p53-independent mechanisms, which lead to the proapoptotic and hypoproliferative phenotype of erythroid cells [2]. Research in the last few years suggested an important role of oxidative stress in DBA. Specifically, increased production of reactive oxygen species (ROS) was observed in shRNA model reproducing DBA, and in cultured primary cells from DBA patients [3]. Moreover, antioxidant treatment was reported to reduce p53 stabilization in RP-deficient cells in vitro [4]. Therefore, it has been suggested that the hypoproliferative phenotype in RP-mutant diseases is associated with increased oxidative stress and DNA damage [2]. Consistently, the activation of DNA damage response (DDR) pathway was detected in RPS19-deficient zebrafish and human fetal liver cells [5].

In certain types of congenital anemia, increased ROS levels are associated with a shortened life span of red blood cells (RBCs) [6,7]. It has also been shown that the accelerated clearance of RBCs in response to oxidative stress is attributed not only to excessive hemolysis, but also to the induction of a programmed cell death of erythrocytes, called eryptosis [6,8]. Although elevated levels of reduced glutathione (GSH), an essential antioxidant, were detected in DBA erythrocytes [9]; the extent of oxidative stress and its possible impact on RBCs' properties has not been investigated in detail.

Here, we aimed to examine the extent of oxidative stress in DBA using patients' samples and Rpl5- and Rps19-deficient cellular models. We observed that the antioxidant defense in Rp-mutant cells is insufficient to cope with increased ROS production, leading to oxidative DNA damage, apoptosis or senescence of erythroid precursors and altered characteristics of mature erythrocytes. This state was accompanied by the induction of proinflammatory cytokines which appeared to play an important role in DBA pathobiology. Despite some known differences in the molecular mechanisms involved in RPL5- and RPS19-mutant DBA [1,3], the above-mentioned features are common to both ribosomal protein deficiencies with a certain degree of variation in the intensity of the pathological phenotype.

2. Results

2.1. Characterization of Patients' Cohort and Assessment of RBCs' Oxidative Stress

The cohort consisted of Czech and Slovak DBA patients (n = 24) [10]. The majority of them (22/24) harbored a mutation in one of the RPs (Table 1). Only erythrocytes of patients in disease remission (n = 12) or hematologically stable on corticosteroids (n = 7) were used for the assessment of RBCs' oxidative stress and enzyme activities. Patients with the most severe anemia receiving repeated erythrocyte transfusions were not included because the donor erythrocytes, likely representing the major fraction in the sample, would yield skewed results. Nevertheless, bone marrow samples from these patients (n = 5, P20–P24) were used for immunohistochemical (IHC) staining. The list of tests performed using samples from individual patients can be found in Table 1.

Initially, peripheral blood erythrocytes were stained with 2′,7′-dichlorofluorescein diacetate (DCF-DA) and the DCF-dependent fluorescence intensity, which is proportional to ROS levels, was measured by flow cytometry. As shown in Figure 1A, significantly increased levels of ROS were observed in DBA erythrocytes (mean fluorescence intensity—MFI: 13.2 ± 3.2) compared to control erythrocytes (MFI: 7.8 ± 1.2), confirming the presence of oxidative stress. Subsequently, the levels of reduced glutathione (GSH) and the activities of enzymes of the pentose phosphate pathway and antioxidant defense were assessed in leukocyte- and platelet-free erythrocytes lysates (Figure 1B). GSH levels, expressed as μmol/g hemoglobin (Hb), were significantly elevated in DBA erythrocytes (7.7 ± 2.73) compared to controls (4.98 ± 0.66). Concomitantly, significantly increased activities (U/g Hb) of glucose-6-phosphate dehydrogenase (G6PD; 7.07 ± 1.94 vs. 5.51 ± 0.83), gluconolactone dehydrogenase (GLD; 8.87 ± 1.48 vs. 5.5 ± 0.9), and glutathione peroxidase (GPx; 11.55 ± 3.33 vs. 8.69 ± 1.53) in DBA erythrocytes, compared to control erythrocytes, showed induction of antioxidant defense.

Table 1. DBA patients included in the study.

Pct	Sex	Age	Gene Mutation	Treatment	ROS	Annexin V	ELISA Inflamm. Cytokines	GSH Measurements	RBC Enzyme Activities	In Vitro Cultivation	IHC
P1	F	*54*	*RPL5*	R	+	+	+	+	NA	+	NA
P2 *	M	*32*	*RPL5*	R	+	+	+	+	+	+	+
P3	F	*42*	*RPL5*	R	NA	+	+	+	+	NA	NA
P4 *	M	*33*	*RPL5*	R	+	+	+	+	+	+	+
P5	M	*20*	*RPS19*	S	+	+	+	NA	NA	NA	NA
P6	F	*19*	*RPS19*	S	+	+	+	NA	NA	NA	NA
P7	M	*40*	*RPS19*	R	+	+	+	+	+	+	NA
P8	M	*46*	*RPS19*	S, Leu	+	NA	+	+	+	+	NA
P9 *	M	*28*	*RPS19*	R	+	+	NA	+	+	NA	+
P10	F	*46*	*RPL11*	R	+	NA	+	NA	NA	NA	NA
P11	M	*8*	*RPL11*	S	+	+	NA	NA	NA	NA	NA
P12	F	*36*	*RPL11*	R	+	NA	NA	NA	NA	NA	NA
P13	F	*54*	*RPL11*	S	+	NA	NA	NA	NA	NA	NA
P14	F	*27*	*RPS7*	R	+	+	+	+	+	NA	NA
P15	F	*41*	*RPS7*	R	+	+	NA	+	+	NA	NA
P16	F	*54*	*RPS7*	R	+	+	NA	+	+	NA	NA
P17	F	*33*	*delRPL35a*	S	+	+	NA	NA	NA	NA	NA
P18	M	*34*	*No mut*	S	+	+	NA	+	+	NA	+
P19	M	*10*	*No mut*	R	+	+	+	NA	NA	NA	NA
P20	F	*30*	*RPS19*	T, DRX	NA	NA	NA	NA	NA	NA	+
P21	F	*34*	*RPS19*	T, DRX	NA	NA	NA	NA	NA	NA	+
P22	M	*26*	*RPS26*	T, DRX	NA	NA	NA	NA	NA	NA	+
P23	M	*21*	*RPS26*	T, DRX	NA	NA	NA	NA	NA	NA	+
P24 §	F	*31*§	*RPL11*	T, DRX	NA	NA	NA	NA	NA	NA	+

Pct—patient, ROS—reactive oxygen species, GSH—reduced glutathione, RBCs—red blood cells, IHC—immunohistochemistry, R—remission, S—steroids, Leu—leucine, T—transfusions, DRX—deferasirox, No mut—no mutation identified, NA—not analyzed, * P2, P4, and P9 recently developed myelodysplastic syndrome (MDS) with bicytopenia in peripheral blood and two or three-lineage dysplasia in the bone marrow with less than 5% of blasts. RPL5-mutant patients (P2 and P4) had no typical MDS changes by flow cytometry, cytogenetics or sequencing; an *ASXL1* mutation was detected in RPS19-mutant patient (P9) [10]. The material from these patients (including bone marrow biopsy) was collected before MDS development. § P24 deceased of triple-negative breast cancer in 2019.

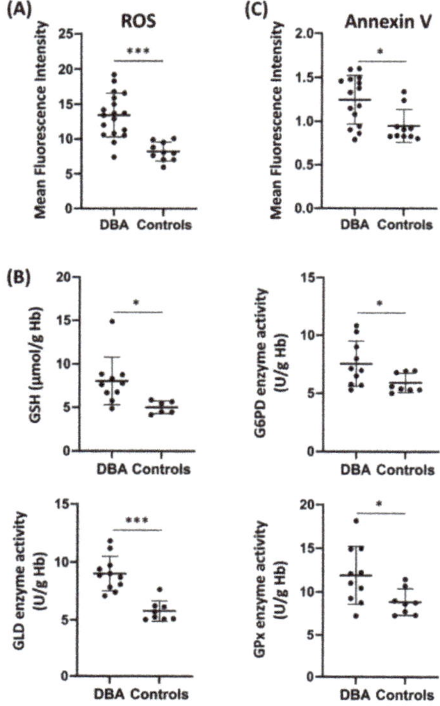

Figure 1. Oxidative stress, antioxidant defense, and Annexin V binding in Diamond-Blackfan anemia (DBA) patients' erythrocytes. (**A**) Significantly elevated intracellular reactive oxygen species (ROS) content in DBA erythrocytes (n = 18) compared to control erythrocytes (n = 10). For a positive control, erythrocytes were exposed to 2 mM hydrogen peroxide (H_2O_2) for 10 min before 2′,7′-dichlorofluorescein diacetate (DCF-DA)-labeling (positive control mean fluorescence intensity (MFI): 230 ± 83). (**B**) Significantly increased reduced glutathione (GSH) levels (given as µmol/g hemoglobin—Hb) and enzymes activities of glucose-6-phosphate dehydrogenase (G6PD), gluconolactone dehydrogenase (GLD) and glutathione peroxidase (GPx) (expressed as U/g Hb) in DBA erythrocytes (n = at least 10) compared to controls (n = at least 6). Specific enzyme activity was calculated using the Lambert-Beer law. (**C**) Enhanced Annexin V binding to the membrane of DBA erythrocytes (n = 15) compared to controls (n = 10). Individual values in Panels (**A–C**) are presented in a dot plot depicting mean with error bars showing standard deviations (SDs). Graphs were created and p values calculated using GraphPad Prism 8 Software (La Jolla, CA, USA, www.graphpad.com); * $p < 0.05$, *** $p < 0.001$.

Thereafter, exposure of phosphatidylserines (PS) on the erythrocyte membrane, as one of the markers of eryptosis [6], was assessed by flow cytometric analysis of Annexin V binding. DBA erythrocytes showed significantly increased Annexin V binding (MFI: 1.30 ± 0.34) when compared to control erythrocytes (MFI: 0.95 ± 0.21) (Figure 1C). This confirms increased PS exposure on the membrane of DBA erythrocytes and suggests that excessive ROS production exceeds the capacity of scavenging mechanisms in DBA erythrocytes, resulting in erythrocyte membrane alterations which may cause their enhanced recognition and destruction by reticuloendothelial macrophages [7,11].

2.2. Creation and Validation of DBA Cellular Model

To investigate the extent of oxidative damage in DBA erythroid lineage in more detail, CRISPR/Cas9 technology was used to create Rpl5- and Rps19-deficient murine erythroleukemia (MEL) cells (a detailed description of the procedure can be found in Supplementary Materials and Figure S1A,B). *RPL5* and *RPS19* are the two most frequently mutated and studied genes associated with DBA [1]. Two clones for each gene with reduced levels of Rpl5 and Rps19 (Figure S1C,D) mimicking the Rp haploinsufficiency showed by DBA patients, were selected for further analyses. MEL cells are arrested at the proerythroblast stage and can be chemically induced to erythroid differentiation [12]. Both uninduced cells and cells induced to erythroid differentiation by dimethylsulfoxid (DMSO) for 96 h were evaluated in this study.

First, the typical DBA characteristics were observed in the created Rpl5- and Rps19-deficient cells. This included significantly decreased proliferation capacity, as measured by thiazolyl blue tetrazolium bromide (MTT) test at 48 and 72 h of cultivation (Figure S2A), and higher percentage of nucleated cells undergoing apoptosis detected by terminal deoxynucleotidyl transferase dUTP nick end labeling (TUNEL) assay on cytospin slides (Figure S2B) when compared to control cells. Immunoblot analysis of p53 phosphorylation revealed increased level of p53 activation in Rpl5- and Rps19-deficient cells (Figure S2C). More profound alterations in all above-mentioned assays were detected for Rpl5-deficient cells compared to Rps19-deficient cells, in agreement with a more severe phenotype associated with RPL5 deficiency [3]. Upon induction of erythroid differentiation, the fold-change in the mRNA expression of GATA-binding factor 1 (*Gata1*), a critical regulator of erythroblast maturation, was substantially lower in Rpl5- and Rps19-deficient cells (1.2-fold and 1.7-fold, respectively) when compared to control cells (2.4-fold) (Figure S3). Altogether, these data are consistent with impaired erythroid differentiation [1,13,14] and augmented erythroblasts apoptosis [1,14] that we previously reported in the bone marrow of DBA patients [15].

2.3. Oxidative DNA Damage in DBA Cellular Model

After the basic characterization of MEL clones, CellROX Green reagent was used to determine intracellular ROS content. Significantly increased ROS levels were detected in uninduced Rpl5-deficient (MFI: 1.24 ± 0.16) and Rps19-deficient (MFI: 1.75 ± 0.39) cells when compared to control cells (MFI: 0.71 ± 0.11), and also in both Rp-deficient cells induced to erythroid differentiation (Rpl5-deficient, MFI: 2.14 ± 0.35; Rps19-deficient, MFI: 1.30 ± 0.17; controls, MFI: 0.76 ± 0.40) (Figure 2A). To assess the antioxidant capacity of Rpl5- and Rps19-deficient cells, expression analysis of critical antioxidant enzyme: superoxide dismutase 1 (SOD1), SOD2, and catalase (CAT) was evaluated. As shown in Figure 2B, predominantly *Sod1* and *Cat* mRNA are downregulated in Rpl5- and Rps19-deficient cells compared to control cells; the difference is more pronounced in Rp-deficient cells induced to erythroid differentiation. Further analysis of antioxidant defense enzymes activities (G6PD, GLD, and GPx) revealed that none of the enzymes showed upregulated activity (Figure S4) which would balance the elevated ROS levels in Rpl5- and Rps19-deficient cells. Altogether, this suggests insufficient capacity of Rpl5- and Rps19-deficient cells to cope with the greater ROS accumulation.

In order to assess the damaging effect of elevated ROS on DNA, immunocytochemical (ICC) analysis of an oxidative DNA lesion 8-oxoguanine (8-oxoG) [16] was performed. This ICC staining revealed elevated 8-oxoG levels on cytospin slides of Rpl5- and Rps19-deficient cells than in control cells (Figure 2C). As a result of ROS production, DNA damage may occur, resulting in DDR, which includes phosphorylation of Ser-139 residue of the histone variant H2AX (γ-H2AX) [16]. Indeed, a significantly elevated γ-H2AX signal, detected by increased fluorescence intensity, was observed for uninduced Rpl5-deficient cells (MFI: 13.05 ± 1.7) compared to control cells (MFI: 5.53 ± 0.49) (Figure 2D); the MFI for γ-irradiated control cells that served as a positive control was 18.6 ± 2.4. On the other hand, γ-H2AX in Rps19-deficient cells was less significantly induced (MFI: 7.59 ± 0.57), suggesting a lesser extent of DNA damage. Because γ-H2AX levels dramatically increase during late-stage erythropoiesis [17], MEL cells induced to erythroid differentiation were not evaluated.

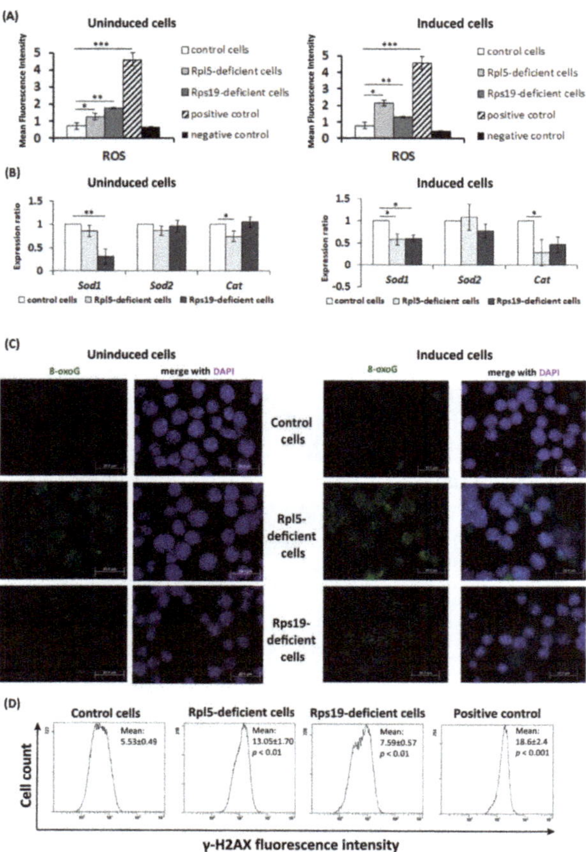

Figure 2. Oxidative stress and DNA damage in Rpl5- and Rps19-deficient murine erythroleukemia (MEL) cells. (**A**) Elevated reactive oxygen species (ROS) levels in uninduced and induced Rpl5- and Rps19-deficient cells compared to control cells. Values are given as mean ± standard deviation (SD). The results represent the mean of four independent experiments. p values were calculated using the Student t-test; * $p < 0.05$, ** $p < 0.01$, *** $p < 0.001$. For a positive control, MEL cells were exposed to 200 μM tert-butyl hydroperoxide (kit component) 1 h before staining with the ROS detection reagent; negative control—unstained cells. (**B**) Decreased relative mRNA expression of superoxide dismutase 1 (*Sod1)* and catalase (*Cat)* (normalized to beta-actin) in uninduced and induced Rpl5- and Rps19-deficient cells compared to controls. p values were calculated using the REST© 2020 software (Technical University Munich, Germany): * $p < 0.05$, ** $p < 0.01$; T bars designate standard error of the mean (SEM). (**C**) Higher nuclear and perinuclear 8-oxoguanine (8-oxoG) positivity (green color) for Rpl5- and Rps19-deficient cells compared to control cells. Cell nuclei were counterstained with 4′,6-diamidino-2-phenylindole dihydrochloride (DAPI, blue color). A positive control for 8-oxoG staining is shown in Figure S5. Immunostained cells were analyzed with an Olympus BX 51 fluorescence microscope (Olympus), original magnification 1000×. Digital images were acquired with an Olympus DP 50 camera driven by DP controller software (provided by Olympus, Tokyo, Japan). Images were cropped, assembled, and labeled using Adobe Photoshop software (Adobe Systems, San Jose, CA, USA). (**D**) Increased phosphorylation of Ser-139 residue of the histone variant H2AX (γ-H2AX) in uninduced Rpl5- and Rps19-deficient cells compared to controls. Values are given as mean ± SDs. Representative histograms of the assay, repeated 3 times, are shown. p values were calculated using the Student t-test; ** $p < 0.01$, *** $p < 0.001$. Positive control: γ-rays irradiated MEL cells (mean fluorescence intensity—MFI: 18.6 ± 2.4). MFI for unstained control cells: 0.86 ± 0.13.

2.4. Activation of DDR Signaling and DNA Repair in DBA Cellular Model

To assess the consequent activation of DDR signaling in response to the observed DNA damage, phosphorylation of Chk1 (p-Chk1) [18] and ATM (p-ATM) [19] was analyzed by immunoblot analysis and ICC staining, respectively. Increased levels of p-Chk1 at S345 in Rpl5-deficient cells and to a lesser degree also in Rps19-deficient cells, compared to controls (Figure 3A), confirmed the induction of ATR-Chk1 signaling. On the other hand, the staining against p-ATM at S1981 revealed predominantly cytoplasmic positivity in Rpl5- and Rps19-deficient cells (Figure 3B), which likely reflects ATM response to ROS [20]. Faint nuclear staining, reflecting a certain threshold level of endogenous DNA damage, was detected only in Rpl5- deficient cells (Figure 3B). These data are in agreement with differences in γ-H2AX fluorescence intensity observed between Rpl5- and Rps19-deficient cells.

Subsequently, selected mRNA expression markers of activated DNA repair were tested. This included an ATP-dependent DNA helicase Q4 (encoded by the *Recql4* gene) involved in homologous recombination (HR), nonhomologous end joining (NHEJ), nucleotide excision repair (NER) and base excision repair (BER) [21,22]; 8-Oxoguanine glycosylase (encoded by the *Ogg1* gene), a marker of BER responding to the presence of 8-oxoG lesions [23]; and a DNA-dependent protein kinase, catalytic subunit (encoded by the *Prkdc* gene) participating in NHEJ [24]. Uninduced Rpl5-deficient cells showed statistically increased mRNA expression of all analyzed markers (Figure 3C), with *Recql4* and *Prkdc* also being significantly upregulated in induced Rpl5-deficient cells. For Rps19-deficient cells, only the expression of *Ogg1* mRNA in uninduced cells and *Recql4* mRNA in cells induced to erythroid differentiation was statistically increased compared to the controls (Figure 3C). The expression of DNA repair genes appears to be more strongly induced in Rpl5-deficient cells than in Rps19-deficient cells, corresponding to a higher rate of DNA damage in Rpl5-deficient cells. As all analyzed markers are reported to be Gata1 targets [25], the degree of their upregulation may be restrained by reduced *Gata1* expression in induced Rpl5- and Rps19-deficient cells (Figure S3) and consequently limit the activation of DNA repair.

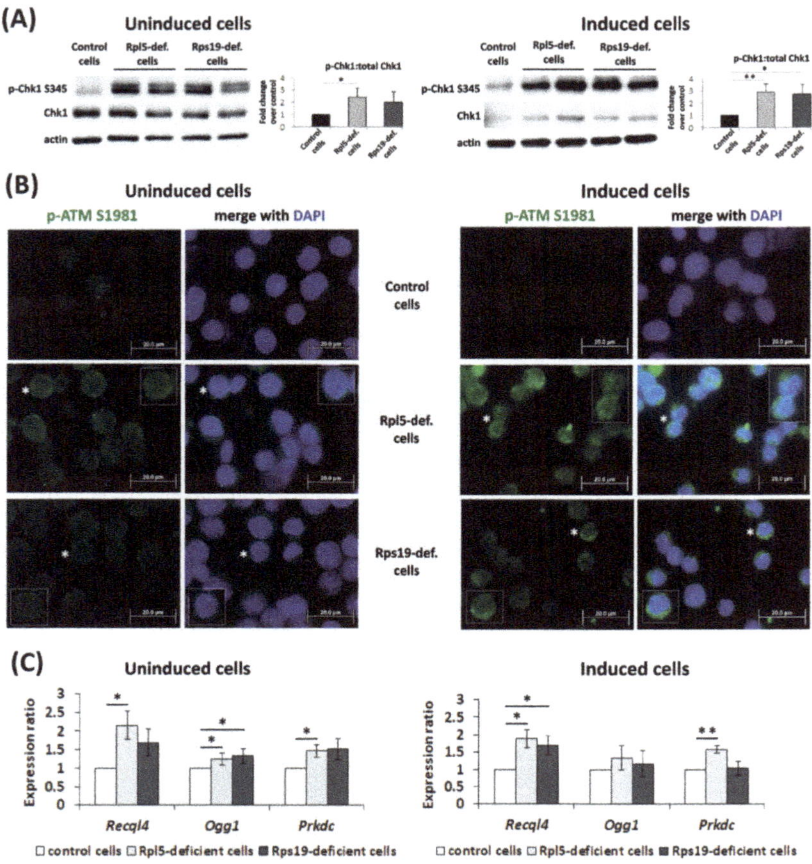

Figure 3. DNA damage response (DDR) signaling and DNA repair in Rpl5- and Rps19-deficient murine erythroleukemia (MEL) cells. (**A**) Increased levels of Chk1 phosphorylation at S345 (p-Chk1 S345) were detected in uninduced and induced Rpl5- and Rps19-deficient cells than in controls. A representative immunoblot is shown. p-Chk1 was normalized to total Chk1 protein using the G:BOX-CHEMI-XX9 imaging system (Syngene, Cambridge, UK). Data in the bar graph showed a fold change in p-Chk1: total Chk1 ratio over control cells and are expressed as means ± standard errors of the mean (SEM) from 3 independent experiments. * $p < 0.05$ and ** $p < 0.01$ versus control cells. (**B**) Higher cytoplasmic positivity for phosphorylated ATM (p-ATM at S1981, green color) was observed in uninduced and induced Rpl5- and Rps19-deficient cells compared to control cells. Induced Rpl5-deficient cells showed, in addition, nuclear p-ATM S1981 staining. Cells nuclei were counterstained with 4′,6-diamidino-2-phenylindole dihydrochloride (DAPI) (blue color). The asterisks indicate cells shown in the insets. Immunostained cells were analyzed with an Olympus BX 51 fluorescence microscope (Olympus), original magnification 1000×. Digital images were acquired with an Olympus DP 50 camera driven by DP controller software (provided by Olympus, Tokyo, Japan). Images were cropped, assembled, and labeled using Adobe Photoshop software (Adobe Systems, San Jose, CA, USA). (**C**) Markers of DNA repair (*Recql4*, *Ogg1*, and *Prkdc*) showed variable increase in mRNA expression (normalized to beta-actin) in uninduced and induced Rpl5- and Rps19-deficient cells compared to controls. p values were calculated using the REST© 2020 software (Technical University Munich, Germany): * $p < 0.05$, ** $p < 0.01$; T bars designate SEM.

2.5. Elevated Inflammatory Cytokines and Senescence in DBA Cellular Model

It is known that oxidative damage may be caused by exposure of cells to inflammatory cytokines [26]. In this context, upregulation of a proinflammatory cytokine tumor necrosis factor-alpha (TNF-α) was previously reported in RPS19-deficient hematopoietic progenitors and rps19-deficient zebrafish [27]. Indeed, upregulated expression of *TNF-α, interleukin 6* (*IL6*), and *IL1b* was detected by real-time PCR in Rpl5- and Rps19-deficient cells, both uninduced and induced, compared to control cells (Figure 4A). This was associated with significantly increased phosphorylation of p38 kinase (measured by flow cytometry) (Figure 4B) or induced expression of mitogen-activated protein kinase kinase kinase 8 (*Map3k8* kinase) mRNA (determined by real-time PCR) (Figure 4C) in Rpl5- and Rps19-deficient cells, respectively. Both of these molecules are known to play an essential role in the production of proinflammatory cytokines [28,29]. Proinflammatory cytokines may contribute to (as well as may result from) the induction of cellular senescence [30]. Significantly elevated senescence-associated β-galactosidase (SA-β-gal) activity was detected in both uninduced and induced Rpl5- and Rps19-deficient cells compared to control cells using the cellular senescence assay kit (Figure 4D).

To assess the involvement of cell-intrinsic production of cytokines and other secreted factors in the pathologies observed in Rpl5- and Rps19-deficient cells, we tested if a conditioned medium from mutant cells can induce cell-nonautonomous responses (bystander effects) in control cells [31]. Uninduced control cells were cultured in conditioned medium harvested from Rpl5- and Rps19-deficient cells as described in Figure S6; conditioned medium harvested from control cells served as a negative control. Immunoblot analysis of corresponding protein lysates showed activation of p53 (phospho-p53) in control cells cultured in conditioned medium harvested from Rpl5- and Rps19-deficient cells (Figure 5A). Simultaneously, propidium iodide (PI) staining revealed accumulation of control cells in the S-phase of cell cycle and an increased proportion of the sub G1 fraction (Figure 5B). These results demonstrate that cell-autonomous production of secreted factors/cytokines by Rpl5- or Rps19-deficient cells may inhibit cell cycle progression and activate p53 checkpoint and as such potentiate cell damage.

Because inflammatory cytokines appeared to be important contributors to DNA damage in Rp-deficient cells in our experiments, next, the effect of pomalidomide, a known TNF-α, IL1b and IL6 inhibitor [32], was tested. Addition of pomalidomide to the culture medium diminished 8-oxoG-positivity in Rpl5- and Rps19-deficient cells (Figure S7A,B) and concomitantly mitigated p53 activation in uninduced Rpl5- and Rps19-deficient cells (Figure S7C). The observed effect of pomalidomide on Rpl5- and Rps19-deficient cells can be attributed to the inhibition of *TNF-α, IL1b*, and *IL6* expression (Figure S8A) rather than changes in cell cycle progression [33] (Figure S8B).

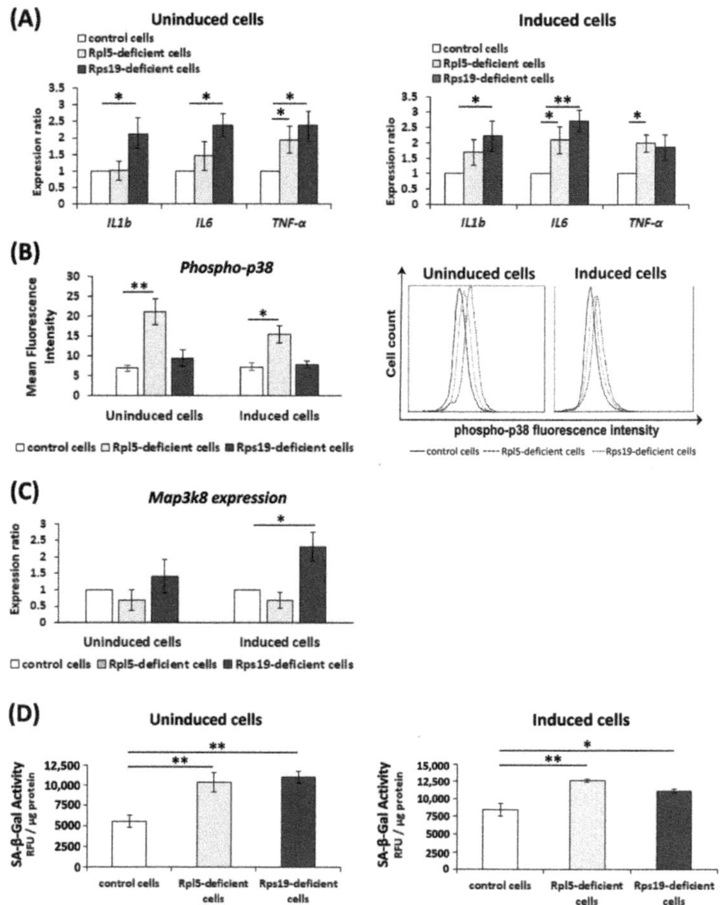

Figure 4. Proinflammatory cytokines expression, signaling pathways, and senescence in Rpl5- and Rps19-deficient murine erythroleukemia (MEL) cells. (**A**) Increased relative mRNA expression of interleukin 1b (*IL1b*), *IL6*, and tumor necrosis factor-alpha (*TNF-α*) (normalized to beta-actin) in uninduced and induced Rpl5- and Rps19-deficient cells compared to controls. *p* values were calculated using the REST© 2020 software (Technical University Munich, Germany): * $p < 0.05$, ** $p < 0.01$; T bars designate standard error of the mean (SEM). (**B**) Significantly elevated phosphorylation of p38 kinase in uninduced and induced Rpl5-deficient cells compared to control cells; mean fluorescence intensity (MFI) of uninduced cells: 21.2 ± 3.2 for Rpl5-deficient cells vs. 6.9 ± 0.8 for control cells; MFI of induced cells: 15.5 ± 2.5 for Rpl5-deficient cells vs. 7.3 ± 0.9 for control cells. MFI for Rps19-deficient cells (uninduced: 9.5 ± 2.1 and induced: 7.8 ± 0.8) was comparable to controls. Values are given as mean ± standard deviation (SD). *p* values were calculated using the Student *t*-test; * $p < 0.05$, ** $p < 0.01$. (**C**) Significantly increased expression of *Map3k8* mRNA (normalized to beta-actin) in induced Rps19-deficient cells. *p* values were calculated using the REST© 2020 software (Technical University Munich, Germany): * $p < 0.05$; T bars designate SEM. (**D**) Significantly augmented SA-β-gal activity (relative fluorescence unit—RFU/μg protein) in uninduced and induced Rpl5-deficient (10,330 ± 1227 and 12,626 ± 221) and Rps19-deficient (10,983 ± 703 and 11,136 ± 307) cells compared to control cells (5566 ± 735 and 8435 ± 862). The absorbance was read at 360 nm (excitation)/ 465 nm (emission) on fluorescence reader (GENios, Tecan, Männedorf, Switzerland). Values are given as mean ± SD; the assay was repeated four times. The values were normalized to total protein levels; * $p < 0.05$, ** $p < 0.01$.

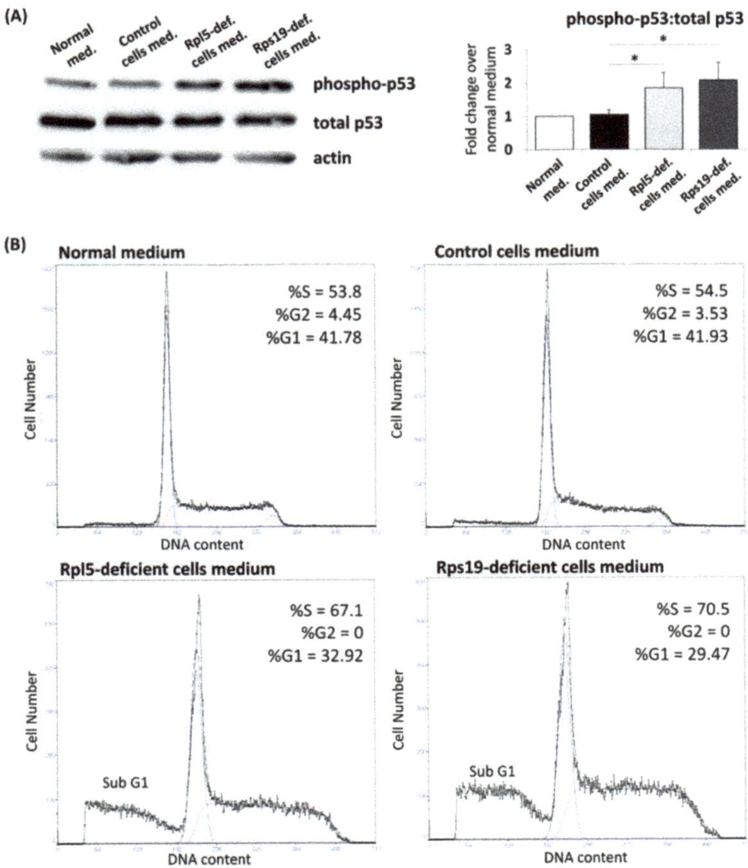

Figure 5. Effect of conditioned medium harvested from Rpl5- and Rps19-deficient cells on p53 activation and cell cycle of control cells. (**A**) Control cells cultivated in conditioned medium harvested from Rpl5- and Rps19-deficient cells showed activation of p53 (phospho-p53). In contrast, p53 phosphorylation in control cells cultivated in normal culture medium and those cultivated in conditioned medium harvested from control cells was comparable. A representative immunoblot is shown. Data in the bar graph showed a fold change in phospho-p53:total p53 ratio over cells cultured in normal cell culture medium and are expressed as means ± standard errors of the mean from three independent experiments; * $p < 0.05$ versus conditioned control cells medium. G:BOX-CHEMI-XX9 imaging system (Syngene, Cambridge, UK) was used for densitometry. (**B**) Control cells cultivated in conditioned medium harvested from Rpl5- and Rps19-deficient cells showed accumulation in the S phase and reduction in the G2/M phase of cell cycle. An increase in sub G1 fraction can also be noted. There was no difference between the cell cycle distribution of controls cells cultivated in normal culture medium and those cultivated in conditioned medium harvested from control cells. Similar data were obtained in three independent experiments.

2.6. Oxidative DNA Damage, Elevated Inflammatory Cytokines, and Activated DDR Signaling in DBA Patients

In order to validate the data obtained in DBA cellular models, erythroblast cell cultures derived from mononuclear cells (MNCs) isolated from selected RPL5- and RPS19-mutant patients (Table 1) were established as we previously described [34]. Subsequent ICC analyses revealed increased numbers of 8-oxoG-positive RPL5- and RPS19-mutant erythroblasts compared to control erythroblasts (Figure 6A).

In addition, phosphorylated ATM at S1981 (p-ATM S1981), localized primarily in the cytoplasm, was observed in RPL5- and RPS19-mutant erythroblasts, but not in controls (Figure 6B).

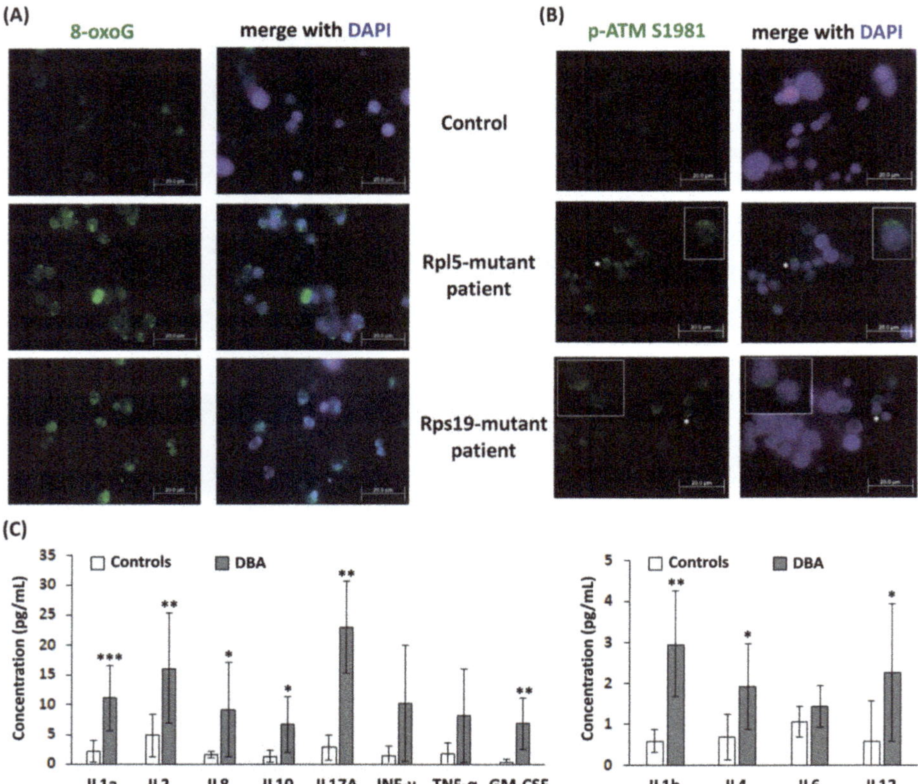

Figure 6. Oxidative DNA damage, p-ATM, and inflammatory cytokines in Diamond-Blackfan anemia (DBA) patients' samples. (**A**) Increased numbers of 8-oxoguanine (8-oxoG)-positive (green color) RPL5-mutant erythroblasts (patient P2: 18.9%, patient P4: 5.7%) and RPS19-mutant erythroblasts (patient P7: 5.9%, patient P8: 6.9%) compared to control erythroblasts (n = 3, mean: 0.5 ± 0.2%). (**B**) Phosphorylated ATM at S1981 (p-ATM S1981, green color) in RPL5- and RPS19-mutant erythroblasts in comparison to control erythroblasts; representative images obtained for patient P1 and P7, and one control are shown. Cells nuclei were counterstained with 4′,6-diamidino-2-phenylindole dihydrochloride (DAPI) (blue color). The asterisks indicate cells shown in the insets. The slides were analyzed with an Olympus BX 51 fluorescence microscope (Olympus), original magnification 1000×. Digital images were acquired with an Olympus DP 50 camera driven by DP controller software (provided by Olympus, Tokyo, Japan). Images were cropped, assembled, and labeled using Adobe Photoshop software (Adobe Systems, San Jose, CA, USA). (**C**) Elevated cytokines in DBA patients' serum (n = 11) compared to controls (n = 11). Values are given as mean ± standard deviation (SD). p values were calculated using the Student t-test; * $p < 0.05$, ** $p < 0.01$, and *** $p < 0.001$.

Next, elevated levels of several cytokines were detected in the serum of DBA patients (n = 11). In particular, IL1a, IL1b, IL2, IL4, IL8, IL10, IL12, IL17a, and granulocyte-macrophage colony-stimulating factor (GM-CSF) were significantly increased compared to controls (C). Importantly, eight out of eleven analyzed patients were in disease remission. This confirms induction of inflammatory cytokine signature in DBA patients in vivo.

Finally, IHC analyses of DBA patients' bone marrow samples ($n = 9$, Table 1) revealed elevated p53 positivity and apoptosis, which were not limited to erythroid cells (Figure S9). Moreover, sporadic 8-oxoG-positive bone marrow cells belonging to granulocyte cell lineage were detected in three patients (P2, P9, and P23) (Figure S10). In agreement, the p-ATM S1981 immunoreactivity detected in the cytoplasm of rare cells in patients P2 and P23 (Figure S11) indicated ATM phosphorylation in response to elevated ROS. However, in the RPS19-mutant patient P9, the number of p-ATM S1981 positive cells dramatically increased and the p-ATM S1981 signal became detectable not only in the cytoplasm but also in the cell nuclei, reflecting accumulation of DNA damage and activation of DNA damage signaling [35]. Importantly, patients P2 and P9 recently developed MDS. Altogether, this suggests that cells of different lineages (other than erythroid) may be vulnerable to DNA damage due to inherited RP-haploinsufficiency. Simultaneously, however, noncell autonomous (microenvironment-dependent) inflammatory stress may fuel oxidative damage of RP-deficient hematopoietic cells (as proposed by the experiment with conditioned medium shown in Figure 5).

3. Discussion

In this study, we used Rpl5- and Rps19-deficient cellular models and DBA patients' samples to show that the ROS generation in Rp-mutant cells overpowers their antioxidant capacity, leading to oxidative DNA damage in erythroid precursors and altered characteristics of mature erythrocytes. In addition, our results imply inflammatory cytokines as mediators associated with oxidative stress in DBA cells. The above-mentioned features are shared to both ribosomal protein deficiencies with a certain degree of variation in the intensity of pathological phenotype.

Two aspects of DBA pathobiology have been evaluated in our study. The first one involved the assessment of consequences of oxidative stress on the properties of DBA RBCs. Mature erythrocytes are highly susceptible to oxidative damage and, therefore, possess several mechanisms, involving both nonenzymatic antioxidants and enzymatic antioxidants, to avoid excessive ROS formation [36]. It was documented that elevated ROS levels induce PS exposure to the outer surface of the erythrocyte membrane leading to enhanced clearance of erythrocytes by macrophages and reduced erythrocyte lifespan [6]. Excessive eryptosis has been described for multiple clinical conditions [8]. Here, we observed induced exposure of PS to the outer surface of the DBA erythrocyte membrane together with elevated ROS and despite the upregulation of antioxidant defense (Figure 1). This might indicate enhanced recognition of DBA erythrocytes by macrophages in vivo. As severe anemia is associated with tissue hypoxia, and ROS generation in hypoxia often exceeds the antioxidant buffering system of erythrocytes [37], profound hypoxemia may limit the antioxidant capacity of DBA erythrocytes. Moreover, studies on animal models and clinical evidence have indicated that erythrocytes are also sensitive to the presence of inflammatory cytokines, even at low levels of chronic inflammation [38]. In particular, IL8, one of the cytokines elevated in the serum of our DBA patients (Figure 6C), was shown to induce pathological changes to the erythrocyte membrane typical for eryptosis [38]. Based on these results, we propose that DBA erythrocytes have limited ROS buffering capacity which makes them more vulnerable to induced stress. Enhanced oxidative stress in DBA erythrocytes, for example during infections, may reinforce erythrophagocytosis and thus contribute to worsening of anemia. Supportive antioxidant treatment might therefore be beneficial in these conditions.

The second aspect addressed here was the extent of oxidative damage in developing DBA erythroblasts. An earlier study proposed the involvement of dysbalanced globin-heme synthesis in excessive ROS formation, predominantly in RPL5- and RPL11-deficient cells [3]. Using the CRISPR/Cas9-created Rpl5- and Rps19-deficient MEL cells, we observed insufficient capacity of both Rpl5- and Rps19-deficient cells to cope with higher ROS production (Figure 2 and Figure S4). This resulted in oxidative DNA damage, detected by increased 8-oxoG and γ-H2AX accumulation (Figure 2), and in the activation of ATR-Chk1 pathway (Figure 3A). This is consistent with previous reports showing downregulation of ROS scavengers in rpl11-mutant zebrafish and RPS19-deficient cells [39,40] and induction of γ-H2AX and activation of DDR in an rps19-deficient zebrafish model [5]. Nevertheless, the staining against p-ATM at S1981 revealed predominantly cytoplasmic p-ATM immunoreactivity in both Rpl5- and Rps19-deficient MEL cells (Figure 3B). Weak nuclear positivity, detected only in Rpl5-deficient cells, indicated low levels of DNA damage. Importantly, these data were confirmed in DBA patients' erythroblasts derived in in vitro cultures where increased 8-oxoG positivity (Figure 6A) and cytoplasmic p-ATM immunoreactivity (Figure 6B) were detected. Thus, ATM in Rp-deficient cells exerts its actions mainly from the cytoplasm where it responds to the overproduction of ROS [20]. The cytoplasmic p-ATM was shown to regulate autophagy in order to maintain redox homeostasis [41]. Consistently, the induction of autophagy was previously observed in the cells with the knock-down of *RPS19* [4].

We also showed that in response to DNA damage, Rpl5- and Rps19-deficient cells activate DNA repair signaling molecules (Figure 3C). Nevertheless, the activation might be insufficient to completely prevent genomic instability. In agreement, leukemia-associated somatic RP-mutants induce oxidative stress and excessive DNA damage [42,43]. Increased risk of cancer, including both solid tumors and hematological malignancies (primarily MDS and acute myeloid leukemia), is reported in DBA [1]. Even though DBA primarily presents with the erythroid phenotype, the RP-haploinsufficiency is inherited to every cell in the body and the ongoing nucleolar and ribosomal stress may presensitize RP-haploinsufficient cells to DNA damage. Indeed, nucleolar stress can lead to cell cycle arrest and/or apoptosis in a p53-dependent manner [44,45]. Consistently, cells showing p53-positivity, apoptosis, 8-oxoG foci, activation of ATM kinase and belonging to the granulocyte lineage, were detected in the bone marrow of DBA patients from our cohort (Figures S9–S11). Importantly, three of the analyzed patients recently progressed to MDS.

Finally, our study underscored proinflammatory cytokines as important contributors to oxidative damage in Rpl5- and Rps19-deficient cells. The induction of proinflammatory cytokines in Rpl5- and Rps19-deficient cells in vitro (Figure 4A–C) and in the serum of DBA patients in vivo (Figure 6C) is consistent with previous studies reporting on a chronic subclinical inflammatory microenvironment in DBA bone marrow [46]. It is also in agreement with the upregulation of interferon (INF) signaling, inflammatory cytokines and mediators, and the complement system in rps19- and rpl11-deficient zebrafish [47]. This increased production of proinflammatory cytokines may either directly (via regulation of gene expression) or indirectly (via induction of cellular senescence) result from the activation of p53 in response to ribosomal stress (Figure 7) [48,49]. A positive feedback loop between p53 and proinflammatory cytokines can also be expected (Figure 7) [27,48,49]. Indeed, inhibition of TNF-α, IL1b, and IL6 in Rpl5- and Rps19-deficient cells upon pomalidomide treatment resulted in the reduction of oxidative DNA damage and inactivation of p53 (Figure S7). Elevated inflammatory cytokines may reinforce cellular senescence [30] observed in Rpl5- and Rps19-deficient cells (Figure 4D) and senescence may be potentiated by excessive oxidative stress (Figure 7) [50]. The induction of senescence is not necessarily conflicting with apoptosis of Rpl5- and Rps19-deficient cells (Figure S2B). Cells with activated DDR and exposed to inflammation and ROS may undergo senescence or apoptosis [51], depending on the cellular context and DNA damage signaling thresholds (Figure 7). Both these cellular phenotypes may, therefore, contribute to bone marrow failure in DBA patients in vivo. Altogether, our data indicate that anti-inflammatory treatment might relieve pathological DBA features associated with oxidative damage and activated p53.

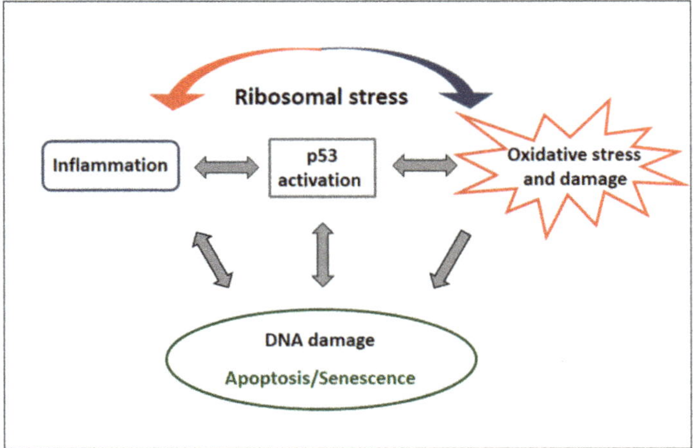

Figure 7. Interplay of deregulated pathways in Diamond-Blackfan anemia (DBA). Ribosomal protein (RP)-haploinsufficiency induces ribosomal stress leading to the activation of p53, induction of inflammatory signaling, and oxidative stress/damage. Multiple feedback loops between DNA damage response (DDR)/p53 activation and inflammatory cytokine production and between p53 activation and cell fate decisions (apoptosis/senescence) is expected. Similarly, a crosstalk between inflammatory cytokines and oxidative damage is presumed. Cells with activated DDR/p53 and exposed to inflammation and reactive oxygen species (ROS) may undergo senescence or apoptosis depending on the cellular context and DDR signaling thresholds. Damaged/senescent cells may further reinforce inflammatory cytokine production. The detailed hierarchical nature of cooperation between each deregulated pathway in DBA pathobiology needs to be defined.

Although a defective microenvironment is not considered to be the major cause of DBA [1], our data strongly support the hypothesis that RP-deficient erythroid cells contribute to and/or induce the production of proinflammatory cytokines, leading to the formation of an inflammatory bone marrow microenvironment. Accordingly, many cytokines, including IL6 and IL8, are among the hub genes reported in RPS19 mutant DBA [52]. The seeming discrepancy between induced *TNF-α* gene expression and signaling in DBA models [27,46] and in our Rp-deficient cells (Figure 4A), and rather normal levels of TNF-α in the serum of our DBA patients (Figure 6C), may reflect the fact that circulating cytokine levels not necessarily correlate with cellular cytokine production [53]. Indeed, most cytokines act in a local microenvironment where they have autocrine and paracrine functions. Increased levels of proinflammatory cytokines (e.g., TNF-α, IL1b, INF-γ) are known to suppress erythropoiesis [54–56]. Moreover, our experiment with conditioned medium harvested from Rpl5- and Rps19-deficient cell cultures (Figure 5) suggested that factors secreted by RP-deficient cells could induce/potentiate checkpoint signaling and cell death in a cell-nonautonomous fashion [57]. Our in vivo data showing oxidative damage and p53 activation in nonerythroid cells in the bone marrow from DBA patients (Figures S9–S11) further support this hypothesis. Consistently, analysis of a mouse model of Fanconi anemia, another inherited bone marrow failure syndrome with defective DNA repair, showed that TNF-α exposure creates an environment for clonal selection of somatically mutated preleukemic stem cells, thus leading to leukemogenesis [58]. The link between oxidative damage, inflammatory cytokines, and preleukemia risk in DBA deserves further comprehensive analyses.

In conclusion, our study extends the concept of a complex interplay of multiple mechanisms converging to DBA development and contributing to the high heterogeneity of clinical symptoms and response to therapy observed in DBA patients. We propose that defective ribosomal biogenesis is associated with the induction of inflammatory cytokine signature and oxidative damage. Further research is needed to conclusively elucidate the hierarchy of all deregulated pathways in DBA.

Nevertheless, the obtained results indicate that therapeutic interventions targeting elevated ROS and/or inflammatory cytokines could alleviate the DBA phenotype in vivo.

4. Materials and Methods

4.1. Patient Samples

DBA patients (n = 24) from a recently updated Czech and Slovak DBA registry [10] were included in the study. The patients' basic characteristics can be found in Table 1. The control group consisted of age- and sex-matched healthy individuals. All patients´ and controls´ samples were obtained with an informed consent. The study was conducted in accordance with the Declaration of Helsinki and approved by the Ethics Committee of University Hospital Olomouc on 17 June 2015 (Project identification code 16-32105A).

4.1.1. Determination of ROS

Erythrocytes from DBA patients and healthy controls were incubated with 0.4 mM DCF-DA (Sigma-Aldrich, Darmstadt, Germany) for 15 min at 37 °C according to Amer et al. [59] DCF-dependent intensity of fluorescence was measured by FACS Calibur (BD Bioscience, Franklin Lakes, NJ, USA).

4.1.2. Erythrocyte Annexin V Binding

Annexin V/FITC kit was purchased from BD Biosciences (Franklin Lakes, NJ, USA) and the assay performed as previously described [11]. Fluorescence intensity was measured by FACS Calibur.

4.1.3. Glutathione Measurements and Erythrocyte Enzyme Activity

GSH was determined using a quantification kit for oxidized and reduced glutathione (Sigma-Aldrich, Darmstadt, Germany) according to manufacturer´s instructions. Activity of enzymes involved in the pentose phosphate pathway and oxidative defense: G6PD, GLD, and GPx was determined according to the methods recommended by the International Committee for Standardization in Haematology [60], as we previously described [61,62]. Leukocyte- and platelet-free erythrocyte lysates were used, and the absorbance was measured by spectrophotometer (Infinite 200 Nanoquant; Tecan, Männedorf, Switzerland). All chemicals and purified enzymes were purchased from Sigma-Aldrich (Darmstadt, Germany).

4.1.4. Erythroblast Cell Culture and ICC Analyses

MNCs were isolated from the whole peripheral blood using density centrifugation and cultivated according to previously published protocols [63,64], with minor changes. For the first seven days, 1×10^6/mL MNCs were cultured in StemPro™-34 SFM medium (ThermoFisher Scientific, Waltham, MA, USA) containing L-glutamine (2 mM, ThermoFisher Scientific, Waltham, MA, USA), 1X cytokine cocktail StemSpan™ CC110 (StemCell Technologies, Vancouver, BC, Canada), recombinant human erythropoietin (EPO, 2 U/mL; StemCell Technologies, Vancouver, BC, Canada), and the synthetic glucocorticoid dexamethasone (Dex, 1 μM; Sigma-Aldrich, Darmstadt, Germany). To induce erythroid differentiation, proliferating erythroblasts were cultured in StemPro™-34 SFM medium supplemented with L-glutamine (2 mM), human stem cell factor (50 ng/mL, StemCell Technologies, Vancouver, BC, Canada), human insulin like growth factor-1 (50 ng/mL, StemCell Technologies, Vancouver, BC, Canada), EPO (10 U/mL), and holo-transferrin (1 mg/mL, Sigma-Aldrich, Darmstadt, Germany). The cultures were maintained at 37 °C in 5% CO_2/95% air atmosphere with a medium changed every two days.

Differentiated cells (day 14) were cytospined on glass slides and fixed with 3% paraformaldehyde (PHA) and methanol solution. After permeabilization (0.1% Tween in PBS for 10 min), the slides were incubated with primary antibodies: 8-oxoG (clone 483.15, Santa Cruz Biotechnology, Dallas, TX, USA) or ATM pS1981 (clone 7C10D8, Rockland Immunochemicals, Pottstown, PA, USA) for 1 h. After washing, Alexa Fluor®488-conjugated secondary antibodies were used (both from ThermoFisher Scientific,

Waltham, MA, USA). Cells nuclei were stained with 0.001% DAPI (Sigma-Aldrich, Darmstadt, Germany); cells were visualized using fluorescence microscopy.

4.1.5. Determination of Inflammatory Cytokines Levels

Human Inflammatory Cytokines Multi-Analyte ELISArray™ Kit was used according to manufacturer's instructions (Qiagen, Venlo, The Netherlands).

4.1.6. Immunohistochemistry on Bone Marrow Samples

IHC staining was performed using formalin-fixed and paraffin-embedded bone marrow biopsy samples as we previously described [15,65] with the use of following antibodies: p53 (clone 7F5, Cell Signaling Technologies, Danvers, MA, USA), 8-oxoG (clone 2Q2311, Abcam, Cambridge, UK), and ATM pS1981 (clone 7C10D8, Rockland Immunochemicals, Pottstown, PA, USA) and an EnVision+ Dual Link Detection System (HRP and DAB+ as a visualization chromogen; both DAKO/Agilent, Santa Clara, CA, USA). The alkaline phosphatase (AP) in situ cell death detection kit (Roche Applied Science, Mannheim, Germany) was used according to the manufacturer's instructions as we previously described [15]. The slides were analyzed by light microscopy.

4.2. Cell Lines

Rpl5- and Rps19-deficient MEL were prepared by CRISPR/Cas9 technology. Detailed information on cell transfection and protocol for genotyping of individual clones is given in Supplementary Materials. The cells were maintained in DMEM medium containing 10% fetal bovine serum (FBS; ThermoFisher Scientific, Waltham, MA, USA). Erythroid differentiation was induced with 1.8% DMSO for 96 h [12]. In selected experiments, pomalidomide (10 μM, Sigma-Aldrich, Darmstadt, Germany) was added to the culture medium [66].

4.2.1. Proliferation Assay and Apoptosis

Rpl5- and Rps19-deficient MEL clones were plated on a 96-well plate. The thiazolyl blue tetrazolium bromide (MTT) proliferation assay was performed according to manufacturer's instructions (Sigma-Aldrich, Darmstadt, Germany), as we previously described [67]. The fluorescein in situ cell death detection kit (Roche Applied Science, Mannheim, Germany) was used for apoptosis detection on cytospined slides of MEL cells as we previously described [68]; for more details see Supplementary Materials.

4.2.2. Immunoblotting

MEL cells were lysed on ice in Radio-Immunoprecipitation Assay (RIPA) lysis buffer (ThermoFisher Scientific, Waltham, MA, USA) with 100 μM Na orthovanadate, 100 μM (phenylmethylsulfonyl fluoride) PMSF, and a cocktail of protease inhibitors (all from Sigma-Aldrich, Darmstadt, Germany). The following primary antibodies were used for immunoblotting: RPS19 (Abcam, Cambridge, UK), RPL5 (Abcam, Cambridge, UK), p53 (Cell Signaling Technologies, Danvers, MA, USA), phospho-p53 (Ser15, Cell Signaling Technologies, Danvers, MA, USA), Chk1 (clone G-4, Santa Cruz Biotechnology, Dallas, TX, USA), phospho-Chk1 (Ser345, clone 133D3, Cell Signaling Technologies, Danvers, MA, USA), tubulin (Cell Signaling Technologies, Danvers, MA, USA), and actin (Sigma-Aldrich, Darmstadt, Germany). The Western blots were analyzed by chemiluminescent detection method using SuperSignal™ West Dura Extended Duration Substrate (ThermoFisher Scientific, Waltham, MA, USA). The bands were detected by traditional x-ray film system or G:BOX-CHEMI-XX9 imaging system (Syngene, Cambridge, UK). ImageJ was used for densitometry of protein expression evaluated by traditional x-ray film system [69].

4.2.3. Real-Time PCR Assay

RNA was isolated using TRI Reagent (Sigma-Aldrich, Darmstadt, Germany) and reverse-transcribed using SuperScript® VILO™ cDNA Synthesis Kit (ThermoFisher Scientific, Waltham, MA, USA) or Transcriptor First Strand cDNA Synthesis Kit (Roche Applied Science, Mannheim, Germany) according to manufacturers' instructions. Real-time polymerase chain reaction was performed in triplicates on a LightCycler 480 (Roche Applied Science, Mannheim, Germany) using TaqMan® Gene Expression Master Mix (ThermoFisher Scientific, Waltham, MA, USA) or LightCycler® 480 Probes Master Mix. For real-time PCR using the UPL probes (Roche Applied Science, Mannheim, Germany) cDNA was treated with Turbo DNA-free kit (ThermoFisher Scientific, Waltham, MA, USA). The list of TaqMan® Gene Expression probes (ThermoFisher Scientific, Waltham, MA, USA) and UPL probes and primers can be found in Supplementary Material and Methods. The data were normalized to the expression of beta-actin and to mRNA levels of control cells. The statistical significance of relative expression changes of target mRNA levels was analyzed using REST© 2020 software (Technical University Munich, Germany) [70].

4.2.4. Flow Cytometry Analysis of γ-H2AX and Phosphorylated p38

MEL cells were fixed with 3% PHA and permeabilized with ice-cold 70% ethanol. After the washing and blocking step (with 0.5% BSA/PBS), the cells were incubated with Alexa Fluor®488-conjugated antibody against phospho-histone H2AX (Ser139; Cell Signaling Technologies, Danvers, MA, USA) or primary antiphospho-p38 antibody (Cell Signaling Technologies, Danvers, MA, USA) for 1 h. For phospho-p38 detection, FITC-conjugated secondary antibody (BD Biosciences, Franklin Lakes, NJ, USA) was used for another hour. The intensity of fluorescence was measured by FACS Calibur. To induce γ-H2AX, control MEL cells were irradiated with γ-rays (4 Gy, 30 min) before staining.

4.2.5. ROS Measurement

MEL cells were stained with 0.5 μM CellROX® Green reagent (ThermoFisher Scientific, Waltham, MA, USA) for 45 min at 37 °C in the dark, washed with PBS, and fixed with 4% formaldehyde (ThermoFisher Scientific, Waltham, MA, USA) in PBS. The intensity of fluorescence was measured by FACS Calibur.

4.2.6. ICC of 8-OxoG and ATM pS1981

ICC on cytospin slides of MEL cells was performed as described for differentiated erythroblasts (Section 4.1.4).

4.2.7. Cellular Senescence Activity Assay

MEL cells were lysed in ice-cold lysis buffer (Enzo Life Sciences, Farmingdale, NY, USA) containing protease inhibitors for 15 min on ice. The measurement of SA-β-gal activity was performed according to manufacturer's instructions (ENZ-kit129, Enzo Life Sciences Farmingdale, NY, USA) using a fluorescence reader (GENios, Tecan, Männedorf, Switzerland).

4.2.8. Cell Cycle Analysis

The cells were harvested, fixed (ice-cold 70% ethanol), permeabilized (1% BSA/0.5% Tween-20), and stained with PI (Sigma-Aldrich, Darmstadt, Germany) for 30 min. Cell cycle was measured by flow cytometry (FACS Calibur) as previously described [67] and the cell cycle distribution was analyzed by MultiCycle AV software (Phoenix Flow System, San Diego, CA, USA).

4.3. Statistical Analyses

Student's *t*-test was used to determine the statistical significance of the results. p values less than 0.05 were considered statistically significant. Statistical analyses were conducted using Origin

6.1 software (OriginLab Corporation, Northampton, MA, USA). Enzyme activity graphs were created, and the corresponding *p* values calculated using GraphPad Prism 8 Software (GraphPad Software Inc., La Jolla, CA, USA).

Supplementary Materials: The following are available online at http://www.mdpi.com/1422-0067/21/24/9652/s1. Figure S1: Creation of Rpl5- and Rps19-deficient MEL cells; Figure S2: Characterization of Rpl5- and Rps19-deficient MEL cells; Figure S3: Relative expression of *Gata1*; Figure S4: Activities of ROS scavenging enzymes in Rpl5- and Rps19-deficient cells; Figure S5: Positive controls for 8-oxoG staining; Figure S6: Effect of conditioned medium harvested from Rpl5- and Rps19-deficient cells on p53 activation and cell cycle of control cells; Figure S7: Immunocytochemistry staining for 8-oxoG and immunoblot analysis for phospho-p53 protein in Rpl5- and Rps19-deficient cells; the effect of pomalidomide; Figure S8: Effect of pomalidomide on the expression of inflammatory cytokines and cell cycle progression in uninduced Rpl5- and Rps19-deficient cells; Figure S9: Immunohistochemical staining of bone marrow specimens; Figure S10: Immunohistochemistry for 8-oxoG in the bone marrow specimens; Figure S11. Immunohistochemistry for pATM in the bone marrow specimens.

Author Contributions: K.K., O.J., P.K. and J.G. performed experiments. L.L. designed and prepared the plasmids. K.K., O.J., and M.H. analyzed results, K.K and O.J. made the figures; M.H. designed the research and wrote the manuscript; D.P. collected patients' samples and analyzed the clinical data. V.D. contributed to study design and edited the manuscript. All authors have read and agreed to the published version of the manuscript.

Funding: This work was supported by the grant of The Ministry of Health of the Czech Republic (AZV 16-32105A), The Ministry of Education, Youth and Sports of the Czech Republic (8F20005), and by the Internal grant of Palacký University (IGA_LF_2020_005).

Acknowledgments: We thank Zuzana Saxova for the assistance with IHC, Renata Mojzikova for the assistance with GSH measurements, and Patrik Flodr (Dept. of Pathology, Faculty of Medicine and Dentistry, Palacký University Olomouc) for the assistance with histopathological evaluation of bone marrow samples.

Conflicts of Interest: The authors declare no conflict of interest.

References

1. Da Costa, L.; Narla, A.; Mohandas, N. An update on the pathogenesis and diagnosis of Diamond-Blackfan anemia. *F1000Research* **2018**, *7*. [CrossRef]
2. Kampen, K.R.; Sulima, S.O.; Vereecke, S.; De Keersmaecker, K. Hallmarks of ribosomopathies. *Nucleic Acids Res.* **2020**, *48*, 1013–1028. [CrossRef]
3. Rio, S.; Gastou, M.; Karboul, N.; Derman, R.; Suriyun, T.; Manceau, H.; Leblanc, T.; El Benna, J.; Schmitt, C.; Azouzi, S.; et al. Regulation of globin-heme balance in Diamond-Blackfan anemia by HSP70/GATA1. *Blood* **2019**, *133*, 1358–1370. [CrossRef] [PubMed]
4. Heijnen, H.F.; van Wijk, R.; Pereboom, T.C.; Goos, Y.J.; Seinen, C.W.; van Oirschot, B.A.; van Dooren, R.; Gastou, M.; Giles, R.H.; van Solinge, W.; et al. Ribosomal protein mutations induce autophagy through S6 kinase inhibition of the insulin pathway. *PLoS Genet.* **2014**, *10*, e1004371. [CrossRef] [PubMed]
5. Danilova, N.; Bibikova, E.; Covey, T.M.; Nathanson, D.; Dimitrova, E.; Konto, Y.; Lindgren, A.; Glader, B.; Radu, C.G.; Sakamoto, K.M.; et al. The role of the DNA damage response in zebrafish and cellular models of Diamond Blackfan anemia. *Dis. Model. Mech.* **2014**, *7*, 895–905. [CrossRef] [PubMed]
6. Lang, E.; Qadri, S.M.; Lang, F. Killing me softly—Suicidal erythrocyte death. *Int. J. Biochem. Cell Biol.* **2012**, *44*, 1236–1243. [CrossRef]
7. Zidova, Z.; Kapralova, K.; Koralkova, P.; Mojzikova, R.; Dolezal, D.; Divoky, V.; Horvathova, M. DMT1-mutant Erythrocytes Have Shortened Life Span, Accelerated Glycolysis and Increased Oxidative Stress. *Cell Physiol. Biochem.* **2014**, *34*, 2221–2231. [CrossRef]
8. Lang, F.; Abed, M.; Lang, E.; Föller, M. Oxidative Stress and Suicidal Erythrocyte Death. *Antioxid. Redox Signal.* **2014**, *21*, 138–153. [CrossRef]
9. Utsugisawa, T.; Uchiyama, T.; Toki, T.; Ogura, H.; Aoki, T.; Hamaguchi, I.; Ishiguro, A.; Ohara, A.; Kojima, S.; Ohga, S.; et al. Erythrocyte glutathione is a novel biomarker of Diamond-Blackfan anemia. *Blood Cells Mol. Dis.* **2016**, *59*, 31–36. [CrossRef]
10. Volejnikova, J.; Vojta, P.; Urbankova, H.; Mojzíkova, R.; Horvathova, M.; Hochova, I.; Cermak, J.; Blatny, J.; Sukova, M.; Bubanska, E.; et al. Czech and Slovak Diamond-Blackfan Anemia (DBA) Registry Update: Clinical Data and Novel Causative Genetic Lesions. *Blood Cells Mol. Dis.* **2020**, *81*, 102380. [CrossRef]
11. Kempe, D.S.; Lang, P.A.; Duranton, C.; Akel, A.; Lang, K.S.; Huber, S.M.; Wieder, T.; Lang, F. Enhanced programmed cell death of iron-deficient erythrocytes. *FASEB J.* **2006**, *20*, 368–370. [CrossRef] [PubMed]

12. Antoniou, M. Induction of Erythroid-Specific Expression in Murine Erythroleukemia (MEL) Cell Lines. *Methods Mol. Biol.* **1991**, *7*, 421–434. [PubMed]
13. Ohene-Abuakwa, Y.; Orfali, K.A.; Marius, C.; Ball, S.E. Two-phase Culture in Diamond Blackfan Anemia: Localization of Erythroid Defect. *Blood* **2005**, *105*, 838–846. [CrossRef] [PubMed]
14. Lipton, J.M.; Kudisch, M.; Gross, R.; Nathan, D.G. Defective erythroid progenitor differentiation system in congenital hypoplastic (Diamond-Blackfan) anemia. *Blood* **1986**, *67*, 962–968. [CrossRef]
15. Pospisilova, D.; Holub, D.; Zidova, Z.; Sulovska, L.; Houda, J.; Mihal, V.; Hadacova, I.; Radova, L.; Dzubak, P.; Hajduch, M.; et al. Hepcidin levels in Diamond-Blackfan anemia reflect erythropoietic activity and transfusion dependency. *Haematologica* **2014**, *99*, e118–e121. [CrossRef]
16. Srinivas, U.S.; Tan, B.W.Q.; Vellayappan, B.A.; Jeyasekharan, A.D. ROS and the DNA damage response in cancer. *Redox Biol.* **2019**, *25*, 101084. [CrossRef]
17. Zhao, B.; Tan, T.L.; Mei, Y.; Yang, J.; Yu, Y.; Verma, A.; Liang, Y.; Gao, J.; Ji, P. H2AX deficiency is associated with erythroid dysplasia and compromised haematopoietic stem cell function. *Sci. Rep.* **2016**, *6*, 19589. [CrossRef]
18. Bartek, J.; Lukas, J. Chk1 and Chk2 kinases in checkpoint control and cancer. *Cancer Cell* **2003**, *3*, 421–429. [CrossRef]
19. Bartek, J.; Lukas, J. DNA damage checkpoints: From initiation to recovery or adaptation. *Curr. Opin. Cell Biol.* **2007**, *19*, 238–245. [CrossRef]
20. Zhang, J.; Tripathi, D.N.; Jing, J.; Alexander, A.; Kim, J.; Powell, R.T.; Dere, R.; Tait-Mulder, J.; Lee, J.-H.; Paull, T.T.; et al. ATM functions at the peroxisome to induce pexophagy in response to ROS. *Nat. Cell Biol.* **2015**, *17*, 1259–1269. [CrossRef]
21. Lu, H.; Shamanna, R.A.; Keijzers, G.; Anand, R.; Rasmussen, L.J.; Cejka, P.; Croteau, D.L.; Bohr, V.A. RECQL4 Promotes DNA End Resection in Repair of DNA Double-Strand Breaks. *Cell Rep.* **2016**, *16*, 161–173. [CrossRef] [PubMed]
22. Shamanna, R.A.; Singh, D.K.; Lu, H.; Mirey, G.; Keijzers, G.; Salles, B.; Croteau, D.L.; Bohr, V.A. RECQ helicase RECQL4 participates in non-homologous end joining and interacts with the Ku complex. *Carcinogenesis* **2014**, *35*, 2415–2424. [CrossRef] [PubMed]
23. Maynard, S.; Schurman, S.H.; Harboe, C.; de Souza-Pinto, N.C.; Bohr, V.A. Base excision repair of oxidative DNA damage and association with cancer and aging. *Carcinogenesis* **2009**, *30*, 2–10. [CrossRef] [PubMed]
24. Chang, H.H.Y.; Pannunzio, N.R.; Adachi, N.; Lieber, M.R. Non-homologous DNA end joining and alternative pathways to double-strand break repair. *Nat. Rev. Mol. Cell Biol.* **2017**, *18*, 495–506. [CrossRef] [PubMed]
25. ENCODE Project Consortium. The ENCODE (ENCyclopedia of DNA Elements) Project. *Science* **2004**, *306*, 636–640. [CrossRef]
26. Kidane, D.; Chae, W.J.; Czochor, J.; Eckert, K.A.; Glazer, P.M.; Bothwell, A.L.M.; Sweasy, J.B. Interplay between DNA repair and inflammation, and the link to cancer. *Crit. Rev. Biochem. Mol. Biol.* **2014**, *49*, 116–139. [CrossRef]
27. Bibikova, E.; Youn, M.Y.; Danilova, N.; Ono-Uruga, Y.; Konto-Ghiorghi, Y.; Ochoa, R.; Narla, A.; Glader, B.; Lin, S.; Sakamoto, K.M. TNF-mediated inflammation represses GATA1 and activates p38 MAP kinase in RPS19-deficient hematopoietic progenitors. *Blood* **2014**, *124*, 3791–3798. [CrossRef]
28. Schieven, G.L. The p38alpha kinase plays a central role in inflammation. *Curr. Top. Med. Chem.* **2009**, *9*, 1038–1048. [CrossRef]
29. Xu, D.; Matsumoto, M.L.; McKenzie, B.S.; Zarrin, A.A. TPL2 kinase action and control of inflammation. *Pharmacol. Res.* **2018**, *129*, 188–193. [CrossRef]
30. Freund, A.; Orjalo, A.V.; Desprez, P.Y.; Campisi, J. Inflammatory networks during cellular senescence: Causes and consequences. *Trends Mol. Med.* **2010**, *16*, 238–246. [CrossRef]
31. Coppé, J.P.; Patil, C.K.; Rodier, F.; Sun, Y.; Muñoz, D.P.; Goldstein, J.; Nelson, P.S.; Desprez, P.Y.; Campisi, J. Senescence-associated secretory phenotypes reveal cell-nonautonomous functions of oncogenic RAS and the p53 tumor suppressor. *PLoS Biol.* **2008**, *6*, 2853–2868. [CrossRef]
32. Corral, L.G.; Haslett, P.A.; Muller, G.W.; Chen, R.; Wong, L.M.; Ocampo, C.J.; Patterson, R.T.; Stirling, D.I.; Kaplan, G. Differential Cytokine Modulation and T Cell Activation by Two Distinct Classes of Thalidomide Analogues That Are Potent Inhibitors of TNF-alpha. *J. Immunol.* **1999**, *163*, 380–386. [PubMed]

33. Escoubet-Lozach, L.; Lin, I.L.; Jensen-Pergakes, K.; Brady, H.A.; Gandhi, A.K.; Schafer, P.H.; Muller, G.W.; Worland, P.J.; Chan, K.W.H.; Verhelle, D. Pomalidomide and Lenalidomide Induce p21 WAF-1 Expression in Both Lymphoma and Multiple Myeloma Through a LSD1-mediated Epigenetic Mechanism. *Cancer Res.* **2009**, *69*, 7347–7356. [CrossRef] [PubMed]
34. Horos, R.; Ijspeert, H.; Pospisilova, D.; Sendtner, R.; Andrieu-Soler, C.; Taskesen, E.; Nieradka, A.; Cmejla, R.; Sendtner, M.; Touw, I.P.; et al. Ribosomal Deficiencies in Diamond-Blackfan Anemia Impair Translation of Transcripts Essential for Differentiation of Murine and Human Erythroblasts. *Blood* **2012**, *119*, 262–272. [CrossRef] [PubMed]
35. Alexander, A.; Walker, C.L. Differential localization of ATM is correlated with activation of distinct downstream signaling pathways. *Cell Cycle* **2010**, *9*, 3685–3686. [CrossRef] [PubMed]
36. Mohanty, J.G.; Nagababu, E.; Rifkind, J.M. Red blood cell oxidative stress impairs oxygen delivery and induces red blood cell aging. *Front. Physiol.* **2014**, *5*, 84. [CrossRef] [PubMed]
37. Rogers, S.C.; Said, A.; Corcuera, D.; McLaughlin, D.; Kell, P.; Doctor, A. Hypoxia limits antioxidant capacity in red blood cells by altering glycolytic pathway dominance. *FASEB J.* **2009**, *23*, 3159–3170. [CrossRef]
38. Bester, J.; Pretorius, E. Effects of IL-1β, IL-6 and IL-8 on erythrocytes, platelets and clot viscoelasticity. *Sci. Rep.* **2016**, *6*, 32188. [CrossRef]
39. Aspesi, A.; Pavesi, E.; Robotti, E. Dissecting the Transcriptional Phenotype of Ribosomal Protein Deficiency: Implications for Diamond-Blackfan Anemia. *Gene* **2014**, *545*, 282–289. [CrossRef]
40. Danilova, N.; Sakamoto, K.M.; Lin, S. Ribosomal Protein L11 Mutation in Zebrafish Leads to Haematopoietic and Metabolic Defects. *Br. J. Haematol.* **2011**, *152*, 217–228. [CrossRef]
41. Alexander, A.; Cai, S.; Kim, J.; Nanez, A.; Sahin, M.; MacLean, K.H.; Inoki, K.; Guan, K.; Shen, J.; Person, M.D.; et al. ATM signals to TSC2 in the cytoplasm to regulate mTORC1 in response to ROS. *Proc. Natl. Acad. Sci. USA* **2010**, *107*, 4153–4158. [CrossRef] [PubMed]
42. Kampen, K.R.; Sulima, S.O.; Vereecke, S.; De Keersmaecker, K. The Ribosomal RPL10 R98S Mutation Drives IRES-dependent BCL-2 Translation in T-ALL. *Leukemia* **2019**, *33*, 319–332. [CrossRef] [PubMed]
43. Sulima, S.O.; Kampen, K.R.; Vereecke, S.; Pepe, D.; Fancello, L.; Verbeeck, J.; Dinman, J.D.; De Keersmaecker, K. Ribosomal Lesions Promote Oncogenic Mutagenesis. *Cancer Res.* **2019**, *79*, 320–327. [CrossRef] [PubMed]
44. Lindstrom, M.S.; Jin, A.; Deisenroth, C.; White Wolf, G.; Zhang, Y. Cancer-associated mutations in the MDM2 zinc finger domain disrupt ribosomal protein interaction and attenuate MDM2-induced p53 degradation. *Mol. Cell Biol.* **2007**, *27*, 1056–1068. [CrossRef] [PubMed]
45. Pestov, D.G.; Strezoska, Z.; Lau, L.F. Evidence of p53-dependent cross-talk between ribosome biogenesis and the cell cycle: Effects of nucleolar protein Bop1 on G(1)/S transition. *Mol. Cell Biol.* **2001**, *21*, 4246–4255. [CrossRef]
46. Matsui, K.; Giri, N.; Alter, B.P.; Pinto, L.A. Cytokine production by bone marrow mononuclear cells in inherited bone marrow failure syndromes. *Br. J. Haematol.* **2013**, *163*, 81–92. [CrossRef]
47. Danilova, N.; Wilkes, M.; Bibikova, E.; Youn, M.Y.; Sakamoto, K.M.; Lin, S. Innate immune system activation in zebrafish and cellular models of Diamond Blackfan Anemia. *Sci. Rep.* **2018**, *8*, 5165. [CrossRef]
48. Cooks, T.; Harris, C.C.; Oren, M. Caught in the cross fire: p53 in inflammation. *Carcinogenesis* **2014**, *35*, 1680–1690. [CrossRef]
49. Gudkov, A.V.; Gurova, K.V.; Komarova, E.A. Inflammation and p53. A Tale of Two Stresses. *Genes Cancer* **2011**, *2*, 503–516. [CrossRef]
50. McHugh, D.; Gil, J. Senescence and aging: Causes, consequences, and therapeutic avenues. *J. Cell Biol.* **2018**, *217*, 65–77. [CrossRef]
51. Childs, B.G.; Baker, D.J.; Kirkland, J.L.; Campisi, J.; van Deursen, J.M. Senescence and apoptosis: Dueling or complementary cell fates? *EMBO Rep.* **2014**, *15*, 1139–1153. [CrossRef] [PubMed]
52. Khan, A.; Ali, A.; Junaid, M.; Liu, C.; Kaushik, A.C.; Cho, W.C.S.; Wei, D.Q. Identification of novel drug targets for diamond-blackfan anemia based on RPS19 gene mutation using protein-protein interaction network. *BMC Syst. Biol.* **2018**, *12*, 39. [CrossRef] [PubMed]
53. Jason, J.; Archibald, L.K.; Nwanyanwu, O.C.; Byrd, M.G.; Kazembe, P.N.; Dobbie, H.; Jarvis, W.R. Comparison of serum and cell-specific cytokines in humans. *Clin. Diagn. Lab. Immunol.* **2001**, *8*, 1097–1103. [CrossRef] [PubMed]
54. Johnson, R.A.; Waddelow, T.A.; Caro, J.; Oliff, A.; Roodman, G.D. Chronic exposure to tumor necrosis factor in vivo preferentially inhibits erythropoiesis in nude mice. *Blood* **1989**, *74*, 130–138. [CrossRef]

55. Means, R.T., Jr.; Dessypris, E.N.; Krantz, S.B. Inhibition of human erythroid colony-forming units by interleukin-1 is mediated by gamma interferon. *J. Cell Physiol.* **1992**, *150*, 59–64. [CrossRef] [PubMed]
56. Dai, C.H.; Price, J.O.; Brunner, T.; Krantz, S.B. Fas ligand is present in human erythroid colony-forming cells and interacts with Fas induced by interferon γ to produce erythroid cell apoptosis. *Blood* **1998**, *91*, 1235–1242. [CrossRef]
57. Hubackova, S.; Krejcikova, K.; Bartek, J.; Hodny, Z. IL1- and TGFβ-Nox4 signaling, oxidative stress and DNA damage response are shared features of replicative, oncogene-induced, and drug-induced paracrine 'bystander senescence'. *Aging* **2012**, *4*, 932–951. [CrossRef]
58. Li, J.; Sejas, D.P.; Zhang, X.; Qiu, Y.; Nattamai, K.J.; Rani, R.; Rathbun, K.R.; Geiger, H.; Williams, D.A.; Bagby, G.C.; et al. TNF-alpha induces leukemic clonal evolution ex vivo in Fanconi anemia group C murine stem cells. *J. Clin. Investig.* **2007**, *117*, 3283–3295. [CrossRef]
59. Amer, J.; Dana, M.; Fibach, E. The antioxidant effect of erythropoietin on thalassemic blood cells. *Anemia* **2010**, *2010*, 978710. [CrossRef]
60. Beutler, E.; Blume, K.G.; Kaplan, J.C.; Löhr, G.W.; Ramot, B.; Valentine, W.N. International Committee for Standardization in Haematology: Recommended methods for red-cell enzyme analysis. *Br. J. Haematol.* **1977**, *35*, 331–340. [CrossRef]
61. Mojzikova, R.; Dolezel, P.; Pavlicek, J.; Mlejnek, P.; Pospisilova, D.; Divoky, V. Partial glutathione reductase deficiency as a cause of diverse clinical manifestations in a family with unstable hemoglobin (Hemoglobin Haná, β63(E7) His-Asn). *Blood Cells Mol. Dis.* **2010**, *45*, 219–222. [CrossRef] [PubMed]
62. Mojzikova, R.; Koralkova, P.; Holub, D.; Zidova, Z.; Pospisilova, D.; Cermak, J.; Striezencova Laluhova, Z.; Indrak, K.; Sukova, M.; Partschova, M.; et al. Iron status in patients with pyruvate kinase deficiency: Neonatal hyperferritinaemia associated with a novel frameshift deletion in the PKLR gene (p.Arg518fs), and low hepcidin to ferritin ratios. *Br. J. Haematol.* **2014**, *165*, 556–563. [CrossRef] [PubMed]
63. Gaikwad, A.; Nussenzveig, R.; Liu, E.; Gottshalk, S.; Chang, K.; Prchal, J.T. In vitro expansion of erythroid progenitors from polycythemia vera patients leads to decrease in JAK2 V617F allele. *Exp. Hematol.* **2007**, *35*, 587–595. [CrossRef] [PubMed]
64. Leberbauer, C.; Boulme, F.; Unfried, G.; Huber, J.; Beug, H.; Mullner, E.W. Different steroids co-regulate long-term expansion versus terminal differentiation in primary human erythroid progenitors. *Blood* **2005**, *105*, 85–94. [CrossRef] [PubMed]
65. Stetka, J.; Vyhlidalova, P.; Lanikova, L.; Koralkova, P.; Gursky, J.; Hlusi, A.; Flodr, P.; Hubackova, S.; Bartek, J.; Hodny, Z.; et al. Addiction to DUSP1 protects JAK2V617F-driven polycythemia vera progenitors against inflammatory stress and DNA damage, allowing chronic proliferation. *Oncogene* **2019**, *38*, 5627–5642. [CrossRef] [PubMed]
66. Muller, G.W.; Chen, R.; Huang, S.Y.; Corral, L.G.; Wong, L.M.; Patterson, R.T.; Chen, Y.; Kaplan, G.; Stirling, D.I. Amino-substituted thalidomide analogs: Potent inhibitors of TNF-α production. *Bioorg. Med. Chem. Lett.* **1999**, *9*, 1625–1630. [CrossRef]
67. Koledova, Z.; Kafkova, L.R.; Calabkova, L.; Krystof, V.; Dolezel, P.; Divoky, V. Cdk2 inhibition prolongs G1 phase progression in mouse embryonic stem cells. *Stem Cells Dev.* **2010**, *19*, 181–194. [CrossRef] [PubMed]
68. Horvathova, M.; Kapralova, K.; Zidova, Z.; Dolezal, D.; Pospisilova, D.; Divoky, V. Erythropoietin-driven signaling ameliorates the survival defect of DMT1-mutant erythroid progenitors and erythroblasts. *Haematologica* **2012**, *97*, 1480–1488. [CrossRef] [PubMed]
69. Schneider, C.A.; Rasband, W.S.; Eliceiri, K.W. NIH Image to ImageJ: 25 years of image analysis. *Nat. Methods* **2012**, *9*, 671–675. [CrossRef] [PubMed]
70. Pfaffl, M.W.; Horgan, G.W.; Dempfle, L. Relative expression software tool (REST) for group-wise comparison and statistical analysis of relative expression results in real-time PCR. *Nucleic Acids Res.* **2002**, *30*, e36. [CrossRef] [PubMed]

International Journal of
Molecular Sciences

Article

Modulatory Effects of Silymarin on Benzo[a]pyrene-Induced Hepatotoxicity

Seung-Cheol Jee, Min Kim and Jung-Suk Sung *

Department of Life Science, Dongguk University-Seoul, Biomedi Campus, 32 Dongguk-ro, Ilsandong-gu, Goyang-si, Gyeonggi-do 10326, Korea; markjee@naver.com (S.-C.J.); pipikimmin@naver.com (M.K.)
* Correspondence: sungjs@dongguk.edu; Tel.: +82-31-961-5132; Fax: +82-31-961-5108

Received: 13 March 2020; Accepted: 28 March 2020; Published: 30 March 2020

Abstract: Benzo[a]pyrene (B[a]P), a polycyclic aromatic hydrocarbon, is a group 1 carcinogen that introduces mutagenic DNA adducts into the genome. In this study, we investigated the molecular mechanisms underlying the involvement of silymarin in the reduction of DNA adduct formation by B[a]P-7,8-dihydrodiol-9,10-epoxide (BPDE), induced by B[a]P. B[a]P exhibited toxicity in HepG2 cells, whereas co-treatment of the cells with B[a]P and silymarin reduced the formation of BPDE-DNA adducts, thereby increasing cell viability. Determination of the level of major B[a]P metabolites in the treated cells showed that BPDE levels were reduced by silymarin. Nuclear factor erythroid 2-related factor 2 (Nrf2) and pregnane X receptor (PXR) were found to be involved in the activation of detoxifying genes against B[a]P-mediated toxicity. Silymarin did not increase the expression of these major transcription factors, but greatly facilitated their nuclear translocation. In this manner, treatment of HepG2 cells with silymarin modulated detoxification enzymes through NRF2 and PXR to eliminate B[a]P metabolites. Knockdown of Nrf2 abolished the preventive effect of silymarin on BPDE-DNA adduct formation, indicating that activation of the Nrf2 pathway plays a key role in preventing B[a]P-induced genotoxicity. Our results suggest that silymarin has anti-genotoxic effects, as it prevents BPDE-DNA adduct formation by modulating the Nrf2 and PXR signaling pathways.

Keywords: benzo[a]pyrene; BPDE-DNA adduct; silymarin; detoxification; Nrf2; PXR

1. Introduction

Benzo[a]pyrene (B[a]P) is a ubiquitous environmental pollutant produced during incomplete combustion from sources such as diesel engine exhaust, cigarette smoke, and industrial activities. B[a]P is also produced during certain types of food processing such as grilling and broiling [1]. It is classified by the International Agency for Research on Cancer (IARC) as a group I carcinogen [2,3]. Low-dose B[a]P is constantly absorbed into the body through the inhalation of polluted air, as well as the consumption of charbroiled food. Prolonged exposure to B[a]P accelerates metastasis and angiogenesis in the liver and induces cancer in the liver, lungs, skin, cervix, and the gastrointestinal tract (colorectal and stomach) [4–7]. Furthermore, polycyclic aromatic hydrocarbons such as B[a]P have been shown to possess very strong bioaccumulation characteristics in animal studies [8–10]. After exposure to B[a]P, cellular cytochrome P450 (CYP) metabolizes B[a]P to B[a]P-7,8-dihydrodiol-9,10-epoxide (BPDE), which interacts with DNA to form carcinogenic BPDE-DNA adducts in vitro and in vivo [11–13].

In the body, defense systems against xenobiotics such as BPDE include the activation of detoxifying phase II and III enzymes to prevent further cellular damage; specifically, glutathione S-transferases (GSTs) [14], nicotinamide adenine dinucleotide phosphate (NAD(P)H): quinone oxidoreductase 1 (NQO1), sulfotransferases (SULTs), and multidrug resistance-associated proteins (ABCCs) [15–17]. Previous studies showed that the genotoxicity of B[a]P was reduced by phase II enzymes conjugated with B[a]P metabolites and that B[a]P was excreted by ABCCs before BPDE-DNA adducts could form [18–20]. Nuclear factor erythroid 2-related factor 2 (Nrf2) and pregnane X receptor (PXR) are important transcriptional factors that regulate the expression of anti-genotoxic phase II detoxification enzymes and phase III transporters [18,19,21–23]. Generally, Nrf2 protein interacts with kelch-like ECH-associated protein 1 (Keap1) to form a dimer [24]. Exposure to endogenous activators such as reactive oxygen species (ROS) or exogenous agents such as electrophilic xenobiotics induces dissociation of the Nrf2 and Keap1 dimer, resulting in Nrf2 degradation by the proteasome and Nrf2 entry into the nucleus. After translocation into the nucleus, small musculoaponeurotic fibrosarcoma (Maf) proteins and other transcription factors help to activate and stabilize Nrf2, leading to the induction of phase II detoxifying enzymes [25,26].

Silymarin, a natural flavonoid, is a constituent of milk thistle (*Silybum marianum*). It is a metabolic regulator known to have anti-oxidant, anti-inflammatory, anti-cancer, anti-mutagenic, anti-bacterial, and anti-virus effects [27–30]. Furthermore, silymarin has multiple pharmacological activities and its components include silibinin, silydianine, and silychristin [27]. Recent studies showed that silymarin can reduce B[a]P-induced toxicity by reducing ROS formation in a rat model and can attenuate colorectal, liver, and lung cancer [28,31–33]. Another report showed that silymarin also exerts anti-oxidant effects by up-regulating NQO1 and heme oxygenase 1 genes through Nrf2 modulation [33]. However, B[a]P induces DNA damage through BPDE-DNA adduct formation as well as ROS formation, and the mechanism of action through which silymarin reduces the formation of BPDE-DNA adducts has not yet been clearly elucidated. In this study, we show the effect of silymarin on the reduction of genotoxicity through the regulation of B[a]P metabolites and the reduction of BPDE-DNA adduct formation via the modulation of Nrf2 and PXR signaling pathways.

2. Results

2.1. *Attenuation of B[a]P-Induced Cytotoxicity by Silymarin*

Regulation of the dose-dependent production of BPDE in human hepatocytes varies depending on the cell line used for BPDE-DNA adduct research [34]. Therefore, we used the well-characterized HepG2 cell line to study the potential preventive effects of silymarin on BPDE formation, thereby reducing the cell toxicity induced by B[a]P. The toxicity of B[a]P and silymarin on HepG2 cells was evaluated using cell viability assays. B[a]P induced cell death in a dose-dependent manner, whereas silymarin was non-toxic (up to 40 μM for 48 h) compared to no treatment (Figure 1A,B). To evaluate the protective effects of silymarin on B[a]P-induced cytotoxicity, B[a]P was co-applied to HepG2 cells with various concentrations of silymarin. Silymarin restored up to 90% of the cell viability in a dose-dependent manner by reducing B[a]P-induced cytotoxicity. This implies that silymarin has a protective effect against B[a]P-induced cytotoxicity (Figure 1C).

The potential protective effect of silymarin against B[a]P-induced genotoxicity was evaluated. BPDE-DNA adduct formation was measured with a BPDE-DNA adduct enzyme-linked immunosorbent assay (ELISA) kit after treatment with B[a]P (10 μM), silymarin (40 μM), and B[a]P (10 μM) co-administered with silymarin (40 μM). The results showed that B[a]P treatment alone increased the BPDE-DNA adduct level compared to that of the untreated control group. In contrast, B[a]P co-treatment with silymarin significantly decreased the BPDE-DNA adduct level compared to B[a]P treatment alone (Figure 1D). These results suggest that silymarin exerts an anti-genotoxic effect by reducing the formation of BPDE-DNA adducts.

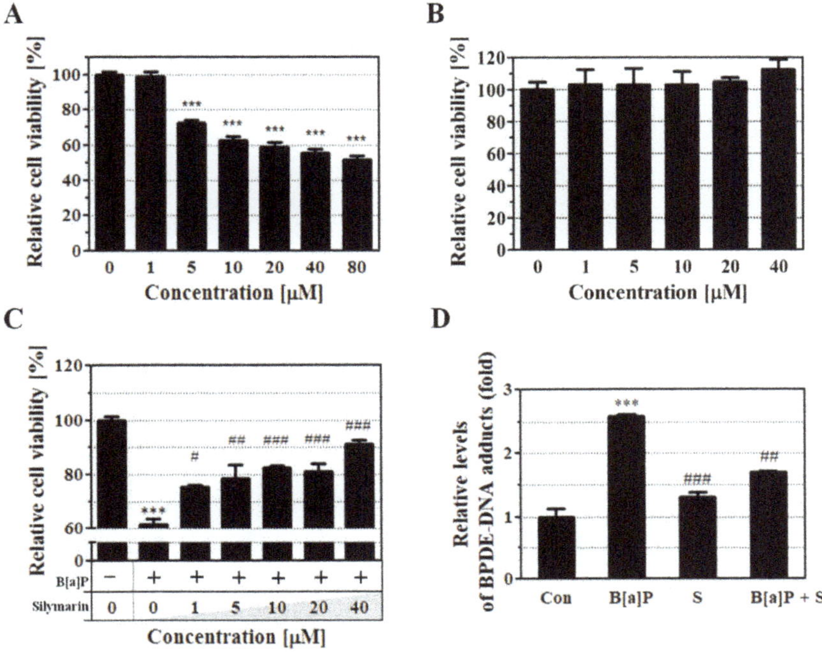

Figure 1. HepG2 cell viability was evaluated by cell viability assay. (**A**,**B**) HepG2 cells were treated with benzo[a]pyrene (B[a]P) or silymarin at various concentrations for 48 h. (**C**) B[a]P-induced cytotoxicity was reduced in cells treated with various concentrations of silymarin for 48 h. (**D**) The inhibitory effect of silymarin on B[a]P-7,8-dihydrodiol-9,10-epoxide (BPDE)-DNA adduct formation was measured by enzyme-linked immunosorbent assay (ELISA). HepG2 cells were treated with B[a]P (10 µM) in the presence or absence of silymarin (40 µM) for 48 h. All treatment group values were significantly different in comparison to the controls (*** $p < 0.001$) and to B[a]P (# $p < 0.05$, ## $p < 0.01$, and ### $p < 0.001$) in Tukey's multiple comparison test.

2.2. Reduction of Intracellular B[a]P Metabolites by Silymarin

B[a]P is sequentially metabolized to B[a]P-7,8-dihydrodiol and BPDE; BPDE induces genotoxicity by introducing mutagenic adducts into guanine [12]. The amount of B[a]P and its metabolites B[a]P-7,8-dihydrodiol and BPDE in the treated HepG2 cells were measured using a high performance liquid chromatography (HPLC) system. We found that the calculated amount of BPDE rapidly increased when the cells were treated with B[a]P alone. In contrast, BPDE levels decreased following B[a]P co-treatment with silymarin compared to the levels with B[a]P treatment alone, but the calculated amounts of B[a]P and B[a]P-7,8-dihydrodiol increased (Figure 2). These results suggest that silymarin reduces B[a]P-induced genotoxicity through the excretion of BPDE and inhibition of BPDE-DNA adduct formation.

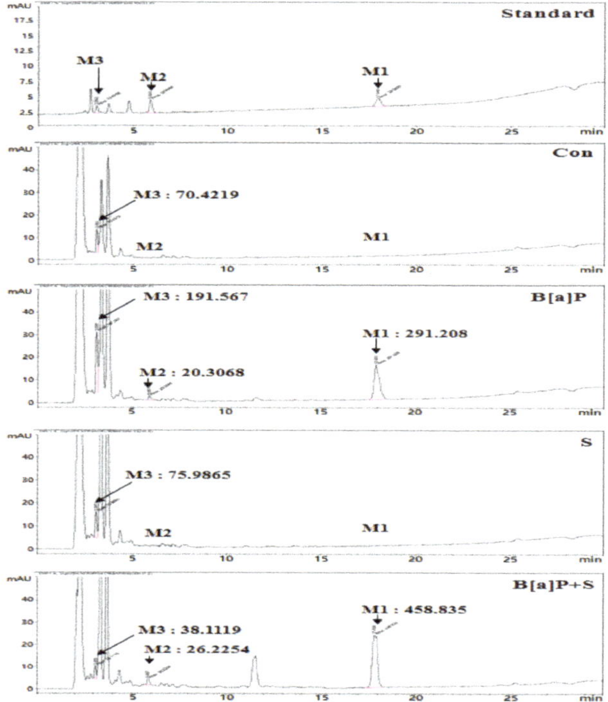

Figure 2. HepG2 cells were incubated with B[a]P (10 μM) and co-treated with silymarin (40 μM) for 48 h. The typical intracellular metabolites of B[a]P were measured by high performance liquid chromatography (HPLC). M1, B[a]P; M2, B[a]P-7,8-dihydrodiol; M3, BPDE.

2.3. Modulatory Effect of Silymarin on the Expression of Phase I, II, and III Enzymes

GST enzymes are important for phase II detoxification processes and for the activation of glutathione (GSH) conjugated with BPDE, which is eliminated by ABCCs [18,35]. ABCC1 is involved in the function of efflux pumps, which contribute to GSH-conjugated with xenobiotic (GS-X) excretion. GSH-BPDE conjugates in the cell are eliminated through excretion to the basolateral side by ABCC1 [20,36,37]. In the liver, GSH-BPDE conjugates are transferred to the blood from luminal surfaces through basolateral efflux by ABCC1, and then BPDE is terminally eliminated from the body [37]. The results of B[a]P application showed that B[a]P induced CYP1A1 expression but reduced GST and ABCC1 expression; however, B[a]P co-treatment with silymarin recovered the expression of GST and ABCC1 (Figure 3A–C). The activation of CYP1A1 and GSTs was evaluated using an activity assay kit. Greater activation of GSTs was observed with B[a]P co-administered with silymarin than with B[a]P treatment alone, whereas the activation of CYP1A1 decreased (Figure 3D,E). These data suggest that silymarin reduces B[a]P metabolism and induces BPDE excretion by regulating phase I, II, and III metabolizing enzymes.

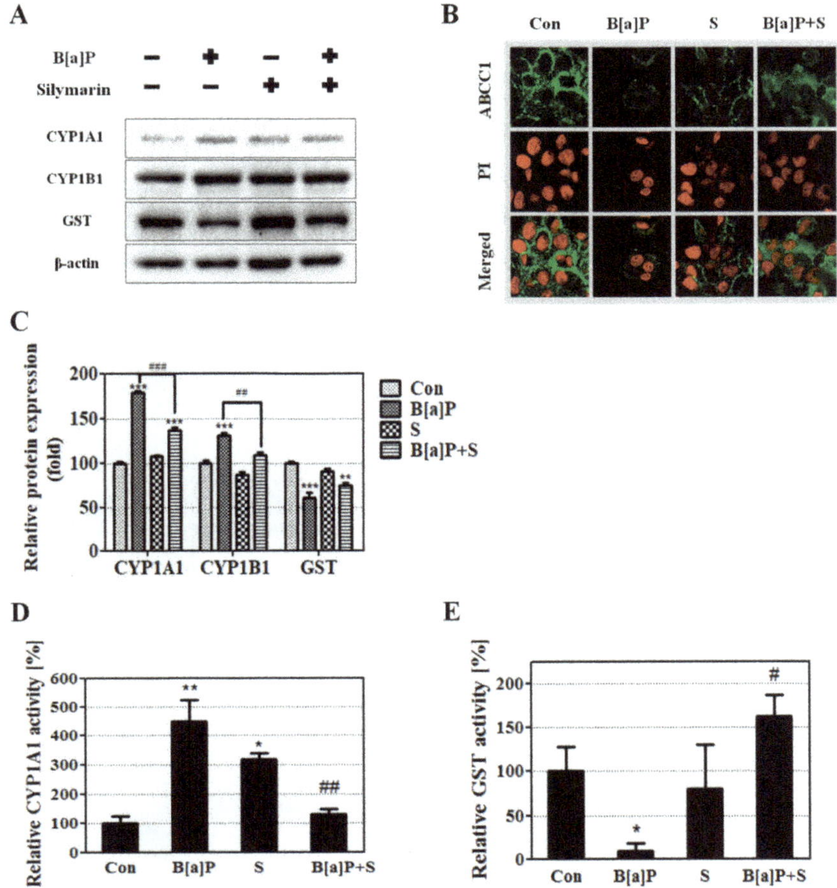

Figure 3. Expression of phase I, II, and III enzymes was induced by incubation for 48 h with B[a]P (10 μM) and co-treatment with silymarin (40 μM). (**A**) Cell lysates were prepared and the level of detoxifying enzymes was measured by Western blot; (**B**) The expression of multidrug resistance-associated protein 1 (ABCC1) was measured using immunocytochemistry. (**C**) Quantitative protein expression was calculated. (**D**,**E**) cellular cytochrome P450 (CYP)1A1 and glutathione S-transferase (GST) were activated and the relative levels of activation were calculated. All values are expressed as mean ± standard error of mean (SEM) (n = 3). The relative activation levels of CYP1A1 and GST were significantly different compared to those of the controls (* $p < 0.05$, ** $p < 0.01$, *** $p < 0.001$) and B[a]P (# $p < 0.05$, ## $p < 0.01$, ### $p < 0.001$) in Tukey's multiple comparison test.

2.4. *Stimulation of Nuclear Translocation of Nrf2 and PXR by Silymarin*

Nrf2 and PXR are well-known as inducers of phase II detoxification enzyme and phase III transporter expression [17,33,38–40]. To elucidate the pathways involved in the silymarin-mediated suppression of BPDE metabolites, the effects of Nrf2 and PXR translocation were evaluated using immunofluorescence staining. The results showed that the protein expression of Nrf2 decreased following B[a]P co-treatment with silymarin, whereas PXR expression was not significantly altered compared to the levels observed with B[a]P treatment alone (Figure 4A,B) However, B[a]P co-treatment with silymarin significantly increased the translocation of Nrf2 and PXR compared to B[a]P alone (Figure 4C,D). These results indicated that silymarin induces the translocation of Nrf2 and PXR, which are mainly involved in pathways related to the elimination of BPDE metabolites.

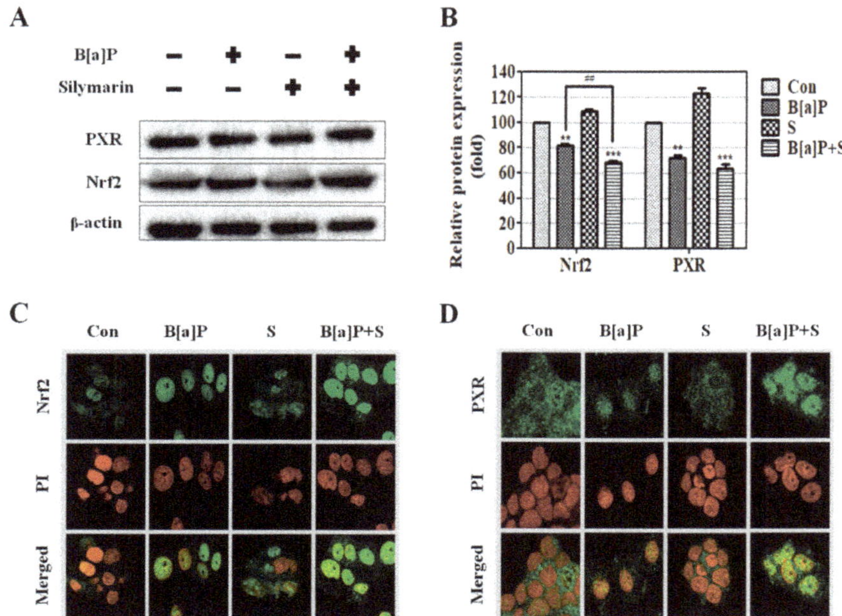

Figure 4. HepG2 cells were treated with B[a]P (10 μM), silymarin (40 μM), and both B[a]P (10 μM) and silymarin (40 μM). (**A**) Nuclear factor erythroid 2-related factor 2 (Nrf2) and pregnane X receptor (PXR) expression was measured using Western blot, and (**B**) the quantitative protein levels were calculated. (**C**,**D**) The translocation levels were measured by immunocytochemistry. DNA was detected by propidium iodide (PI) staining (red).

2.5. Reduction of BPDE-DNA Adduct by Silymarin is Dependent on Nrf2

Previous results showed that silymarin reduces BPDE-DNA adduct formation through the Nrf2 signaling pathway. We confirmed that the preventive effect of Nrf2 was inhibited by knockdown of Nrf2 expression using specific small interfering RNA (siRNA). Depletion of Nrf2 was confirmed by Western blot (Figure 5A). Co-treatment of HepG2 cells with B[a]P and silymarin resulted in the reduction of BPDE-DNA adduct formation with control siRNA (siCon). However, this effect of silymarin on BPDE-DNA adduct formation was eliminated by Nrf2 knockdown (Figure 5B). These results confirmed that BPDE-DNA adduct formation was mainly reduced by silymarin via the up-regulation of Nrf2 translocation.

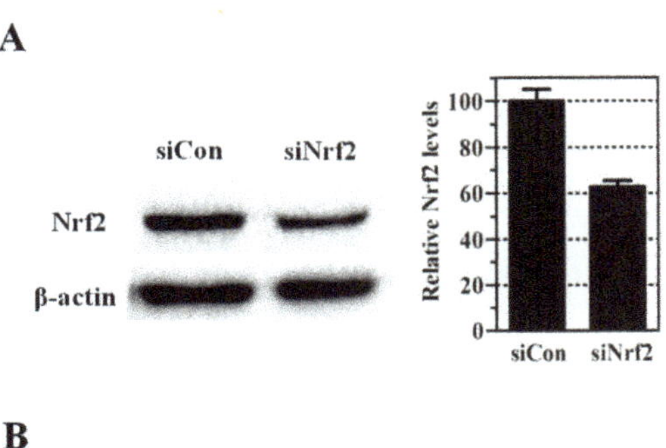

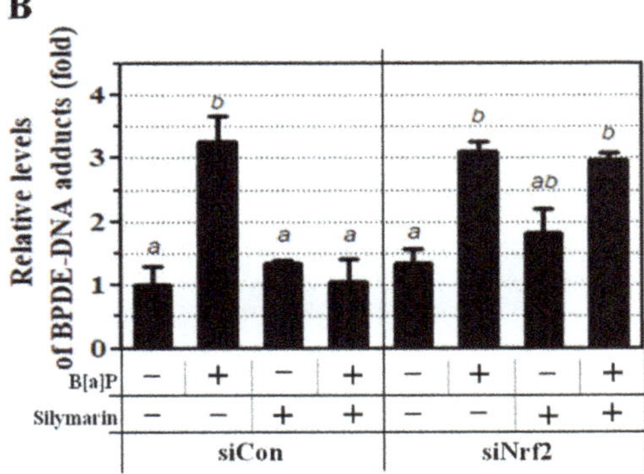

Figure 5. Knockdown of HepG2 cells was performed using Nrf2-small interfering RNA (siRNA) (siNrf2) and control siRNA (siCon). (**A**) Knockdown of cells was confirmed by Western blot. (**B**) Transfected HepG2 cells were exposed to B[a]P for 48 h. BPDE-DNA adduct formation was measured by enzyme-linked immunosorbent assay (ELISA). Each letter indicates significantly ($p < 0.05$) different values (a, b, ab—mean values with the same letters for each level are not significantly different).

3. Discussion

B[a]P is a group I carcinogen that is widely produced in cigarette smoke, charbroiled food, and other sources. It activates several types of cancer, mainly by causing DNA damage through BPDE-DNA adduct and ROS formation [41,42]. Unfortunately, we are constantly exposed to low-doses of B[a]P, which induces cancer in the body through its conversion to the BPDE metabolite [43,44]. Research on the protective effects of natural compounds against B[a]P-induced damage is currently ongoing to identify methods to prevent cancer [45–49].

Silymarin, a natural compound derived from milk thistle (*Silybum marianum*) is widely used in health supplements, medicines, and medical supplies related to liver disease and cancer, and has no reported side-effects [50–52]. Previous studies on silymarin focused on its role as an anti-oxidant and its anti-inflammatory properties against B[a]P-induced ROS formation via modulating detoxifying enzymes in vivo [28]. However, B[a]P-induced genotoxicity is mainly caused by damage from BPDE-DNA adducts as well as ROS-based DNA damage. Additionally, the metabolic counteractive effects of silymarin on B[a]P-induced toxicity have not yet been studied. In this study, we provide evidence that silymarin reduces B[a]P-induced genotoxicity by modulating phase detoxification enzymes and eliminating B[a]P metabolites via the Nrf2 and PXR signaling pathways. Our results show that co-application of B[a]P with silymarin recovers the cell viability levels (Figure 1C). Previous reports showed that silymarin possesses anti-oxidant effects and attenuates B[a]P-induced ROS damage [27,53]. However, we focused on the attenuation of BPDE-DNA adduct formation by silymarin in this study. Our results provide new insight that novel mechanisms of silymarin modulate the attenuation of BPDE, facilitating anti-genotoxic effects against B[a]P. BPDE is the ultimate metabolite of B[a]P and is produced through genotoxic interactions with DNA to induce BPDE-DNA adduct formation. The present study confirms that silymarin inhibits BPDE-DNA adduct formation compared to the effects of B[a]P alone (Figure 1D). This result indicates that silymarin attenuates B[a]P-induced genotoxicity by inhibiting BPDE-DNA adduct formation.

To confirm the mechanism of BPDE-DNA adduct formation by silymarin, we evaluated the amount of B[a]P metabolites produced. Silymarin co-treatment with B[a]P resulted in greater production of B[a]P and B[a]P-7,8-dihydrodiol than B[a]P alone. However, the amount of BPDE was markedly decreased (Figure 2), consistent with the formation of fewer BPDE-DNA adducts as the amount of BPDE decreased (Figure 1D). Additionally, the results suggest that silymarin reduces B[a]P metabolism by inhibiting the transition of B[a]P to BPDE. A previous report indicated that direct exposure of cells to BPDE resulted in instant formation of DNA adducts, whereas B[a]P-exposed cells required multiple enzymatic steps for B[a]P conversion to BPDE [54]. This result indicates that the inhibition of BPDE formation is important to prevent B[a]P-induced genotoxicity by modulating the necessary enzymatic steps. Therefore, we hypothesized that the inhibition of B[a]P conversion to BPDE will be regulated by enzymatic steps.

CYP1A1 is considered as playing a key role in B[a]P metabolism involving both BPDE formation and reduction [55]. Further, abnormal activation of phase I enzymes may have an adverse effect by inducing toxicity and causing cancer in the body [56]. Our results showed that co-treatment with silymarin reduced CYP1A1 and CYP1B1 protein level compared to B[a]P treatment alone (Figure 3A,C). Moreover, silymarin reduced CYP1A1 activity to levels similar to those of the control group (Figure 3D). These results indicated that CYP1A1 and 1B1 contribute to the inhibition of both B[a]P metabolism and BPDE generation by silymarin.

Xenobiotics such as B[a]P are mainly converted to water-soluble metabolites that are easily eliminated from the body [57]. These defense systems are modulated by phase detoxification enzymes such as GSTs, SULTs, uridine 5′-diphospho-glucuronosyltransferases (UGTs), and ABCCs that induce the excretion of toxicants through urine [58]. Previous studies showed that B[a]P-induced genotoxicity was reduced by the modulation of phase detoxifying enzymes before BPDE-DNA adduct formation [17,47]. GSTs are typical phase II detoxifying enzymes and are important in reducing B[a]P-induced DNA damage by inhibiting the formation of DNA adducts and 8-oxo-G [59]. Additionally, GSTs are known to promote GSH conjugation with BPDE and inhibit BPDE-DNA adduct formation [60]. Our results showed that the expression and activity of GSTs were promoted by silymarin compared to the observations with B[a]P alone (Figure 3A,C,E). Reduced BPDE production is caused by two factors: (1) inhibition of B[a]P conversion to BPDE by the regulation of CYP1A1 and CYP1B1, and (2) elimination of BPDE metabolites via the induction of BPDE conjugation with GSH through the modulation of GST.

GSH conjugated with metabolites enhances the water solubility of B[a]P metabolites, facilitating the excretion of BPDE by ABCCs [61]. Previous studies showed that BPDE was detoxified when conjugated with GSH and excreted by ABCCs [20,62]. Specifically, ABCC1 and ABCC2 are mainly involved in the excretion of GSH-BPDE conjugates, with ABCC2 mediating the apical excretion and ABCC1 mediating the basolateral excretion [20]. We confirmed that ABCC1 mRNA expression, but not ABCC2 mRNA expression, is regulated by silymarin based on microarray data (data not shown). Another study also showed that the prevention of xenobiotic toxicity by ABCC1 is important in several types of tissues and resulted in the direct excretion of GS-X [63,64]. Our results demonstrated that the level of ABCC1 is reduced by B[a]P treatment and restored by co-treatment with silymarin (Figure 3B). These results indicate that the mechanism underlying the reduction of B[a]P-induced genotoxicity involves the activation of GST to enhance GSH conjugation with BPDE, excretion of BPDE into the blood from the liver by ABCC1, and elimination from the body through urine. As a consequence, silymarin induces the up-regulation of GST and ABCC1, facilitating BPDE conjugation with GSH and excretion of the conjugate, and inhibits the conversion of BPDE from B[a]P by regulating CYP1A1 and

CYP1B1. Therefore, these results confirm that silymarin reduces B[a]P-induced genotoxicity through the inhibition of BPDE-DNA adduct formation.

Aryl hydrocarbon receptor (AhR), one of major transcriptional factor of phase I enzymes, stimulates expression of CYP enzymes by binding with xenobiotic response element (XRE) [65,66] and induces to B[a]P metabolism. We confirmed that B[a]P transition to BPDE was decreased (Figure 2) whereas CYP1A1 and 1B1 mRNA level were increased by silymarin. Our results indicate that the anti-genotoxic effect of silymarin is due to the activity of CYP enzymes, not mRNA expression level. Nrf2 and PXR are major transcriptional factors of phase II detoxification enzymes and phase III transporters, and they are reported to modulate GSTs, NQO1, UGTs, and ABCCs [67–69]. A previous study demonstrated that Nrf2 plays a critical role in drug metabolism by regulating phase I, II, and III enzymes in the liver [70]. In addition to this, Nrf2 is important for maintaining GSH and regulating the expression of phase I, II, and III enzymes, thereby facilitating the elimination of xenobiotics [71]. Moreover, nuclear translocation of Nrf2 interactions with antioxidant response element (ARE) induces phase detoxifying enzymes [71]. Our results show that the expression of Nrf2 by B[a]P is slightly decreased by co-exposure with silymarin, whereas nuclear translocation of Nrf2 is increased significantly (Figure 4A–C). Accordingly, our results suggest that silymarin induces phase detoxifying enzymes via Nrf2. A previous study showed that PXR regulated the detoxifying enzymes against B[a]P and reduced toxicity in the liver [22]. This report indicates that PXR is also related to the anti-genotoxic effect of silymarin. Therefore, these results suggest that increased Nrf2 and PXR translocation regulates phase detoxifying enzymes such as GSTs and ABCC1, which then facilitate the elimination of B[a]P metabolites.

A previous study showed that the inhibition of Nrf2 increased B[a]P toxicity and decreased the expression of several GSTs [72]. To confirm that silymarin regulates the Nrf2 signaling pathway, which is related to the attenuation of BPDE-DNA adduct formation, we demonstrated that knockdown of Nrf2 expression by RNA interference abolishes the inhibition of BPDE-DNA adduct formation by silymarin (Figure 5B). This indicates that the induction of Nrf2 translocation by silymarin plays a key regulatory role against B[a]P-induced genotoxicity. Previous studies supported that Nrf2 was closely related to the induction of GSTs and ABCC1 through the regulation of Nrf2 expression [73,74]. These reports support our findings that silymarin modulates Nrf2, which causes B[a]P detoxification through the regulation of phase II and III enzymes. These results confirm that silymarin exerts anti-genotoxic effects against B[a]P-induced toxicity through the up-regulation of Nrf2 and PXR translocation (Figure 6).

Previous reports indicated that silymarin attenuated DNA damage by reducing ROS damage. However, BPDE also causes DNA damage by forming BPDE-DNA adducts, which induce toxic and pathologic cellular changes. Our results clearly show that the signaling pathways inhibit B[a]P conversion to BPDE by modulating phase I enzymes and accelerate BPDE conjugation with GSH by regulating phase detoxification enzymes that counteract B[a]P metabolites. In this manner, silymarin can reduce B[a]P-induced genotoxicity by reducing DNA damage through the inhibition of BPDE interactions with DNA.

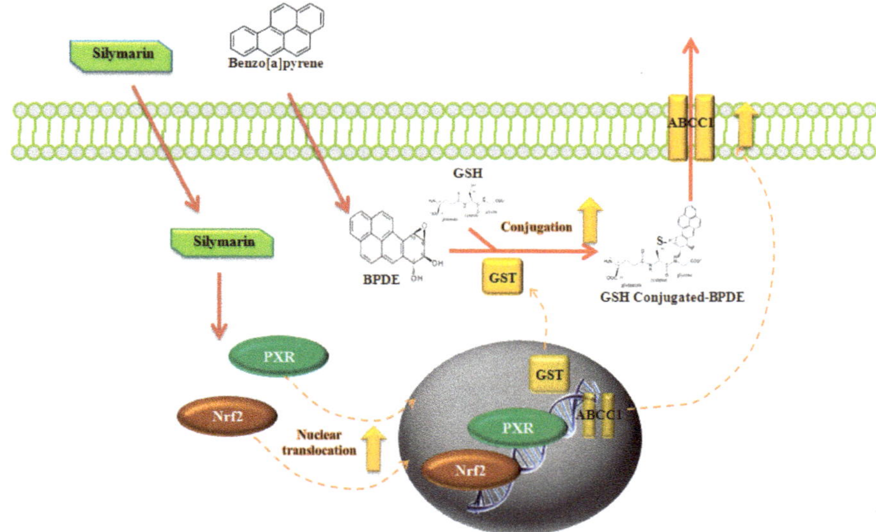

Figure 6. Schematic of reduction of B[a]P-induced genotoxicity by silymarin through Nrf2 and PXR. Silymarin induces Nrf2 and PXR translocation and can regulate the expression of GSTs and ABCC1, enhancing GSH conjugation with BPDE to facilitate the excretion process.

4. Methods

4.1. Chemicals and Reagents

Silymarin, B[a]P, dimethyl sulfoxide (DMSO), 4,6-diamidino-2-phenylindole dihydrochloride (DAPI), and Triton X-100 were purchased from Sigma-Aldrich Chemical (St. Louis, MO, USA). Minimum essential medium (MEM), fetal bovine serum (FBS), penicillin/streptomycin, trypsin-ethylenediaminetetraacetic acid (EDTA), and sodium pyruvate were purchased from Welgene (Daegu, Korea). Phosphate-buffered saline was purchased from Biosesang (Seongnam, Korea). Fluorescent mounting medium was purchased from Dako (Carpinteria, CA, USA). Antibodies (anti-Nrf2, PXR, GST, CYP1A1, ABCC1, and β-actin), horseradish peroxidase-conjugated anti-rabbit immunoglobulin G (IgG) and siRNA transfection reagent, siRNA transfection medium, control siRNA, and Nrf2-siRNA were purchased from Santa Cruz Biotechnology (Santa Cruz, CA, USA). Alexa 488-conjugated anti-rabbit secondary antibody was purchased from Cell Signaling Technology (Beverly, MA, USA).

4.2. Cell Culture and Treatment

The HepG2 cell line was purchased from the American Type Culture Collection (Manassas, VA, USA). HepG2 cells were cultured in a 100 mm^2 cell culture dish with MEM containing 10% FBS, 100 U/mL penicillin, 100 μg/mL streptomycin, and 1 mM sodium pyruvate at 37°C in a humidified atmosphere of 5% CO_2. To determine the effects of the treatment conditions, the cells were incubated with various concentrations of B[a]P and silymarin in MEM for 48 h. The cells were subjected to further analysis or harvested to prepare cell-free extracts.

4.3. Cell Viability Assay

We performed cell viability assays to evaluate the general cytotoxicity of B[a]P and silymarin on HepG2 cells. HepG2 cells with a density of 1 x 10^4 cells/well with MEM were seeded in 96-well plates with B[a]P (0, 1, 5, 10, 20, 40 μM) or silymarin (0, 1, 5, 10, 20, 40 μM) for 48 h. EZ-CYTOX reagent

(DOGEN, Daejeon, Korea) was added to each well and the cells were incubated for 2 h. Absorbance measurement at 450 nm was carried out by using a microplate reader (Molecular Devices, San Jose, CA, USA) and the cell viability levels of the B[a]P and silymarin treatment groups were evaluated compared to those of the non-treatment groups.

4.4. BPDE-DNA Adduct Formation Analysis

HepG2 cells were seeded with MEM in 6-well plates and were treated with 10 μM B[a]P in the presence or absence of 40 μM silymarin for 48 h. DNA was extracted from the HepG2 cells at the end of the treatment period using the QIAamp DNA Mini Kit (Qiagen, Stanford, CA, USA) according to the manufacturer's instructions. The isolated DNA was analyzed for BPDE-DNA adduct formation using a BPDE-DNA adduct ELISA kit (Cell Biolabs, San Diego, CA, USA) according to the manufacturer's instructions. The relative BPDE-DNA adduct levels were measured using a microplate reader with absorbance at 450 nm.

4.5. Metabolite Extraction and HPLC Analysis Conditions

To extract metabolites, cells treated with B[a]P and/or silymarin in 100 mm^2 cell culture dishes for 48 h were dissolved with ethyl acetate and homogenized. The dissolved cells were evaporated using a vacuum centrifuge and re-dissolved with 50% acetonitrile/0.1% acetic acid. After this, 30 μL was injected into an Agilent HPLC Hewlett Packard 1100 series (Hewlett-Packard, Palo Alto, CA, USA). Chromatography was performed using a Kinetex C18 column (4.6 mm X 250 mm, 5 μm, Phenomenex, Torrance, CA, USA) at 25 °C with a flow rate of 1.0 mL/min. HPLC separation was performed using the following linear gradient: 25 to 30 min for 100% acetonitrile in 0.1% acetic acid, and 0 to 40 min for 50% acetonitrile in 0.1% acetic acid (with solvent A, 0.1% acetic acid in distilled water and solvent B, 0.1% acetic acid/acetonitrile in distilled water). For each sample, the retention time and fragmentation patterns were obtained from reference standards of B[a]P, B[a]P-7,8,dihydrodiol, and BPDE, which were purchased from the MRIGlobal Chemical Carcinogen Repository (Kansas City, MO, USA).

4.6. CYP1A1 Activity Assay

HepG2 cells were seeded at a density of 1×10^4 cells into each well of 96-well plates with MEM and were treated with 10 μM B[a]P in the absence or presence of 40 μM silymarin for 48 h. CYP1A1 activity was measured using a CYP1A1 activity assay kit (Promega, Madison, WI, USA) according to the manufacturer's instructions. The luminescence level of CYP1A1 was measured using an Infinite 200 PRO multi-well plate reader (Tecan, Mannedorf, Switzerland).

4.7. GST Activity Assay

GST activity was measured using a GST activity assay kit (Cayman, Ann Arbor, MI, USA) according to the manufacturer's instructions. HepG2 cells were seeded with MEM in 96-well plates at a density of 1×10^4 cells in each well and were treated with 10 μM B[a]P in the absence or presence of 40 μM silymarin for 48 h. The relative activation of GST was measured by using an Infinite 200 PRO multi-well plate reader (Tecan, Mannedorf, Switzerland).

4.8. Immunofluorescence Staining

HepG2 cells were seeded on coverslips in a 6-well cell culture plate. The cells were fixed with 4% formaldehyde for 15 min and treated with 0.25% Triton X-100 after the formaldehyde was discarded. The cells were sequentially incubated with primary antibody for 1 h and Alexa 488-conjugated anti-rat secondary antibody for 1 h. The cells with attached antibody were stained with DAPI. After DAPI staining, fluorescence mounting medium was applied to a glass slide on which was placed a coverslip with the attached cells. The fluorescence images were obtained by confocal microscopy (Olympus, Tokyo, Japan) and quantitative analysis was performed using ImageJ software (Bethesda, MD, USA).

4.9. Western Blot Analysis

Total cell lysates were prepared in radioimmunoprecipitation assay (RIPA) buffer (50 mM Tris-HCl pH 7.4, 150 mM NaCl, 1% Nonidet P-40, 0.25% sodium deoxycholate) containing protease inhibitor cocktail, phosphatase inhibitor cocktail 2, and phosphatase inhibitor cocktail 3 (Sigma-Aldrich, St. Louis, MO, USA). Total cell proteins (30 μg) were separated by 10% sodium dodecyl sulfate-polyacrylamide gel electrophoresis, transferred to a polyvinylidene difluoride membrane (Millipore, Billerica, MA, USA), and hybridized with their respective primary antibodies. The membrane was incubated with secondary antibodies and the immunoreactive proteins bound to the antibodies were detected with enhanced chemiluminescent (ECL) Plus Western blotting detection reagents (Amersham Bioscience, Buckinghamshire, United Kingdom). Images were acquired using a Bio-Rad ChemiDoc XRS (Hercules, CA, USA) and quantified by Quantity One Image Software (Hercules, CA, USA).

4.10. Statistical Analysis

Each experiment was repeated at least three times. All data were expressed as means ± standard error of the mean (SEM). Significant differences among groups were determined by using one-way ANOVA with Tukey's multiple comparison test. Statistical significance was considered as $p < 0.05$.

5. Conclusions

In summary, our results show that the modulation of B[a]P detoxification reduces intracellular B[a]P metabolites, and thereby prevents the formation of BPDE-DNA adducts. Silymarin reduces BPDE by attenuating B[a]P conversion to BPDE and accelerates BPDE detoxification and elimination by modulating phase I, II, and III enzymes via Nrf2 and PXR pathways. In particular, knockdown of the Nrf2 gene abolishes the attenuation of BPDE-DNA adduct formation by silymarin. We confirmed that the Nrf2 signaling pathway is mainly related to the inhibition of BPDE-DNA adduct formation by silymarin. These results suggest that silymarin has anti-genotoxicity properties against B[a]P through the inhibition of BPDE-DNA adduct formation.

Author Contributions: S.-C.J. designed the study, performed the experiments, analyzed the data, and wrote the manuscript. M.K. performed the experiments and analyzed the data. J.-S.S. designed the study and contributed to the writing of the manuscript. All authors have read and agreed to the published version of the manuscript

Funding: This research was supported by a grant (14162MFDS072) from the Ministry of Food and Drug Safety (MFDS, Korea, 2017) and the Dongguk University Research Fund of 2019.

Acknowledgments: We would like to thank Jung at the Korea Basic Science Institute (KBSI, Seoul) for acquiring the HPLC data on the B[a]P metabolites.

Conflicts of Interest: The authors declare no conflict of interest.

References

1. Lee, B.M.; Shim, G.A. Dietary exposure estimation of benzo[a]pyrene and cancer risk assessment. *J. Toxicol. Env. Health A* **2007**, *70*, 1391–1394. [CrossRef] [PubMed]
2. Liu, F.; Liu, J.; Chen, Q.; Wang, B.; Cao, Z. Pollution characteristics, ecological risk and sources of polycyclic aromatic hydrocarbons (PAHs) in surface sediment from Tuhai-Majia River system, China. *Procedia Environ. Sci.* **2012**, *13*, 1301–1314. [CrossRef]
3. Straif, K.; Baan, R.; Grosse, Y.; Secretan, B.; El Ghissassi, F.; Cogliano, V. Carcinogenicity of polycyclic aromatic hydrocarbons. *Lancet Oncol.* **2005**, *6*, 931–932. [CrossRef]
4. Some non-heterocyclic polycyclic aromatic hydrocarbons and some related exposures. *Iarc. Monogr. Eval. Carcinog Risks Hum.* **2010**, *92*, 1–853.
5. Ba, Q.; Li, J.; Huang, C.; Qiu, H.; Chu, R.; Zhang, W.; Xie, D.; Wu, Y.; Wang, H. Effects of benzo[a]pyrene exposure on human hepatocellular carcinoma cell angiogenesis, metastasis, and NF-kappaB signaling. *Env. Health Perspect* **2015**, *123*, 246–254. [CrossRef]

6. Sinha, R.; Kulldorff, M.; Gunter, M.J.; Strickland, P.; Rothman, N. Dietary benzo[a]pyrene intake and risk of colorectal adenoma. *Cancer Epidemiol Biomark. Prev.* **2005**, *14*, 2030–2034. [CrossRef] [PubMed]
7. Samir, A.M.; Shaker, D.A.; Fathy, M.M.; Hafez, S.F.; Abdullatif, M.M.; Rashed, L.A.; Alghobary, H.A.F. Urinary and Genetic Biomonitoring of Polycyclic Aromatic Hydrocarbons in Egyptian Coke Oven Workers: Associations between Exposure, Effect, and Carcinogenic Risk Assessment. *Int. J. Occup. Env. Med.* **2019**, *10*, 124–136. [CrossRef] [PubMed]
8. Costa, J.; Ferreira, M.; Rey-Salgueiro, L.; Reis-Henriques, M.A. Comparision of the waterborne and dietary routes of exposure on the effects of Benzo(a)pyrene on biotransformation pathways in Nile tilapia (Oreochromis niloticus). *Chemosphere* **2011**, *84*, 1452–1460. [CrossRef] [PubMed]
9. Fasulo, S.; Marino, S.; Mauceri, A.; Maisano, M.; Giannetto, A.; D'Agata, A.; Parrino, V.; Minutoli, R.; De Domenico, E. A multibiomarker approach in Coris julis living in a natural environment. *Ecotoxicol. Env. Saf.* **2010**, *73*, 1565–1573. [CrossRef]
10. De Domenico, E.; Mauceri, A.; Giordano, D.; Maisano, M.; Gioffre, G.; Natalotto, A.; D'Agata, A.; Ferrante, M.; Brundo, M.V.; Fasulo, S. Effects of "in vivo" exposure to toxic sediments on juveniles of sea bass (Dicentrarchus labrax). *Aquat. Toxicol.* **2011**, *105*, 688–697. [CrossRef]
11. Briede, J.J.; Godschalk, R.W.; Emans, M.T.; De Kok, T.M.; Van Agen, E.; Van Maanen, J.; Van Schooten, F.J.; Kleinjans, J.C. In vitro and in vivo studies on oxygen free radical and DNA adduct formation in rat lung and liver during benzo[a]pyrene metabolism. *Free Radic. Res.* **2004**, *38*, 995–1002. [CrossRef] [PubMed]
12. Phillips, D.H.; Venitt, S. DNA and protein adducts in human tissues resulting from exposure to tobacco smoke. *Int. J. Cancer* **2012**, *131*, 2733–2753. [CrossRef]
13. Baird, W.M.; Hooven, L.A.; Mahadevan, B. Carcinogenic polycyclic aromatic hydrocarbon-DNA adducts and mechanism of action. *Env. Mol. Mutagen* **2005**, *45*, 106–114. [CrossRef] [PubMed]
14. Gundacker, C.; Neesen, J.; Straka, E.; Ellinger, I.; Dolznig, H.; Hengstschlager, M. Genetics of the human placenta: Implications for toxicokinetics. *Arch. Toxicol.* **2016**, *90*, 2563–2581. [CrossRef] [PubMed]
15. Perumal Vijayaraman, K.; Muruganantham, S.; Subramanian, M.; Shunmugiah, K.P.; Kasi, P.D. Silymarin attenuates benzo(a)pyrene induced toxicity by mitigating ROS production, DNA damage and calcium mediated apoptosis in peripheral blood mononuclear cells (PBMC). *Ecotoxicol. Env. Saf.* **2012**, *86*, 79–85. [CrossRef] [PubMed]
16. Guengerich, F.P. Roles of cytochrome P-450 enzymes in chemical carcinogenesis and cancer chemotherapy. *Cancer Res.* **1988**, *48*, 2946–2954. [PubMed]
17. Yuan, L.; Lv, B.; Zha, J.; Wang, W.; Wang, Z. Basal and benzo[a]pyrene-induced expression profile of phase I and II enzymes and ABC transporter mRNA in the early life stage of Chinese rare minnows (Gobiocypris rarus). *Ecotoxicol. Env. Saf.* **2014**, *106*, 86–94. [CrossRef]
18. Poon, P.Y.; Kwok, H.H.; Yue, P.Y.; Yang, M.S.; Mak, N.K.; Wong, C.K.; Wong, R.N. Cytoprotective effect of 20S-Rg3 on benzo[a]pyrene-induced DNA damage. *Drug Metab. Dispos.* **2012**, *40*, 120–129. [CrossRef]
19. Kranz, J.; Hessel, S.; Aretz, J.; Seidel, A.; Petzinger, E.; Geyer, J.; Lampen, A. The role of the efflux carriers Abcg2 and Abcc2 for the hepatobiliary elimination of benzo[a]pyrene and its metabolites in mice. *Chem. Biol. Interact.* **2014**, *224*, 36–41. [CrossRef]
20. Hessel, S.; John, A.; Seidel, A.; Lampen, A. Multidrug resistance-associated proteins are involved in the transport of the glutathione conjugates of the ultimate carcinogen of benzo[a]pyrene in human Caco-2 cells. *Arch. Toxicol.* **2013**, *87*, 269–280. [CrossRef]
21. Xu, C.; Li, C.Y.; Kong, A.N. Induction of phase I, II and III drug metabolism/transport by xenobiotics. *Arch. Pharm. Res.* **2005**, *28*, 249–268. [CrossRef] [PubMed]
22. Naspinski, C.; Gu, X.; Zhou, G.D.; Mertens-Talcott, S.U.; Donnelly, K.C.; Tian, Y. Pregnane X receptor protects HepG2 cells from BaP-induced DNA damage. *Toxicol. Sci.* **2008**, *104*, 67–73. [CrossRef] [PubMed]
23. Kim, K.C.; Kang, K.A.; Zhang, R.; Piao, M.J.; Kim, G.Y.; Kang, M.Y.; Lee, S.J.; Lee, N.H.; Surh, Y.J.; Hyun, J.W. Up-regulation of Nrf2-mediated heme oxygenase-1 expression by eckol, a phlorotannin compound, through activation of Erk and PI3K/Akt. *Int. J. Biochem. Cell. Biol.* **2010**, *42*, 297–305. [CrossRef]
24. Furukawa, M.; Xiong, Y. BTB protein Keap1 targets antioxidant transcription factor Nrf2 for ubiquitination by the Cullin 3-Roc1 ligase. *Mol Cell Biol* **2005**, *25*, 162–171. [CrossRef] [PubMed]
25. Menegon, S.; Columbano, A.; Giordano, S. The Dual Roles of NRF2 in Cancer. *Trends Mol. Med.* **2016**, *22*, 578–593. [CrossRef]

26. Osburn, W.O.; Kensler, T.W. Nrf2 signaling: An adaptive response pathway for protection against environmental toxic insults. *Mutat Res.* **2008**, *659*, 31–39. [CrossRef]
27. Kiruthiga, P.V.; Shafreen, R.B.; Pandian, S.K.; Arun, S.; Govindu, S.; Devi, K.P. Protective effect of silymarin on erythrocyte haemolysate against benzo(a)pyrene and exogenous reactive oxygen species (H2O2) induced oxidative stress. *Chemosphere* **2007**, *68*, 1511–1518. [CrossRef]
28. Kiruthiga, P.V.; Karthikeyan, K.; Archunan, G.; Pandian, S.K.; Devi, K.P. Silymarin prevents benzo(a)pyrene-induced toxicity in Wistar rats by modulating xenobiotic-metabolizing enzymes. *Toxicol. Ind. Health* **2015**, *31*, 523–541. [CrossRef]
29. Kiruthiga, P.V.; Shafreen, R.B.; Pandian, S.K.; Devi, K.P. Silymarin protection against major reactive oxygen species released by environmental toxins: Exogenous H2O2 exposure in erythrocytes. *Basic Clin. Pharm. Toxicol.* **2007**, *100*, 414–419. [CrossRef]
30. Bektur Aykanat, N.E.; Kacar, S.; Karakaya, S.; Sahinturk, V. Silymarin suppresses HepG2 hepatocarcinoma cell progression through downregulation of Slit-2/Robo-1 pathway. *Pharm. Rep.* **2020**, *72*, 199–207. [CrossRef]
31. Eo, H.J.; Park, G.H.; Song, H.M.; Lee, J.W.; Kim, M.K.; Lee, M.H.; Lee, J.R.; Koo, J.S.; Jeong, J.B. Silymarin induces cyclin D1 proteasomal degradation via its phosphorylation of threonine-286 in human colorectal cancer cells. *Int. Immunopharmacol.* **2015**, *24*, 1–6. [CrossRef] [PubMed]
32. Feher, J.; Lengyel, G. Silymarin in the prevention and treatment of liver diseases and primary liver cancer. *Curr. Pharm. Biotechnol.* **2012**, *13*, 210–217. [CrossRef] [PubMed]
33. Podder, B.; Kim, Y.S.; Zerin, T.; Song, H.Y. Antioxidant effect of silymarin on paraquat-induced human lung adenocarcinoma A549 cell line. *Food Chem. Toxicol.* **2012**, *50*, 3206–3214. [CrossRef] [PubMed]
34. Genies, C.; Maitre, A.; Lefebvre, E.; Jullien, A.; Chopard-Lallier, M.; Douki, T. The extreme variety of genotoxic response to benzo[a]pyrene in three different human cell lines from three different organs. *PLoS ONE* **2013**, *8*, e78356. [CrossRef] [PubMed]
35. Liu, D.; Pan, L.; Li, Z.; Cai, Y.; Miao, J. Metabolites analysis, metabolic enzyme activities and bioaccumulation in the clam Ruditapes philippinarum exposed to benzo[a]pyrene. *Ecotoxicol. Env. Saf.* **2014**, *107*, 251–259. [CrossRef]
36. Muller, M.; Meijer, C.; Zaman, G.J.; Borst, P.; Scheper, R.J.; Mulder, N.H.; de Vries, E.G.; Jansen, P.L. Overexpression of the gene encoding the multidrug resistance-associated protein results in increased ATP-dependent glutathione S-conjugate transport. *Proc. Natl. Acad. Sci. USA* **1994**, *91*, 13033–13037. [CrossRef]
37. Kruh, G.D.; Belinsky, M.G. The MRP family of drug efflux pumps. *Oncogene* **2003**, *22*, 7537–7552. [CrossRef]
38. Traber, M.G. Vitamin E, nuclear receptors and xenobiotic metabolism. *Arch. Biochem. Biophys.* **2004**, *423*, 6–11. [CrossRef]
39. Niestroy, J.; Barbara, A.; Herbst, K.; Rode, S.; van Liempt, M.; Roos, P.H. Single and concerted effects of benzo[a]pyrene and flavonoids on the AhR and Nrf2-pathway in the human colon carcinoma cell line Caco-2. *Toxicol* **2011**, *25*, 671–683. [CrossRef]
40. Geillinger, K.E.; Kipp, A.P.; Schink, K.; Roder, P.V.; Spanier, B.; Daniel, H. Nrf2 regulates the expression of the peptide transporter PEPT1 in the human colon carcinoma cell line Caco-2. *Biochim. Biophys. Acta.* **2014**, *1840*, 1747–1754. [CrossRef]
41. Divi, R.L.; Einem Lindeman, T.L.; Shockley, M.E.; Keshava, C.; Weston, A.; Poirier, M.C. Correlation between CYP1A1 transcript, protein level, enzyme activity and DNA adduct formation in normal human mammary epithelial cell strains exposed to benzo[a]pyrene. *Mutagenesis* **2014**, *29*, 409–417. [CrossRef] [PubMed]
42. Patel, B.; Das, S.K.; Patri, M. Neonatal Benzo[a]pyrene Exposure Induces Oxidative Stress and DNA Damage Causing Neurobehavioural Changes during the Early Adolescence Period in Rats. *Dev. Neurosci.* **2016**, *38*, 150–162. [CrossRef] [PubMed]
43. Mantey, J.A.; Rekhadevi, P.V.; Diggs, D.L.; Ramesh, A. Metabolism of benzo(a)pyrene by subcellular fractions of gastrointestinal (GI) tract and liver in Apc(Min) mouse model of colon cancer. *Tumour. Biol.* **2014**, *35*, 4929–4935. [CrossRef] [PubMed]
44. Wang, B.Y.; Wu, S.Y.; Tang, S.C.; Lai, C.H.; Ou, C.C.; Wu, M.F.; Hsiao, Y.M.; Ko, J.L. Benzo[a]pyrene-induced cell cycle progression occurs via ERK-induced Chk1 pathway activation in human lung cancer cells. *Mutat Res.* **2015**, *773*, 1–8. [CrossRef]
45. Kou, X.; Kirberger, M.; Yang, Y.; Chen, N. Natural products for cancer prevention associated with Nrf2–ARE pathway. *Food Sci. Hum. Wellness* **2013**, *2*, 22–28. [CrossRef]

46. Zhu, W.; Cromie, M.M.; Cai, Q.; Lv, T.; Singh, K.; Gao, W. Curcumin and vitamin E protect against adverse effects of benzo[a]pyrene in lung epithelial cells. *PLoS ONE* **2014**, *9*, e92992. [CrossRef]
47. Zhang, T.; Jiang, S.; He, C.; Kimura, Y.; Yamashita, Y.; Ashida, H. Black soybean seed coat polyphenols prevent B(a)P-induced DNA damage through modulating drug-metabolizing enzymes in HepG2 cells and ICR mice. *Mutat Res.* **2013**, *752*, 34–41. [CrossRef]
48. Sikdar, S.; Mukherjee, A.; Khuda-Bukhsh, A.R. Ethanolic Extract of Marsdenia condurango Ameliorates Benzo[a]pyrene-induced Lung Cancer of Rats: Condurango Ameliorates BaP-induced Lung Cancer in Rats. *J. Pharmacopunct.* **2014**, *17*, 7–17. [CrossRef]
49. Kim, K.S.; Kim, N.Y.; Son, J.Y.; Park, J.H.; Lee, S.H.; Kim, H.R.; Kim, B.; Kim, Y.G.; Jeong, H.G.; Lee, B.M.; et al. Curcumin Ameliorates Benzo[a]pyrene-Induced DNA Damages in Stomach Tissues of Sprague-Dawley Rats. *Int. J. Mol. Sci.* **2019**, *20*, 5533. [CrossRef]
50. Nanda, V.; Gupta, V.; Sharma, S.N.; Pasricha, A.; Karmakar, A.K.; Patel, A.; Bhatt, V.M.; Kantroo, B.L.; Kumar, B.; Paul, N.K.; et al. Effect of Liverubin on hepatic biochemical profile in patients of alcoholic liver disease: A retrospective study. *Minerva Med.* **2014**, *105*, 1–8.
51. Marjani, M.; Baghaei, P.; Kazempour Dizaji, M.; Gorji Bayani, P.; Fahimi, F.; Tabarsi, P.; Velayati, A.A. Evaluation of Hepatoprotective Effect of Silymarin Among Under Treatment Tuberculosis Patients: A Randomized Clinical Trial. *Iran J. Pharm. Res.* **2016**, *15*, 247–252. [PubMed]
52. Neha; Jaggi, A.S.; Singh, N. Silymarin and Its Role in Chronic Diseases. *Adv. Exp. Med. Biol.* **2016**, *929*, 25–44. [CrossRef] [PubMed]
53. Das, S.K.; Vasudevan, D.M. Protective effects of silymarin, a milk thistle (Silybium marianum) derivative on ethanol-induced oxidative stress in liver. *Indian J. Biochem. Biophys.* **2006**, *43*, 306–311. [PubMed]
54. Souza, T.; Jennen, D.; van Delft, J.; van Herwijnen, M.; Kyrtoupolos, S.; Kleinjans, J. New insights into BaP-induced toxicity: Role of major metabolites in transcriptomics and contribution to hepatocarcinogenesis. *Arch. Toxicol.* **2016**, *90*, 1449–1458. [CrossRef]
55. Shiizaki, K.; Kawanishi, M.; Yagi, T. Modulation of benzo[a]pyrene-DNA adduct formation by CYP1 inducer and inhibitor. *Genes Env.* **2017**, *39*, 14. [CrossRef]
56. Ashrap, P.; Zheng, G.; Wan, Y.; Li, T.; Hu, W.; Li, W.; Zhang, H.; Zhang, Z.; Hu, J. Discovery of a widespread metabolic pathway within and among phenolic xenobiotics. *Proc. Natl. Acad. Sci. USA* **2017**, *114*, 6062–6067. [CrossRef]
57. Rose, R.L.; Hodgson, E. Metabolism of Toxicants. In *A Textbook of Modern Toxicology*; John Wiley & Sons, Inc.: Hoboken, NJ, USA, 2004; pp. 111–148. [CrossRef]
58. Arbuckle, T.E.; Marro, L.; Davis, K.; Fisher, M.; Ayotte, P.; Belanger, P.; Dumas, P.; LeBlanc, A.; Berube, R.; Gaudreau, E.; et al. Exposure to free and conjugated forms of bisphenol A and triclosan among pregnant women in the MIREC cohort. *Env. Health Perspect.* **2015**, *123*, 277–284. [CrossRef]
59. Cai, Y.; Pan, L.; Miao, J. In vitro study of the effect of metabolism enzymes on benzo(a)pyrene-induced DNA damage in the scallop Chlamys farreri. *Env. Toxicol. Pharm.* **2016**, *42*, 92–98. [CrossRef]
60. Sundberg, K.; Dreij, K.; Seidel, A.; Jernstrom, B. Glutathione conjugation and DNA adduct formation of dibenzo[a,l]pyrene and benzo[a]pyrene diol epoxides in V79 cells stably expressing different human glutathione transferases. *Chem. Res. Toxicol.* **2002**, *15*, 170–179. [CrossRef]
61. Zhu, S.; Li, L.; Thornton, C.; Carvalho, P.; Avery, B.A.; Willett, K.L. Simultaneous determination of benzo[a]pyrene and eight of its metabolites in Fundulus heteroclitus bile using ultra-performance liquid chromatography with mass spectrometry. *J. Chromatogr. B Anal. Technol. Biomed. Life Sci.* **2008**, *863*, 141–149. [CrossRef] [PubMed]
62. John, A.; Hessel, S.; Lampen, A.; Seidel, A. Analysis of GSH Conjugates of Bay- and Fjord-Region Dihydrodiol Epoxides of Benzo[a]pyrene and Dibenzo[a,l]pyrene and their Transport in Enterocyte-like Caco-2 Cells. *Polycycl. Aromat. Compd.* **2012**, *32*, 221–237. [CrossRef]
63. Cole, S.P. Targeting multidrug resistance protein 1 (MRP1, ABCC1): Past, present, and future. *Annu Rev Pharm. Toxicol* **2014**, *54*, 95–117. [CrossRef] [PubMed]
64. Cole, S.P.; Deeley, R.G. Transport of glutathione and glutathione conjugates by MRP1. *Trends Pharm. Sci.* **2006**, *27*, 438–446. [CrossRef] [PubMed]
65. Terashima, J.; Jimma, Y.; Jimma, K.; Hakata, S.; Yachi, M.; Habano, W.; Ozawa, S. The regulation mechanism of AhR activated by benzo[a]pyrene for CYP expression are different between 2D and 3D culture of human lung cancer cells. *Drug Metab. Pharm.* **2018**, *33*, 211–214. [CrossRef] [PubMed]

66. Ye, W.; Chen, R.; Chen, X.; Huang, B.; Lin, R.; Xie, X.; Chen, J.; Jiang, J.; Deng, Y.; Wen, J. AhR regulates the expression of human cytochrome P450 1A1 (CYP1A1) by recruiting Sp1. *Febs. J.* **2019**, *286*, 4215–4231. [CrossRef]
67. Gong, H.; Li, H.D.; Yan, M.; Zhang, B.K.; Jiang, P.; Fan, X.R.; Deng, Y. Effect of licorice on the induction of phase II metabolizing enzymes and phase III transporters and its possible mechanism. *Pharmazie* **2014**, *69*, 894–897.
68. Thimmulappa, R.K.; Mai, K.H.; Srisuma, S.; Kensler, T.W.; Yamamoto, M.; Biswal, S. Identification of Nrf2-regulated genes induced by the chemopreventive agent sulforaphane by oligonucleotide microarray. *Cancer Res.* **2002**, *62*, 5196–5203.
69. Nakata, K.; Tanaka, Y.; Nakano, T.; Adachi, T.; Tanaka, H.; Kaminuma, T.; Ishikawa, T. Nuclear receptor-mediated transcriptional regulation in Phase I, II, and III xenobiotic metabolizing systems. *Drug Metab Pharm.* **2006**, *21*, 437–457. [CrossRef]
70. Wu, K.C.; Cui, J.Y.; Klaassen, C.D. Effect of graded Nrf2 activation on phase-I and -II drug metabolizing enzymes and transporters in mouse liver. *PLoS ONE* **2012**, *7*, e39006. [CrossRef]
71. Tonelli, C.; Chio, I.I.C.; Tuveson, D.A. Transcriptional Regulation by Nrf2. *Antioxid. Redox Signal* **2017**. [CrossRef]
72. Lim, J.; Ortiz, L.; Nakamura, B.N.; Hoang, Y.D.; Banuelos, J.; Flores, V.N.; Chan, J.Y.; Luderer, U. Effects of deletion of the transcription factor Nrf2 and benzo [a]pyrene treatment on ovarian follicles and ovarian surface epithelial cells in mice. *Reprod. Toxicol.* **2015**, *58*, 24–32. [CrossRef] [PubMed]
73. Kwak, M.K.; Itoh, K.; Yamamoto, M.; Kensler, T.W. Enhanced expression of the transcription factor Nrf2 by cancer chemopreventive agents: Role of antioxidant response element-like sequences in the nrf2 promoter. *Mol. Cell. Biol.* **2002**, *22*, 2883–2892. [CrossRef] [PubMed]
74. Hayashi, A.; Suzuki, H.; Itoh, K.; Yamamoto, M.; Sugiyama, Y. Transcription factor Nrf2 is required for the constitutive and inducible expression of multidrug resistance-associated protein 1 in mouse embryo fibroblasts. *Biochem. Bioph. Res. Co.* **2003**, *310*, 824–829. [CrossRef] [PubMed]

International Journal of
Molecular Sciences

Article

DNA Damage and DNA Damage Response in Chronic Myeloid Leukemia

Henning D. Popp [1,*], Vanessa Kohl [1], Nicole Naumann [1], Johanna Flach [1], Susanne Brendel [1], Helga Kleiner [1], Christel Weiss [2], Wolfgang Seifarth [1], Susanne Saussele [1], Wolf-Karsten Hofmann [1] and Alice Fabarius [1]

1 Department of Hematology and Oncology, Medical Faculty Mannheim, Heidelberg University, 68167 Mannheim, Germany; Vanessa.Kohl@medma.uni-heidelberg.de (V.K.); Nicole.Naumann@medma.uni-heidelberg.de (N.N.); Johanna.Flach@medma.uni-heidelberg.de (J.F.); Susanne.Brendel@medma.uni-heidelberg.de (S.B.); Helga.Kleiner@medma.uni-heidelberg.de (H.K.); Wolfgang.Seifarth@medma.uni-heidelberg.de (W.S.); Susanne.Saussele@medma.uni-heidelberg.de (S.S.); w.k.hofmann@medma.uni-heidelberg.de (W.-K.H.); Alice.Fabarius@medma.uni-heidelberg.de (A.F.)

2 Department of Medical Statistics and Biomathematics, Medical Faculty Mannheim, Heidelberg University, 68167 Mannheim, Germany; Christel.Weiss@medma.uni-heidelberg.de

* Correspondence: henning.popp@medma.uni-heidelberg.de; Tel.: +49-621-383-71307

Received: 26 November 2019; Accepted: 8 February 2020; Published: 11 February 2020

Abstract: DNA damage and alterations in the DNA damage response (DDR) are critical sources of genetic instability that might be involved in BCR-ABL1 kinase-mediated blastic transformation of chronic myeloid leukemia (CML). Here, increased DNA damage is detected by γH2AX foci analysis in peripheral blood mononuclear cells (PBMCs) of de novo untreated chronic phase (CP)-CML patients ($n = 5$; 2.5 γH2AX foci per PBMC ± 0.5) and blast phase (BP)-CML patients ($n = 3$; 4.4 γH2AX foci per PBMC ± 0.7) as well as CP-CML patients with loss of major molecular response (MMR) ($n = 5$; 1.8 γH2AX foci per PBMC ± 0.4) when compared to DNA damage in PBMC of healthy donors ($n = 8$; 1.0 γH2AX foci per PBMC ± 0.1) and CP-CML patients in deep molecular response or MMR ($n = 26$; 1.0 γH2AX foci per PBMC ± 0.1). Progressive activation of erroneous non-homologous end joining (NHEJ) repair mechanisms during blastic transformation in CML is indicated by abundant co-localization of γH2AX/53BP1 foci, while a decline of the DDR is suggested by defective expression of (p-)ATM and (p-)CHK2. In summary, our data provide evidence for the accumulation of DNA damage in the course of CML and suggest ongoing DNA damage, erroneous NHEJ repair mechanisms, and alterations in the DDR as critical mediators of blastic transformation in CML.

Keywords: chronic myeloid leukemia; genetic instability; DNA double-strand breaks; DNA damage response

---

1. Introduction

Chronic myeloid leukemia (CML) is a myeloproliferative neoplasm characterized by increased proliferation of myeloid cells in the bone marrow and expansion of these cells in the peripheral blood [1]. The leukemic clone originates from a hematopoietic stem cell by acquisition of the chromosomal translocation t(9;22)(q34;q11) containing the *BCR-ABL1* fusion gene [1]. The natural course of untreated CML is characterized by an initial chronic phase (CP) that progresses to an accelerated phase (AP) and a terminal blast phase (BP) [2]. During this process, the constitutively activated BCR-ABL1 tyrosine kinase stimulates different oncogenic pathways (e.g., WNT, PI3K/AKT, JAK/STAT, Hedgehog signaling) [3], which drive malignant differentiation (e.g., proliferation, G2/M delay, cell survival) [4]. In addition, BCR-ABL1 kinase-mediated genetic instability (e.g., reactive oxygen species, replication stress, error-prone DNA repair, centrosomal dysfunction) presumably plays a critical role in the blastic transformation of CML [5–8].

Tyrosine kinase inhibitors (TKIs) have revolutionized CML therapy, and induce high rates of deep or major molecular responses (DMRs or MMRs) [9]. However, TKI failure may occur, for example, by acquired point mutations in the BCR-ABL1 tyrosine kinase domain, clonal evolution, or BCR-ABL1 independent pathways resulting in loss of MMR and disease progression [10].

DNA double-strand breaks (DSBs) are serious DNA lesions that may accumulate during the course of CML. In response to DSB, the histone variant H2AX is phosphorylated at Ser139 in a region of several megabase pairs around the DSB, resulting in the formation of discrete γH2AX foci in the nucleus that are detectable by immunofluorescence microscopy [11]. γH2AX recruits additional proteins engaged in chromatin remodeling, DNA repair, and signal transduction [12–15]. One of these proteins is 53BP1 [16], which promotes ATM-dependent checkpoint signaling, regulates DSB repair pathway choice, and tethers DNA ends during non-homologous end joining (NHEJ) [17,18]. Importantly, 53BP1 triggers the repair of DSBs by erroneous NHEJ and microhomology-mediated end joining (MMEJ), which may aggravate genetic instability [19–21]. In this study, γH2AX and 53BP1 foci were analyzed by immunofluorescence microscopy in peripheral blood mononuclear cells (PBMCs) of CML patients at different stages.

The DNA damage response (DDR) is activated upon DNA damage and acts as an "anti-cancer barrier" [22]. In this signaling network, the ATM-CHK2 and ATR-CHK1 axes promote the repair of DSB and single-strand breaks, respectively. In addition, TP53 may induce apoptosis if apoptotic factors overwhelm DNA repair factors [23–25]. However, the frequency of additional chromosomal aberrations (ACAs) is about 5% in CP-CML and increases to about 80% in BP-CML [26,27], making a strong argument for the occurrence of DDR defects in the course of CML.

In summary, genetic instability is most likely involved in blastic transformation of CML. The goal of our project was to analyze mechanisms of genetic instability related to DNA damage, DSB repair, and DDR signaling in CML. For this purpose, immunofluorescence microscopy of γH2AX/53BP1 and Western blotting of (p-)ATM/(p-)CHK2 were applied in PBMC of CML patients at different disease stages in comparison to healthy controls.

2. Results

2.1. γH2AX Foci in PBMCs of Healthy Donors and CML Patients

γH2AX foci were analyzed in PBMCs of healthy donors ($n = 8$, group 1), CP-CML patients in DMR or MMR ($n = 22 + 4$, group 2), CP-CML patients with loss of MMR ($n = 5$, group 3), de novo CP-CML patients ($n = 5$, group 4), and BP-CML patients ($n = 3$, group 5) (Table 1, Figure 1a,b). The number of γH2AX foci varied in each cell of a given sample. γH2AX foci levels were similar in PBMCs of healthy donors (1.0 γH2AX foci per PBMC ± 0.1) and in PBMCs of CP-CML patients in DMR/MMR (1.0 γH2AX foci per PBMC ± 0.1). Importantly, γH2AX foci levels were significantly increased ($p = 0.0003$) in PBMCs of de novo CP-CML patients (2.5 γH2AX foci per PBMC ± 0.5) and BP-CML patients (4.4 γH2AX foci per PBMC ± 0.7) as well as CP-CML patients with loss of MMR (1.8 γH2AX foci per PBMC ± 0.4) when compared to γH2AX foci levels in PBMC of healthy donors (1.0 γH2AX foci per PBMC ± 0.1) and CP-CML patients in DMR or MMR (1.0 γH2AX foci per PBMC ± 0.1) (Figure 1b).

Table 1. Sample characterization of healthy donors and CML patients.

Gp	Pt	Age/Sex	Disease	TKI	γH2AX Foci Per Nucleus ± SEM	BCR/ABL MR	BCR/ABL Transcript	Cytogenetics/FISH
Group 1	HEALTHY#1	49/♂	Healthy	-	0.6 ± 0.1	-	-	-
	HEALTHY#2	62/♀	Healthy	-	0.7 ± 0.1	-	-	-
	HEALTHY#3	43/♂	Healthy	-	0.7 ± 0.1	-	-	-
	HEALTHY#4	NA	Healthy	-	0.9 ± 0.2	-	-	-
	HEALTHY#5	54/♀	Healthy	-	1.2 ± 0.2	-	-	-
	HEALTHY#6	36/♀	Healthy	-	1.2 ± 0.2	-	-	-
	HEALTHY#7	54/♀	Healthy	-	1.3 ± 0.1	-	-	-
	HEALTHY#8	53/♀	Healthy	-	1.6 ± 0.2	-	-	-
Group 2	CP-CML DMR#1	61/♀	CP-CML	DASA	0.1 ± 0.1	MR4.5	e13a2 (b2a2)	46,XX,t(9;22)(q34;q11) [25]
	CP-CML DMR#2	64/♀	CP-CML	Stop	0.4 ± 0.1	MR4.5	e13a2 (b2a2)	NA
	CP-CML DMR#3	74/♀	CP-CML	Stop	0.7 ± 0.2	MR4.5	e13a2/e14a2 (b2a2/b3a2)	NA
	CP-CML DMR#4	48/♂	CP-CML	Stop	0.7 ± 0.2	MR4.5	e14a2 (b3a2)	NA
	CP-CML DMR#5	63/♀	CP-CML	Stop	0.8 ± 0.2	MR4.5	e14a2 (b3a2)	NA
	CP-CML DMR#6	55/♂	CP-CML	DASA	0.8 ± 0.2	MR4	e14a2 (b3a2)	NA
	CP-CML DMR#7	61/♀	CP-CML	Stop	0.8 ± 0.2	MR4.5	e14a2 (b3a2)	NA
	CP-CML DMR#8	33/♀	CP-CML	DASA	0.8 ± 0.2	MR4	e13a2 (b2a2)	NA
	CP-CML DMR#9	58/♂	CP-CML	Stop	0.8 ± 0.2	MR5	e14a2 (b3a2)	NA
	CP-CML DMR#10	72/♂	CP-CML	Stop	0.9 ± 0.2	MR4.5	e14a2 (b3a2)	NA
	CP-CML DMR#11	34/♂	CP-CML	DASA	0.9 ± 0.1	MR4	e13a2 (b2a2)	NA
	CP-CML DMR#12	68/♀	CP-CML	DASA	0.9 ± 0.2	MR4	e13a2 (b2a2)	NA
	CP-CML DMR#13	78/♀	CP-CML	NILO	1.0 ± 0.2	MR4.5	e13a2/e14a2 (b2a2/b3a2)	NA
	CP-CML DMR#14	59/♀	CP-CML	IMA	1.0 ± 0.2	MR4	e13a2 (b2a2)	NA
	CP-CML DMR#15	60/♀	CP-CML	DASA	1.1 ± 0.2	MR4.5	e14a2 (b3a2)	NA
	CP-CML DMR#16	67/♀	CP-CML	Stop	1.1 ± 0.2	MR4.5	e14a2 (b3a2)	NA
	CP-CML DMR#17	76/♀	CP-CML	Stop	1.1 ± 0.2	MR4.5	e13a2 (b2a2)	NA
	CP-CML DMR#18	77/♂	CP-CML	Stop	1.2 ± 0.2	MR4.5	e13a2/e14a2 (b2a2/b3a2)	NA
	CP-CML DMR#19	84/♀	CP-CML	IMA	1.3 ± 0.2	MR4	e13a2 (b2a2)	NA
	CP-CML DMR#20	71/♀	CP-CML	Stop	1.3 ± 0.3	MR5	e14a2 (b3a2)	NA
	CP-CML DMR#21	63/♂	CP-CML	Stop	1.3 ± 0.2	MR4.5	e14a2 (b3a2)	NA
	CP-CML DMR#22	62/♂	CP-CML	DASA	1.7 ± 0.2	MR4.5	e14a2 (b3a2)	NA
	CP-CML MMR#1	70/♀	CP-CML	IMA	0.8 ± 0.2	0.04%	e13a2 (b2a2)	NA
	CP-CML MMR#2	59/♀	CP-CML	Stop	1.2 ± 0.2	0.07%	e13a2/e14a2 (b2a2/b3a2)	NA
	CP-CML MMR#3	77/♀	CP-CML	DASA	1.2 ± 0.2	0.09%	e14a2 (b3a2)	NA
	CP-CML MMR#4	54/♂	CP-CML	DASA	1.3 ± 0.2	0.02%	e14a2 (b3a2)	NA

Table 1. *Cont.*

Gp	Pt	Age/Sex	Disease	TKI	γH2AX Foci Per Nucleus ± SEM	BCR/ABL MR	BCR/ABL Transcript	Cytogenetics/FISH
Group 3	CP-CML loss MMR#1	54/♀	CP-CML	NILO	1.0 ± 0.2	0.18%	e1a2	46,XX,t(9;22)(q34;q11) [25]
	CP-CML loss MMR#2	46/♂	CP-CML	DASA	1.1 ± 0.2	45%	e14a2 (b3a2)	51,XY,+6,+8,+8,+8, t(9;22)(q34;q11),+19 [25]
	CP-CML loss MMR#3	81/♂	CP-CML	Stop	1.3 ± 0.2	0.35%	e13a2/e14a2 (b2a2/b3a2)	NA
	CP-CML loss MMR#4 *	83/♀	CP-CML	Stop	2.4 ± 0.2	1.00%	e14a2 (b3a2)	NA
	CP-CML loss MMR#5	83/♀	CP-CML	Stop	3.0 ± 0.2	30%	e13a2/e14a2 (b2a2/b3a2)	NA
Group 4	CP-CML de novo#1	19/♂	CP-CML	-	1.4 ± 0.2	60%	e13a2 (b2a2)	46,XY,t(9;22)(q34;q11) [25]
	CP-CML de novo#2	79/♀	CP-CML	-	1.6 ± 0.3	63%	e13a2 (b2a2)	46,XX,t(9;22)(q34;q11) [25]
	CP-CML de novo#3	18/♀	CP-CML	-	2.3 ± 0.3	68%	e13a2 (b2a2)	46,XX,t(9;22)(q34;q11) [25]
	CP-CML de novo#4	66/♀	CP-CML	-	3.4 ± 0.2	29%	e14a2 (b3a2)	46,XX,t(9;22)(q34;q11) [25]
	CP-CML de novo#5	53/♀	CP-CML	-	3.7 ± 0.3	68%	e13a2 (b2a2)	46,XX,t(9;22)(q34;q11) [25]
Group 5	BP-CML#1 **	76/♀	BP-CML	Stop	3.4 ± 0.4	32%	e13a2 (b2a2)	NA (rejected)
	BP-CML#2	63/♂	BP-CML	DASA	4.0 ± 0.4	63%	e14a2 (b3a2)	46,XY,inv(3)(q21;q26), t(9;22)(q34;q11) [25]
	BP-CML#3 ***	38/♀	BP-CML	PONA	5.8 ± 0.4	256%	e13a2 (b2a2)	46,XX,t(9;22)(q34;q11) [1] 44,XX,der(3)t(3;9)(p11;q11) t(9;22)(q34;q11),−7,−9, der(13)t(7;13)(q22;q34),der(22) t(9;22)(q34;q11) [5] 46,XX [14]

"Stop" = no TKI treatment. BP-CML, blast phase CML; CML, chronic myeloid leukemia; CP-CML, chronic phase CML; DASA, dasatinib; DMR, deep molecular response; FISH, fluorescence in situ hybridization; Gp, group; IMA, imatinib; MMR, major molecular response; MR, molecular response; NA, not assessed; NILO, nilotinib; PONA, ponatinib; Pt, patient; SEM, standard error of mean; TKI, tyrosine kinase inhibitor; ♀, female; ♂, male; *, V379I detected; **, M351V, E459K detected; ***, T315I detected.

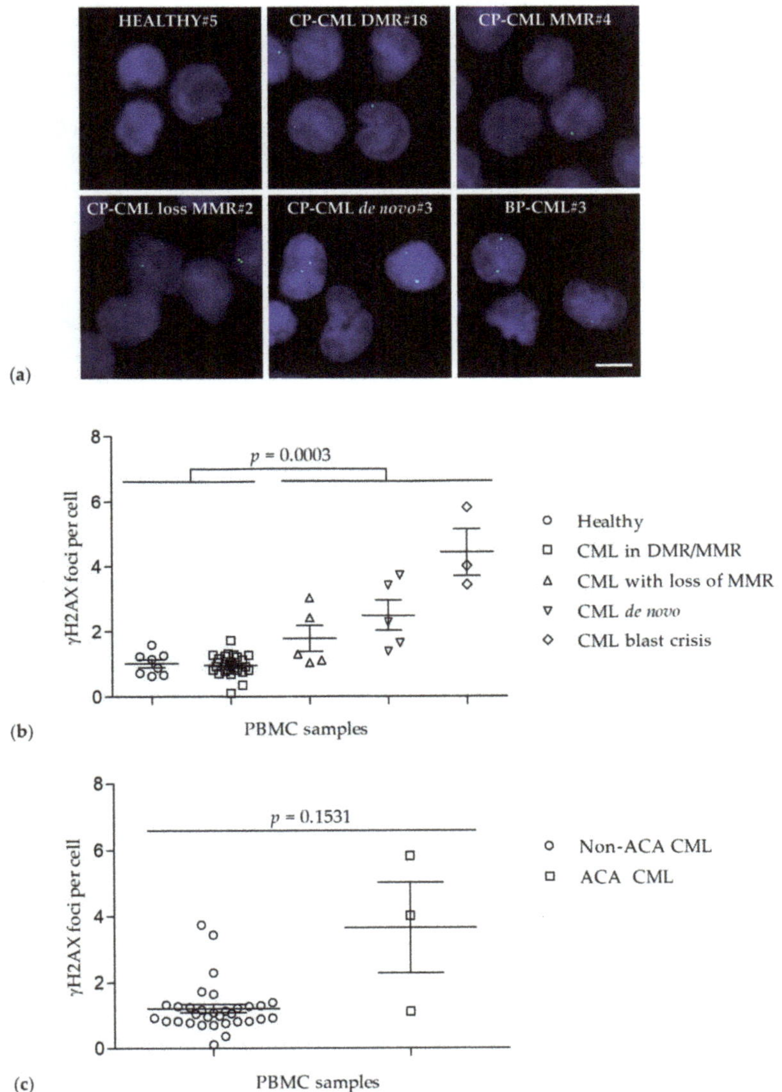

Figure 1. γH2AX foci in peripheral blood mononuclear cells (PBMC) of healthy donors and CML patients. (**a**) Representative recordings of γH2AX foci (green, Alexa 488) in PBMC nuclei (blue, DAPI) of a healthy donor (HEALTHY#5), a CP-CML patient with a DMR (CP-CML DMR#18), a CP-CML patient with a MMR (CP-CML MMR#4), a CP-CML patient with a loss of MMR (CP-CML loss MMR#2), a de novo untreated CP-CML patient (CP-CML de novo#3), and a BP-CML patient (BP-CML#3). Scale bar = 5 µm. (**b**) γH2AX foci counts in PBMCs of healthy donors and in PBMCs of CML patients at different stages. (**c**) γH2AX foci counts in PBMCs of CML patients without additional chromosomal aberrations (Non-ACA CML) and in PBMCs of CML patients with additional chromosomal aberrations (ACA CML).

Furthermore, γH2AX foci levels were correlated with the detection of ACAs in CML (Figure 1c). γH2AX foci levels tended to be increased ($p = 0.1531$) in CML samples with ACAs ($n = 3$; 1 CP-CML sample with loss of MMR, 2 BP-CML samples) as compared to γH2AX foci levels in CML samples without ACAs (n = 32; 26 CP-CML samples in DMR/MMR, 1 CP-CML sample with loss of MMR and 5 de novo CP-CML samples).

2.2. Co-Localization of γH2AX and 53BP1 Foci

γH2AX and 53BP1 foci were analyzed in the PBMCs of healthy donors and all CML groups 2–5 (Figure 2). All samples showed co-localizing γH2AX/53BP1 foci in similar patterns. Notably, the number of γH2AX/53BP1 foci in PBMCs increased across the spectrum from CP-CML towards BP-CML patients, suggesting the promotion of erroneous NHEJ and MMEJ during blastic transformation.

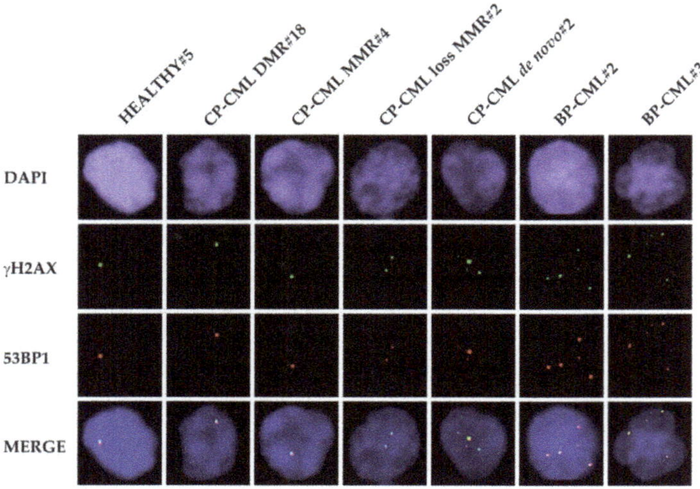

Figure 2. γH2AX and 53BP1 foci in PBMCs of healthy donors and CML patients. Representative recordings show co-localization of γH2AX foci (green, Alexa 488) and 53BP1 foci (red, Cy3) in PBMC nuclei (blue, DAPI) of a healthy donor (HEALTHY#5), a CP-CML patient with DMR (CP-CML DMR#18), a CP-CML patient with MMR (CP-CML MMR#4), a CP-CML patient with loss of MMR (CP-CML loss MMR#2), a de novo untreated CP-CML patient (CP-CML de novo#2), and two BP-CML patients (BP-CML#2 and BP-CML#3).

2.3. DNA Damage Response

Western blotting of (p-)ATM and (p-)CHK2 was performed in PBMCs of healthy donors and all CML groups 2–5 (Figure 3). PBMCs of healthy donors, CP-CML patients with DMR or MMR, and CP-CML patients with loss of MMR demonstrated no or minor expression of p-ATM and variable expression of p-CHK2, which was in accordance with the presence of low levels of DNA damage in these cells as evidenced by γH2AX foci analysis. However, PBMC of de novo CP-CML patients and BP-CML patients demonstrated no expression of p-ATM and predominantly no expression of p-CHK2 (apart from strong expression of p-CHK2 in BP-CML#2), which was in contrast to the presence of relatively high levels of DNA damage in these cells as evidenced by γH2AX foci analysis. The results suggest an alteration of the DDR in transformed CML cells.

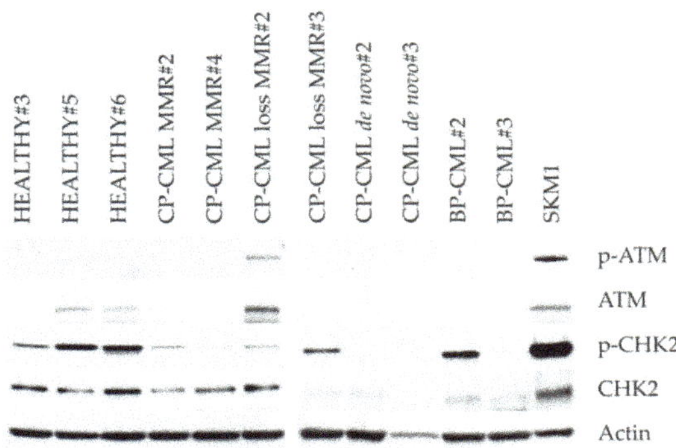

Figure 3. DNA damage response (DDR) in PBMCs of healthy donors and CML patients. Representative Western blots of (p-)ATM and (p-)CHK2 in PBMC of three healthy donors (HEALTHY#3, HEALTHY#5, AND HEALTHY#6), two CP-CML patients in MMR (CP-CML MMR#2 and CP-CML MMR#4), two CP-CML patients with loss of MMR (CP-CML loss MMR#2 and CP-CML loss MMR#3), two de novo untreated CP-CML patients (CP-CML de novo#2 and CP-CML de novo#3), and two BP-CML patients (BP-CML#2 and BP-CML#3). In PBMCs of healthy donors, CP-CML patients in MMR, and CP-CML patients with loss of MMR, no or minor expression of p-ATM and variable expression of p-CHK2 were evident. Notably, in PBMCs of de novo CP-CML patients and BP-CML patients, no expression of p-ATM and predominantly no expression of p-CHK2 were observed (except for strong expression of p-CHK2 in BP-CML#2). SKM-1 acute myeloid leukemia cells served as positive control.

3. Discussion

DSBs continuously occur in each genome [28] and are elementarily involved in the blastic transformation of CML, as evidenced by the formation of t(9;22)(q34;q11) in CP-CML and a high degree of ACA in BP-CML. For the detection of DSBs, the immunofluorescence microscopy of γH2AX and 53BP1 foci was performed in PBMCs of healthy donors, CP-CML patients in DMR/MMR, CP-CML patients with loss of MMR, de novo CP-CML patients, and BP-CML patients. Analyses showed a gain in DSB in the course from CP-CML towards BP-CML, indicating a biological continuum to accumulate DNA damage across different CML stages. Moreover, the intersection of γH2AX foci levels between different CML stages indicated that γH2AX foci were a biological marker, rather than a stage-specific marker. Further, γH2AX foci levels were evaluated in CML patient samples in relation to ACA. The ACA-bearing BP-CML samples demonstrated increased γH2AX foci levels as compared to the non-ACA bearing CP-CML samples. This demonstrates that γH2AX foci and ACA increase concordantly in the process of blastic transformation in CML.

γH2AX and 53BP1 foci were analyzed semi-quantitatively in PBMCs of all healthy donors and CML patients. Analyses revealed similar patterns of γH2AX and 53BP1 foci co-localization in all samples. Furthermore, co-localizing γH2AX/53BP1 foci increased in the course from CP-CML patients towards BP-CML patients. This observation suggests the promotion of erroneous NHEJ and MMEJ as critical mechanisms of blastic transformation in CML. Considering the contribution of erroneous NHEJ and particularly MMEJ to the formation of chromosomal aberrations such as deletions, translocations, inversions, and other complex rearrangements [29], this data might, at least in part, explain the occurrence of ACAs in BP-CML.

γH2AX and 53BP1 foci were evaluated in different CML patients at various stages. One might ask for follow-up data in same CML patients; however, none of the CML patients in DMR or MMR (group 2), who were treated by TKI or were closely monitored in a TKI–free setting, demonstrated loss of MMR during study. Further, CML patients with loss of MMR (group 3) were changed promptly to another TKI according to BCR/ABL kinase domain mutations and patient characteristics. These patients achieved DMR or MMR shortly again, except for one patient who took the TKI irregularly and still demonstrated loss of MMR. Although our study lacks follow-up data on γH2AX and 53BP1 foci in same CML patients, the potential increases in the long-term course in these patients may be similar to the numbers of γH2AX and 53BP1 foci in different CML patients at various stages.

Western blotting of (p-)ATM and (p-)CHK2 was performed in the PBMCs of healthy donors and CML patients. Minor activation of the DDR proteins was detected in the PBMCs of healthy donors, CP-CML patients in DMR or MMR, and CP-CML patients with loss of MMR, which was in accordance with the detection of low levels of DNA damage in these cells, as evidenced by γH2AX foci analysis. However, in the PBMCs of de novo CP-CML patients and BP-CML patients, almost no activation of the DDR proteins was observed, too, despite of the presence of relatively high levels of DNA damage in these cells, as evidenced by γH2AX foci analysis and the presence of ACAs. The absence of p-ATM and the predominant absence of p-CHK2 in the PBMCs of de novo CP-CML and BP-CML patients might be explained by missing or minor expression of ATM and CHK2, respectively. Further, there may also have been defective phosphorylation of ATM and CHK2, aggravating DDR defects. Overall, our results implicate alterations of the DDR in CML cells and correspond well with the known function of the DDR as an "anti-cancer barrier" that becomes activated upon DNA damage in healthy cells and disarranged in cancer cells [22,30].

Our data may generate hypotheses for the development of genetic instability in CML, in which γH2AX and 53BP1 foci accumulate with ongoing DNA damage by intrinsic sources (e.g., reactive oxygen species, replication-stress-induced DNA damage), erroneous DSB repair (e.g., NHEJ and MMEJ), and alterations of the DDR (e.g., sensing and signaling of DNA damage) (Figure 4). Importantly, genetic instability is a common feature in TKI-refractory CML. In consideration of the worse prognosis, the limited effectiveness of chemotherapy and the ineligibility of most of these patients for allogeneic bone marrow transplantation, novel treatment options are an unmet need at present. Here, DNA repair inhibitors (e.g., PARP or APE1 inhibitors) might be beneficial by inducing synthetic lethality in DNA repair defective TKI-refractory CML cells [31] and may constitute possible treatment options.

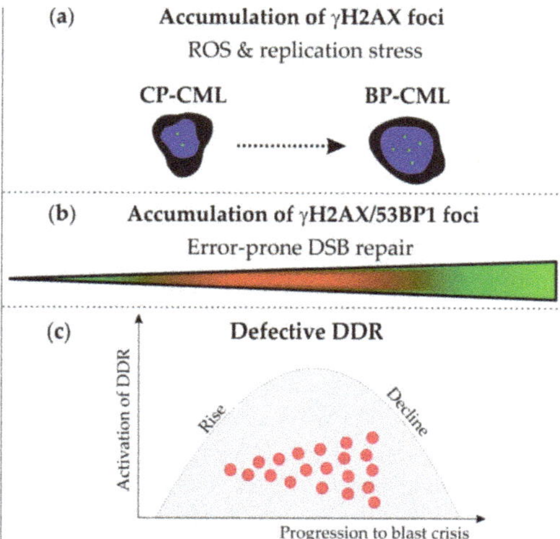

Figure 4. Model of genetic instability in CML. (**a**) Accumulation of γH2AX foci (green) in nuclei (blue) of CP-CML and BP-CML may increase with reactive oxygen species (ROS)- and replication-stress-induced DNA damage. (**b**) Accumulation of γH2AX/53BP1 foci suggests progressive activation of error-prone DNA double-strand break (DSB) repair in the course from CP-CML towards BP-CML. (**c**) After initial activation of the DDR in hematopoietic cells, the DDR might finally decline in transformed CP- and BP-CML cells.

4. Materials and Methods

4.1. Blood Samples

This study was authorized by the Ethics Committee II of the Medical Faculty Mannheim of the Heidelberg University (2018-566N-MA). Written informed consent was given by all participants. Blood samples were collected from 8 healthy donors (3 males, 5 females, mean age: 50 years), 26 CP-CML patients in DMR or MMR (9 males, 17 females, mean age: 63 years), 5 CP-CML patients with loss of MMR (2 males, 3 females, mean age: 69 years), 5 de novo untreated CP-CML patients (1 male, 4 females, mean age: 47 years), and 3 BP-CML patients (1 male, 2 female, mean age: 59 years) (Table 1). Treatment with TKI included imatinib, nilotinib, dasatinib, and bosutinib, respectively. PBMCs were separated from whole-blood samples by Ficoll-Paque density gradient centrifugation (Miltenyi Biotec, Bergisch Gladbach, Germany).

4.2. Cytology, Cytogenetics, and Molecular Analyses

Peripheral blood smears were stained by May-Gruenwald-Giemsa [32]. Fluorescence in situ hybridization (FISH) of *BCR-ABL1*-rearrangement was performed with probes for the detection of t(9;22)(q34;q11) [33]. ACAs were determined by conventional cytogenetics of G-banded chromosomes [34]. The molecular response (MR) of CML was assessed by international scale (IS)-standardized polymerase chain reaction (PCR) of *BCR-ABL1* in all CML samples [35]. For the quantitative real-time PCR reaction (qRT-PCR) of *BCR-ABL1* fusion gene and *GUSB* control gene, a TaqMan detection system (TaqMan 7500 Fast Real-Time PCR System, Thermo Fisher Scientific, Waltham, US) was used for quantification of e13a2/e14a2 (b2a2/b3a2) transcripts and a LightCycler detection system (Roche Applied Science, Penzberg, Germany) for quantification of e1a2 transcripts as previously described [36,37].

4.3. Immunofluorescence Staining of γH2AX and 53BP1

γH2AX and 53BP1 were detected in the PBMCs of healthy donors and CML patients using a mouse monoclonal anti-γH2AX antibody (clone JBW301; Merck Millipore, Darmstadt, Germany) and a polyclonal rabbit anti-53BP1 antibody (NB100-304; Novus Biologicals, Littleton, US), respectively. An Alexa Fluor 488-conjugated goat anti-mouse antibody and an Alexa Fluor 555-conjugated goat anti-rabbit antibody (Thermo Fisher Scientific) were used as previously described [38,39]. At least 50 PBMCs were analyzed for each measurement.

4.4. Western Blotting

Western blotting of (p-)ATM and (p-)CHK2 was conducted in the PBMCs of healthy donors, CP-CML patients in DMR or MMR, CP-CML patients with loss of MMR, de novo untreated CP-CML patients, and BP-CML patients as previously described [39]. (p-)ATM and (p-)CHK2 expressing SKM-1 acute myeloid leukemia cells (Leibniz Institute DSMZ–German Collection of Microorganisms and Cell Cultures GmbH, Braunschweig, Germany) served as positive control.

4.5. Statistics

Statistical calculations were performed using SAS software, release 9.4 (SAS Institute Inc., Cary, NC, USA). Quantitative variables were stated by mean values and standard errors. The number of foci in each sample were modeled by a Poisson regression model using the SAS procedure PROC GENMOD. Furthermore, the SAS "repeated" statement was applied with statistical consideration encompassing several cells, which were counted in each sample. p values < 0.05 were regarded statistically significant.

5. Conclusions

Our study reveals accumulation of DNA damage in the course from CP-CML towards BP-CML. In addition, our data suggest increase of DNA damage, erroneous DSB repair, and alterations of the DDR as critical mediators of blastic transformation in CML. Finally, γH2AX and 53BP1 foci might be useful markers for targeting genetic instability in TKI-refractory CML in future studies.

Author Contributions: Conceptualization, H.D.P.; methodology, H.D.P, V.K., S.B., and H.K.; software, H.D.P., V.K., and C.W.; validation, H.D.P.; formal analysis, H.D.P., V.K., and C.W.; investigation, H.D.P., S.B., and H.K.; resources, W.-K.H.; data curation, H.D.P.; writing—original draft preparation, H.D.P.; writing—review and editing, V.K., J.F., N.N., W.S., S.S., W.-K.H., and A.F.; visualization, H.D.P; supervision, A.F. and W.-K.H.; project administration, H.D.P. and A.F.; funding acquisition, H.D.P. and A.F. All authors have read and agreed to the published version of the manuscript.

Funding: This research was funded by Deutsche José Carreras Leukämie-Stiftung (DJCLS 14 R/2017).

Acknowledgments: We acknowledge financial support by Deutsche Forschungsgemeinschaft within the funding program Open Access Publishing, by the Baden-Württemberg Ministry of Science, Research and the Arts and by Ruprecht-Karls-Universität Heidelberg.

Conflicts of Interest: The authors declare no conflict of interest. The funders had no role in the design of the study; in the collection, analyses, or interpretation of data; in the writing of the manuscript, or in the decision to publish the results.

Abbreviations

AP	Accelerated phase
ACA	Additional chromosomal aberrations
BP	Blast phase
CML	Chronic myeloid leukemia
CP	Chronic phase
DASA	Dasatinib
DDR	DNA damage response
DMR	Deep molecular response
DSB	DNA double-strand breaks
IMA	Imatinib
MMEJ	Microhomology mediated end-joining
MMR	Major molecular response
MR	Molecular response
NHEJ	Non-homologous end joining
NILO	Nilotinib
PBMC	Peripheral blood mononuclear cells
PONA	Ponatinib
ROS	Reactive oxygen species
TKI	Tyrosine kinase inhibitor

References

1. Vardiman, J.W.; Melo, J.V.; Baccarani, M.; Radich, J.P.; Kvasnicka, H.M. Chronic myeloid leukaemia, BCR-ABL1-positive. In *WHO Classification of Tumours of Haematopoietic and Lymphoid Tissues*, 4th ed.; Swerdlow, S.H., Campo, E., Harris, N.L., Jaffe, E.S., Pileri, S.A., Stein, H., Thiele, J., Arber, D.A., Hasserjian, R.P., Le Beau, M.M., Eds.; International Agency for Research on Cancer: Lyon, France, 2017; pp. 30–36.
2. NCCN. Chronic Myeloid Leukemia, Version 1.2019, NCCN Clinical Practice Guidelines in Oncology. Available online: https://jnccn.org/view/journals/jnccn/16/9/article-p1108.xml (accessed on 05 November 2019).
3. Holyoake, T.L.; Vetrie, D. The chronic myeloid leukemia stem cell: Stemming the tide of persistence. *Blood* **2017**, *129*, 1595–1606. [CrossRef] [PubMed]
4. Muvarak, N.; Nagaria, P.; Rassool, F.V. Genomic instability in chronic myeloid leukemia: Targets for therapy? *Curr. Hematol. Malig. Rep.* **2012**, *7*, 94–102. [CrossRef] [PubMed]

5. Nowicki, M.O.; Falinski, R.; Koptyra, M.; Slupianek, A.; Stoklosa, T.; Gloc, E.; Nieborowska-Skorska, M.; Blasiak, J.; Skorski, T. BCR/ABL oncogenic kinase promotes unfaithful repair of the reactive oxygen species-dependent DNA double-strand breaks. *Blood* **2004**, *104*, 3746–3753. [CrossRef] [PubMed]
6. Koptyra, M.; Cramer, K.; Slupianek, A.; Richardson, C.; Skorski, T. BCR/ABL promotes accumulation of chromosomal aberrations induced by oxidative and genotoxic stress. *Leukemia* **2008**, *22*, 1969–1972. [CrossRef] [PubMed]
7. Cramer, K.; Nieborowska-Skorska, M.; Koptyra, M.; Slupianek, A.; Penserga, E.T.; Eaves, C.J.; Aulitzky, W.; Skorski, T. BCR/ABL and other kinases from chronic myeloproliferative disorders stimulate single-strand annealing, an unfaithful DNA double-strand break repair. *Cancer Res.* **2008**, *68*, 6884–6888. [CrossRef]
8. Stoklosa, T.; Poplawski, T.; Koptyra, M.; Nieborowska-Skorska, M.; Basak, G.; Slupianek, A.; Rayevskaya, M.; Seferynska, I.; Herrera, L.; Blasiak, J.; et al. BCR/ABL inhibits mismatch repair to protect from apoptosis and induce point mutations. *Cancer Res.* **2008**, *68*, 2576–2580. [CrossRef]
9. Baccarani, M.; Deininger, M.W.; Rosti, G.; Hochhaus, A.; Soverini, S.; Apperley, J.F.; Cervantes, F.; Clark, R.E.; Cortes, J.E.; Guilhot, F.; et al. European LeukemiaNet recommendations for the management of chronic myeloid leukemia: 2013. *Blood* **2013**, *122*, 872–884. [CrossRef]
10. Patel, A.B.; O'Hare, T.; Deininger, M.W. Mechanisms of Resistance to ABL Kinase Inhibition in Chronic Myeloid Leukemia and the Development of Next Generation ABL Kinase Inhibitors. *Hematol. Oncol. Clin. North. Am.* **2017**, *31*, 589–612. [CrossRef]
11. Rogakou, E.P.; Pilch, D.R.; Orr, A.H.; Ivanova, V.S.; Bonner, W.M. DNA double-stranded breaks induce histone H2AX phosphorylation on serine 139. *J. Biol. Chem.* **1998**, *273*, 5858–5868. [CrossRef]
12. Scully, R.; Xie, A. Double strand break repair functions of histone H2AX. *Mutat Res.* **2013**, *750*, 5–14. [CrossRef]
13. Stucki, M.; Clapperton, J.A.; Mohammad, D.; Yaffe, M.B.; Smerdon, S.J.; Jackson, S.P. MDC1 directly binds phosphorylated histone H2AX to regulate cellular responses to DNA double-strand breaks. *Cell* **2005**, *123*, 1213–1226. [CrossRef] [PubMed]
14. Kobayashi, J.; Tauchi, H.; Sakamoto, S.; Nakamura, A.; Morishima, K.; Matsuura, S.; Kobayashi, T.; Tamai, K.; Tanimoto, K.; Komatsu, K. NBS1 localizes to gamma-H2AX foci through interaction with the FHA/BRCT domain. *Curr. Biol.* **2002**, *12*, 1846–1851. [CrossRef]
15. Paull, T.T.; Rogakou, E.P.; Yamazaki, V.; Kirchgessner, C.U.; Gellert, M.; Bonner, W.M. A critical role for histone H2AX in recruitment of repair factors to nuclear foci after DNA damage. *Curr. Biol.* **2000**, *10*, 886–895. [CrossRef]
16. Ward, I.M.; Minn, K.; Jorda, K.G.; Chen, J. Accumulation of checkpoint protein 53BP1 at DNA breaks involves its binding to phosphorylated histone H2AX. *J. Biol. Chem.* **2003**, *278*, 19579–19582. [CrossRef] [PubMed]
17. Difilippantonio, S.; Gapud, E.; Wong, N.; Huang, C.Y.; Mahowald, G.; Chen, H.T.; Kruhlak, M.J.; Callen, E.; Livak, F.; Nussenzweig, M.C.; et al. 53BP1 facilitates long-range DNA end-joining during V(D)J recombination. *Nature* **2008**, *456*, 529–533. [CrossRef]
18. Lee, J.H.; Paull, T.T. Activation and regulation of ATM kinase activity in response to DNA double-strand breaks. *Oncogene* **2007**, *26*, 7741–7748. [CrossRef]
19. Xiong, X.; Du, Z.; Wang, Y.; Feng, Z.; Fan, P.; Yan, C.; Willers, H.; Zhang, J. 53BP1 promotes microhomology-mediated end-joining in G1-phase cells. *Nucleic Acids Res.* **2015**, *43*, 1659–1670. [CrossRef]
20. Callen, E.; Di Virgilio, M.; Kruhlak, M.J.; Nieto-Soler, M.; Wong, N.; Chen, H.T.; Faryabi, R.B.; Polato, F.; Santos, M.; Starnes, L.M.; et al. 53BP1 mediates productive and mutagenic DNA repair through distinct phosphoprotein interactions. *Cell* **2013**, *153*, 1266–1280. [CrossRef]
21. Chapman, J.R.; Barral, P.; Vannier, J.B.; Borel, V.; Steger, M.; Tomas-Loba, A.; Sartori, A.A.; Adams, I.R.; Batista, F.D.; Boulton, S.J. RIF1 is essential for 53BP1-dependent nonhomologous end joining and suppression of DNA double-strand break resection. *Mol. Cell* **2013**, *49*, 858–871. [CrossRef]
22. Bartkova, J.; Horejsi, Z.; Koed, K.; Kramer, A.; Tort, F.; Zieger, K.; Guldberg, P.; Sehested, M.; Nesland, J.M.; Lukas, C.; et al. DNA damage response as a candidate anti-cancer barrier in early human tumorigenesis. *Nature* **2005**, *434*, 864–870. [CrossRef]
23. Norbury, C.J.; Zhivotovsky, B. DNA damage-induced apoptosis. *Oncogene* **2004**, *23*, 2797–2808. [CrossRef] [PubMed]
24. Roos, W.P.; Kaina, B. DNA damage-induced cell death: From specific DNA lesions to the DNA damage response and apoptosis. *Cancer Lett.* **2013**, *332*, 237–248. [CrossRef] [PubMed]

25. Roos, W.P.; Thomas, A.D.; Kaina, B. DNA damage and the balance between survival and death in cancer biology. *Nat. Rev. Cancer* **2016**, *16*, 20–33. [CrossRef] [PubMed]
26. Mitelman, F. The cytogenetic scenario of chronic myeloid leukemia. *Leuk Lymphoma* **1993**, *11*, 11–15. [CrossRef] [PubMed]
27. Fabarius, A.; Leitner, A.; Hochhaus, A.; Muller, M.C.; Hanfstein, B.; Haferlach, C.; Gohring, G.; Schlegelberger, B.; Jotterand, M.; Reiter, A.; et al. Impact of additional cytogenetic aberrations at diagnosis on prognosis of CML: Long-term observation of 1151 patients from the randomized CML Study IV. *Blood* **2011**, *118*, 6760–6768. [CrossRef] [PubMed]
28. Vilenchik, M.M.; Knudson, A.G. Endogenous DNA double-strand breaks: Production, fidelity of repair, and induction of cancer. *Proc. Natl. Acad. Sci. USA* **2003**, *100*, 12871–12876. [CrossRef]
29. Bunting, S.F.; Nussenzweig, A. End-joining, translocations and cancer. *Nat. Rev. Cancer* **2013**, *13*, 443–454. [CrossRef]
30. Halazonetis, T.D.; Gorgoulis, V.G.; Bartek, J. An oncogene-induced DNA damage model for cancer development. *Science* **2008**, *319*, 1352–1355. [CrossRef]
31. Tobin, L.A.; Robert, C.; Rapoport, A.P.; Gojo, I.; Baer, M.R.; Tomkinson, A.E.; Rassool, F.V. Targeting abnormal DNA double-strand break repair in tyrosine kinase inhibitor-resistant chronic myeloid leukemias. *Oncogene* **2013**, *32*, 1784–1793. [CrossRef]
32. Löffler, H.; Rastetter, J.; Haferlach, T. Light microscopic procedures. In *Atlas of Clinical Hematology*, 6th ed.; Springer: Heidelberg, Germany, 2005; p. 8.
33. MLL. Request for Testing Form. Available online: https://www.mll.com/en.html (accessed on 05 November 2019).
34. Gisselsson, D. Cytogenetic methods. In *Cancer Cytogenetics*, 3rd ed.; Heim, S., Mitelman, F., Eds.; Wiley-Blackwell: Hoboken, NJ, USA, 2009; pp. 9–16.
35. UMM. Request for Testing Form. Available online: https://www.umm.de/iii-medizinische-klinik/wissenschaftliches-labor/ (accessed on 05 November 2019).
36. Emig, M.; Saussele, S.; Wittor, H.; Weisser, A.; Reiter, A.; Willer, A.; Berger, U.; Hehlmann, R.; Cross, N.C.; Hochhaus, A. Accurate and rapid analysis of residual disease in patients with CML using specific fluorescent hybridization probes for real time quantitative RT-PCR. *Leukemia* **1999**, *13*, 1825–1832. [CrossRef]
37. Spiess, B.; Rinaldetti, S.; Naumann, N.; Galuschek, N.; Kossak-Roth, U.; Wuchter, P.; Tarnopolscaia, I.; Rose, D.; Voskanyan, A.; Fabarius, A.; et al. Diagnostic performance of the molecular BCR-ABL1 monitoring system may impact on inclusion of CML patients in stopping trials. *PLoS ONE* **2019**, *14*, e0214305. [CrossRef] [PubMed]
38. Popp, H.D.; Brendel, S.; Hofmann, W.K.; Fabarius, A. Immunofluorescence Microscopy of gammaH2AX and 53BP1 for Analyzing the Formation and Repair of DNA Double-strand Breaks. *J. Vis. Exp.* **2017**. [CrossRef] [PubMed]
39. Popp, H.D.; Naumann, N.; Brendel, S.; Henzler, T.; Weiss, C.; Hofmann, W.K.; Fabarius, A. Increase of DNA damage and alteration of the DNA damage response in myelodysplastic syndromes and acute myeloid leukemias. *Leuk Res.* **2017**, *57*, 112–118. [CrossRef] [PubMed]

International Journal of
Molecular Sciences

Article

Does Neuraxial Anesthesia as General Anesthesia Damage DNA? A Pilot Study in Patients Undergoing Orthopedic Traumatological Surgery

Monika Kucharova [1,2], David Astapenko [2,3], Veronika Zubanova [1,4], Maria Koscakova [1], Rudolf Stetina [2], Zdenek Zadak [2] and Miloslav Hronek [2,5,*]

1 Department of Biophysics and Physical Chemistry, Faculty of Pharmacy, Charles University, Heyrovskeho 1203, 500 05 Hradec Kralove, Czech Republic; kucharova@faf.cuni.cz (M.K.); Veronikazubanova@gmail.com (V.Z.); makoscakova@gmail.com (M.K.)
2 Department of Research and Development, University Hospital in Hradec Kralove, Sokolska 581, 500 05 Hradec Kralove, Czech Republic; david.astapenko@fnhk.cz (D.A.); r.stetina@tiscali.cz (R.S.); zdenek.zadak@fnhk.cz (Z.Z.)
3 Department of Anesthesiology, Resuscitation and Intensive Medicine, University Hospital Hradec Kralove, Sokolska 581, 500 05 Hradec Kralove, Czech Republic
4 Institute for Clinical Biochemistry and Diagnostics, Charles University, Faculty of Medicine in Hradec Kralove and University Hospital Hradec Kralove, Sokolska 581, 500 05 Hradec Kralove, Czech Republic
5 Department of Biological and Medical Sciences, Faculty of Pharmacy, Charles University, Zborovska 2089, 500 03 Hradec Kralove, Czech Republic
* Correspondence: Hronek@faf.cuni.cz; Tel.: +420-495-067

Received: 29 November 2019; Accepted: 18 December 2019; Published: 20 December 2019

Abstract: The human organism is exposed daily to many endogenous and exogenous substances that are the source of oxidative damage. Oxidative damage is one of the most frequent types of cell component damage, leading to oxidation of lipids, proteins, and the DNA molecule. The predominance of these damaging processes may later be responsible for human diseases such as cancer, neurodegenerative disease, or heart failure. Anesthetics undoubtedly belong to the group of substances harming DNA integrity. The goal of this pilot study is to evaluate the range of DNA damage by general and neuraxial spinal anesthesia in two groups of patients undergoing orthopedic traumatological surgery. Each group contained 20 patients, and blood samples were collected before and after anesthesia; the degree of DNA damage was evaluated by the comet assay method. Our results suggest that general anesthesia can cause statistically significant damage to the DNA of patients, whereas neuraxial anesthesia has no negative influence.

Keywords: DNA damage; general anesthesia; neuraxial anesthesia; comet assay

1. Introduction

DNA is continuously exposed to a variety of biological, chemical, and physical agents, which may alter its structure and modify its function. Such exogenous compounds include anesthetic substances commonly used in general anesthesia (GA) or neuraxial anesthesia (NA). These have attracted attention because of concern about their potential genotoxic effect. The aim of anesthesia during surgery is to reduce pain. However, anesthetics cannot completely block autonomic nervous system responses to traumatic stimuli. The autonomic nervous system-dependent responses to traumatic stress remain progressive, resulting in increased oxygen consumption and production of oxygen free radicals, and decreased antioxidant activity. Typically, several different drugs and techniques are combined during anesthesia [1]. Various methods have described DNA structure damage due to anesthesia [2–5], and thus there is evidence that anesthetics change human genetic information. Recent studies have been published concerning the genotoxicity of inhalational anesthesia in patients who have undergone surgery and in personnel who are occupationally exposed to anesthetics. Even a trace concentration of waste anesthetic gases may lead to an increase in genetic damage [6,7].

In GA, inhalation anesthetics are the most widely used. The anesthetic mechanism of volatile agents is complicated, targeting a number of sites including GABA (gamma aminobutyric acid) and NMDA (N-methyl-D-aspartic acid) receptors. Chemically they are small hydrophobic molecules that pass through lipophilic cell membranes. They cause depression of respiration and oxidative phosphorylation in mitochondria. Exposure to volatile anesthetics generates small quantities of reactive oxygen species (ROS) either directly due to interaction with the electron transport chain or indirectly through a signaling cascade in which G-protein-coupled receptors, protein kinases, and mitochondrial ATP-sensitive potassium channels are involved [8].

Among the inhaled anesthetic gases, the halogenated gases isoflurane (ISF) and sevoflurane (SVF) are the most widely used in general anesthesia. ISF was synthesized in 1965 and introduced to clinical practice several years later. According to the study by Corbett from 1976, this anesthetic may cause liver tumors in rats [9]. Even though this claim was not confirmed in other studies [10], results are not entirely uniform. The chemical structure of ISF is similar to that of some non-anesthetic carcinogens, including chloromethyl methyl ether [11]. The advantage of ISF and SVF is their low metabolism rate and low blood–gas partition coefficient, which decreases its induction and recovery times.

Some studies have shown an association between inhalational anesthesia and increased risk of tumor spread [12,13]. In their retrospective study, Wigmore and coauthors found that cancer patients had a worse survival outcome if they received inhalation anesthesia. Inhalation anesthesia inhibits the immune system by diminishing the function of killer cells that protect the organism against the proliferation of cancer cells [14]. It has been published that ISF could promote the growth and migration of glioblastoma cells and increase levels of hypoxia-inducible factors that are overexpressed in a variety of carcinomas and their metastases, and it is supposed that it is a transcriptional regulator of VEGF expression involved in tumor growth [15,16]. A connection with postoperative cognitive impairment and possible development of Alzheimer's disease is suspected rather than proved [17,18].

The need for a local anesthetic with low toxicity has led to the development of numerous compounds. Bupivacaine, levobupivacaine, and ropivacaine are long-acting amide-based local anesthetics most commonly used in clinical practice. The principle of their effect is the prevention of nerve impulse induction primarily in nerve cell membranes by inhibition of voltage-gated Na^+ channels [19]. Regional analgesic techniques deliver local anesthetics to groups of peripheral nerves outside the central nervous system. They block sensation to specific dermatomes. The principle of action of neuraxial techniques is to deliver local anesthetic and analgesic medications directly or indirectly to the spinal cord [20].

Neuraxial blockade includes epidural and subarachnoid (spinal) anesthesia (SA). The difference between them is in the dosage and the place of administration. In this manner, the speed of the onset of the numbness and the speed of the influencing of motoric is given. During epidural anesthesia, spinal nerve root block occurs by injection of a local anesthetic outside the subarachnoid space. The required dose and concentration of the local anesthetic are high. Efficacy starts slowly, and the duration of action is medium to long (6–8 h). It has a wide range of applications including operative anesthesia, obstetric analgesia, and chronic pain management. Spinal anesthesia is the oldest neuraxial technique, first used by August Bier who injected cocaine into the intrathecal space to provide anesthesia in 1898. During SA, the spinal nerve roots are blocked by the injection of local anesthetic to the subarachnoid space. The concentration of local anesthetic is usually high, but the volume required is usually small. Efficacy starts very quickly, sometimes during the injection. The duration of action depends on the substance used: it may be short, medium, or long (minutes to hours). Use of this technique in general is confined to shorter procedures to the lower extremities or pelvis [20,21]. As with any anesthesia, subarachnoid blockade carries the risk of adverse reactions, such as post-puncture headaches, hypotension, bradycardia, hypothermia, urinary retention, nausea, and vomiting [21].

Genotoxic or mutagenic effects of anesthesia have been of interest in many studies. However, the results of the studies do not coincide, and the differences may be influenced by many factors. Different exposure times, interindividual differences in sensitivity dependent on genetic factors, different design of experiment, type and length of surgery, interindividual parameters of patients, and dissimilar methods of statistical analysis all contribute to the differences in analytical results [22]. DNA damage during GA has been well documented, and it delivers a significant burden to the patients [23–25]. Data on NA are less common. There are several studies that prove that local anesthetics cause neurotoxicity and apoptosis by induction of oxidative DNA damage [26–28]. Unfortunately, there is a lack of information about the influence of spinal anesthesia on DNA damage. Which method offers less burden for the organism remains controversial. Therefore, there is a tendency to continually explore and improve its safety for humans [4].

A suitable method for quantification of the degree of DNA damage in human medicine is the comet assay (single cell gel electrophoresis) [29]. The method was developed in a different modification enabling quantification of single-strand DNA breaks (SSB), double-strand DNA breaks, and oxidized pyrimidine and purine bases, and enables quantification of low-level DNA changes in individual eukaryotic cells [30]. Using this method, the DNA changes in peripheral lymphocytes can be evaluated. Comet assay is appropriate for quantification of DNA damage in patients especially after chemotherapy and radiotherapy [31,32].

The primary objective of this pilot study is to evaluate the applicability of the comet assay method for quantification of DNA changes in patients under anesthesia undergoing orthopedic or traumatological lower limb surgery. The secondary aim is to verify the hypothesis that neuraxial anesthesia damages DNA less than general anesthesia. Obtaining such data should contribute to broadening the knowledge on the impact of surgical interventions on oxidative DNA damage, and should be the basis for further research into finding preventive and/ or protective procedures to minimize DNA damage associated with invasive interventional procedures in clinical medicine.

2. Results

2.1. Preoperative Data

Forty-five patients were included in the study, from 40 of whom were obtained complete data. Twenty patients were in the general anesthesia group (GA group) and 20 in the spinal anesthesia group (SA group). The demographic data and results of the preoperative evaluation are in Table 1. The level of statistical significance was $p = 0.05$. Both groups were comparable except for age and height. Patients were separated into four groups according to ASA (American Society of Anesthesia) classification: ASA I represents normal, healthy patients; ASA II includes patients with mild systemic disease; ASA III includes patients with severe systemic disease; ASA IV includes patients with severe systemic disease that is a constant threat to life. The differences between patients in individual ASA grades between the GA group and the SA group were statistically insignificant ($p = 0.49$).

Table 1. Demographic data and results of preoperative evaluation of both groups.

Parameter	GA Group	SA Group	p-Value
Number	20	20	
Woman	9	11	0.96
Man	10	10	0.96
Age [years]	37 (29; 51)	65.5 (53.25; 74.75)	0.0002 *
Height [cm]	175.5 (170; 183)	170 (166; 175)	0.01 *
Weight [kg]	90 (80; 99)	81,2 (73.5; 85)	0.12
BMI	28 (24; 30)	28.4 (26; 30)	0.64
ASA I [%]	13	21	0.55

Table 1. *Cont.*

Parameter	GA Group	SA Group	p-Value
ASA II [%]	54	68	0.36
ASA III [%]	27	11	0.22
ASA IV [%]	6	0	0.25

BMI—body mass index; ASA—American Society of Anesthesia physical status; data are shown as median (Q_1 = first quartile; Q_3 = third quartile); * $p < 0.05$.

Relevant laboratory results prior to anesthesia and preoperative performance data are shown in Table 2. Patients in GA had a statistically significantly higher duration of operation.

Table 2. Results of laboratory tests of preoperative evaluation.

Parameter	GA Group	SA Group	p-Value
MAP [mm Hg]	130 (120; 143)	132 (120; 140)	0.47
Gly [mmol/l]	5.55 (5.33; 5.78)	5.7 (5.18; 6.08)	0.29
Na [mmol/l]	139 (137.8; 140.5)	140 (139; 142)	0.53
K [mmol/l]	4.2 (4; 4.5)	4.6 (4.58; 4.8)	0.54
Cl [mmol/l]	103 (100; 105)	104 (102; 107)	0.27
Anest. Duration [min]	180 (120; 273)	107.5 (91.3; 118.8)	0.001 *

MAP—mean arterial pressure; Gly—glycemia; Na—natremia; K—kalemia; Cl—chloremia; Anest. duration—duration of anesthesia.; data are shown as median (Q_1 = first quartile; Q_3 = third quartile); * $p < 0.05$.

2.2. DNA Damage

The values of DNA damage for both groups are shown in Table 3. The determined parameter was the percentage of DNA in the tail of the comet. In the table are the percentage values of DNA SSB, oxidized pyrimidine bases (ENDO III), and oxidized purine bases (FPG). In the general anesthesia group, the results show a statistically significant difference between the before and after blood samples for all three parameters (for SSB $r = 0.84$. $p < 0.0001$; for ENDO III $r = 0.83$. $p < 0.0001$; for FPG $r = 0.72$. $p < 0.0001$; Table 3, Figure 1).

Table 3. Results of DNA damage in general and spinal anesthesia groups.

	GA		SA	
	Before	**After**	**Before**	**After**
SSB	7.49	10.05 **	4.00	4.18
	(5.09; 9.66)	(6.97; 11.63)	(1.71; 8.80)	(1.91; 7.39)
ENDO III	8.65	11.85 **	5.83	6.60
	(6.12; 10.20)	(8.27; 13.47)	(3.02; 12.78)	(3.42; 12.73)
FPG	7.97	11.75 **	6.72	8.61
	(5.72; 12.04)	(8.38; 15.32)	(3.34; 15.77)	(4.76; 15.60)

Values are expressed as median (Q_1 = first quartile; Q_3 = third quartile); ** $p < 0.0001$.

The results of spinal anesthesia showed statistically insignificant differences between before and after samples for all the monitored parameters (for SSB $p = 0.97$; for ENDO III $p = 0.29$; for FPG $p = 0.41$; Table 3, Figure 2).

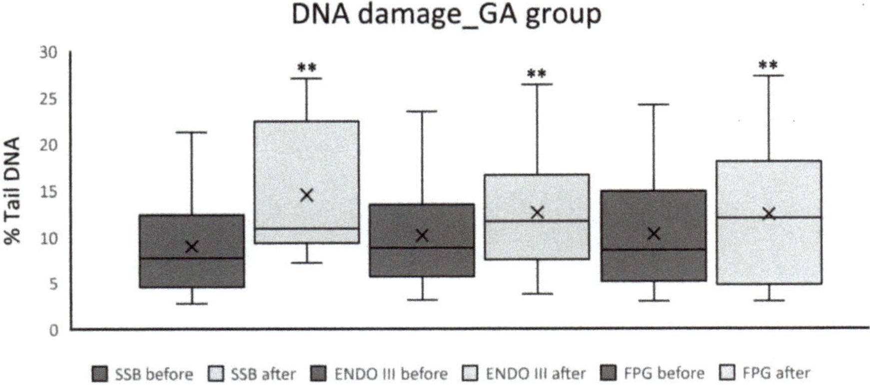

Figure 1. Percentage rail DNA as a measure of SSBs, oxidized pyrimidine bases, and oxidized purine bases in patients underwent general anesthesia (median ± first and third quartile, whiskers are minimal and maximal value), ** $p < 0.0001$.

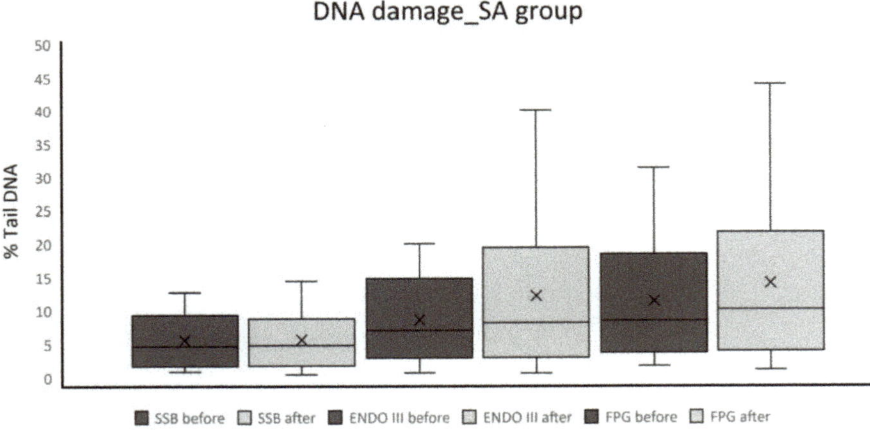

Figure 2. Percentage Tail DNA as a measure of SSBs, oxidized pyrimidine bases, and oxidized purine bases in patients who underwent spinal anesthesia (median ± first and third quartile, whiskers are minimal and maximal value).

2.3. *The Influence of Length of Anesthesia*

The dependence between the duration of anesthesia and SSB breaks in the general anesthesia group came out as statistically significant. This correlation is expressed as linear dependence with $R^2 = 0.36$ (Figure 3). In the spinal anesthesia group, such dependence has not been described (Figure 4).

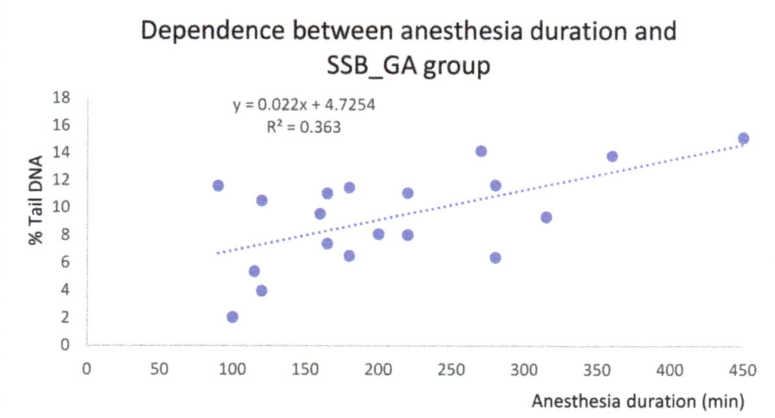

Figure 3. Linear dependence between anesthesia duration and SSB for the general anesthesia group.

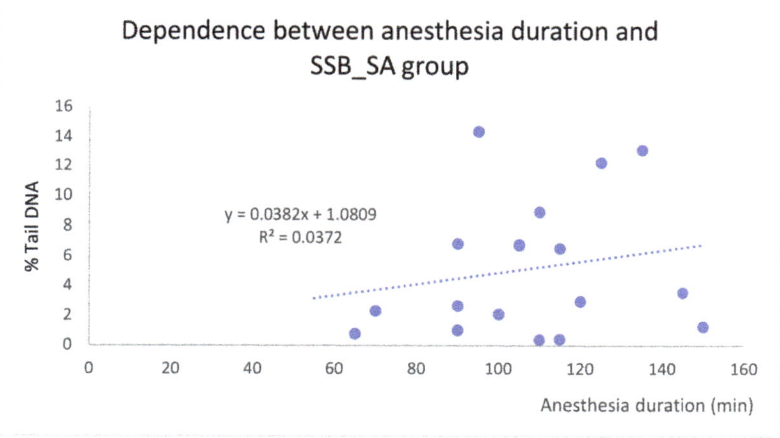

Figure 4. Dependence between anesthesia duration and SSB for the spinal anesthesia group.

3. Discussion

Many drugs and environmental factors can cause DNA damage. Thus, it is important to understand and be able to predict correctly the effect of the DNA damaging agent. The ultimate biological effect of exposure depends on many processes within the cell that affect DNA damage and repair [33]. It is also known that the level of normal oxidative damage is different between individuals according to the condition of antioxidative and stress response systems.

The aim of our pilot study is to describe DNA damage due to the different types of anesthesia during surgery of a similar range. Many authors have addressed the case of general anesthesia. Sardas and co-authors described DNA damage in patients after abdominal surgery under general anesthesia with isoflurane. They discovered a statistically significant difference between the experimental and control group and between states before and after anesthesia [5]. Similar results were obtained by Karybiyik et al. They compared a group of 24 un-premedicated patients with ASA grades 1–2 with a control group consisting of 12 healthy individuals. They used the comet assay method and recorded a significant increase in the mean comet response in blood sampled from patients at 60 and 120 min of anesthesia, and on the first day after anesthesia, in both sevoflurane and isoflurane treated groups [23]. Kadioglu et al. dealt with the influence of sevoflurane on DNA damage in patients indicated for mastectomy. They used two methods—the comet assay and the alkaline halo assay. They described statistically significant damage to DNA by both methods after two-hour surgery [34]. Nogueira and et al. described statistically significant damage the day after minor surgery under general anesthesia maintained with desflurane [35]. Braz with coworkers assessed the DNA damage in patients who had undergone minimally invasive surgery lasting two hours under inhalation of sevoflurane and propofol. No significant difference in DNA damage was observed in either group of patients, and their study is unique [4]. The authors of the cited studies followed the state of the DNA on the several postoperative days. They agree that full DNA repair will occur between the 3rd and 5th postoperative day [5,23,34].

The results of the above studies together with our results raises the questions about the genotoxicity of anesthetics causing iatrogenic damage. Anesthetics, as highly lipophilic substances, do not need any specific transporter to get inside the cell nucleus. It is known that the potential toxic degradation product of SVF is fluoromethyl 2-2 difluoro-1-(trifluoromethyl)vinyl ether (Compound A). The highly reactive nature of Compound A suggests that it may be an alkylating agent and is well known that alkylating agents are electrophilic compounds with affinity towards the nucleophilic parts of macromolecules [23].

Anesthesia as a chemical agent belongs to the group of intracellular agents that cause DNA changes due to its participation in the production of reactive oxygen species (ROS). ROS overproduction leads to oxidative changes on membranes, proteins (also histones), or DNA in the form of DNA breaks. This condition disrupts the homeostasis of the cell and, in combination with the disruption of the mitochondrial transmembrane potential caused by the presence of a pro-oxidant status, could accelerate apoptosis of lymphocytes. We also think that on the other hand, chemical agents could also participate in chemical modification of DNA bases that can be checked by enzymatic modification of the comet assay. These changes lead to mistakes in base pairing and could result in mutagenesis [36].

In our pilot study, two blood samples from each of the 40 patients were processed and the degree of DNA damage was evaluated by the comet assay method. The results showed a statistically significant increase in DNA damage in all three monitored parameters (SSB, and oxidized pyrimidine and purine bases) after surgery in the 20 patients undergoing general anesthesia. In the spinal anesthesia group, no statistically significant differences were found. This study has several limitations. We are aware that both groups of patients were disparate, in the type of surgery resulting from diagnosis, in that the surgery under general anesthesia required more time, and in that some patients underwent repeated surgery (for example, patients after polytrauma). Also, the age of the patients in both groups was statistically significantly different; the patients receiving spinal anesthesia being appreciably older. This was driven by the type of diagnosis, with older patients undergoing knee or hip replacement with total arthroplasty. Our results show that there is a linear dependence between the duration of general anesthesia and the level of single-strand breaks (Figure 3). This dependence has not been demonstrated in spinal anesthesia (Figure 4).

Our aim is to confirm the applicability of comet assay in the research of DNA damage caused by different types of anesthesia. This assay is now widely accepted as a standard method for assessing DNA damage in individual cells and is used in a broad variety of applications including human biomonitoring, genotoxicology, ecological monitoring, and as a tool to investigate DNA damage and repair in different cell types in response to a range of DNA-damaging agents. Because there is no uniform protocol for comet assay, for every lab, the verification of its own protocol is needed. Our protocol was tested and optimized during previous research, where the range for positive and negative results was found [37]. For this experiment, we did not have a positive control yet, but for the next experiment, the samples containing a higher level of oxidative damage should be used as a positive control.

The aim of future experiments should be to expand the patient group to validate the results and verify that these findings will apply to larger experimental groups. It will be necessary to repeat the measurements with larger and more homogeneous groups of patients and collect more samples to have a good statistical strength (to have few age groups, groups according to the duration of anesthesia) where the effect of anesthesia according to higher age or longer duration of anesthesia should be evaluated. It should also be interesting to include some patients with the combination of anesthesia and another oxidative agent (smoking, chemotherapy) to search for the effect of combined agents. A good idea should be the inclusion of other methods for evaluation of biological levels of basic defense or antioxidative system compounds (enzyme activity, protein, or mRNA levels) to compare the interindividual differences in the ability to fight against harmful agents.

Our work has shown that a modified alkaline version of the comet analysis using enzymes is suitable for the quantification of single-strand breaks and oxidized purines and pyrimidines in the DNA of lymphocytes. The advantage of the method is the small number of cells required for analysis, its sensitivity, and the detection of single-cell damage. It is a quick and relatively simple method for assessing oxidative damage. The method is suitable for clinical trials as it does not present a burden on the patient when taking blood samples or by exposure to harmful substances.

This study is a pilot study whose primary objective was to verify the feasibility of the method and with the secondary goal of verifying the hypothesis that neuraxial spinal anesthesia causes less damage to DNA than general anesthesia. Our results and results of the cited studies on patients undergoing surgery and on operating room staff [6,36] raise further questions for the prevention of DNA damage in both patients and operating room staff [38]. Notwithstanding, it is also difficult to assess the impact of surgical trauma on DNA integrity; we may assume it plays a minor role, as DNA damage after surgery was minimal in the SA group.

4. Materials and Methods

4.1. Selection of Patients

This prospective, monocentric, non-randomized study was approved by the Ethics Committee (University Hospital Hradec Králové. reference number 201511 S14P; identification date 22.10.2015.). All patients signed informed consent to participate in the study. The Helsinki Declaration of Patients' Rights was respected. The inclusion criteria were consent to the study and signing of an informed consent form, age over 18, traumatological major surgery of the lower limb or pelvis under general anesthesia, or orthopedic major joint replacement lower limb surgery under neuraxial anesthesia. The exclusion criteria were acute nature of the procedure, a history of cancer with chemotherapy or radiotherapy in the last 12 months, immunosuppressive therapy in the last 12 months, active smoker (more than 1 cigarette/day), a CT scan in the last week before surgery, and patient refusal to participate. Forty-five patients participated in the study and complete data were obtained from 40 of them. The age of patients in the GA group was between 18 and 66 years (40 ± 15.6 years), with weight in the range 60 to 120 kg (88.5 ± 170.3 kg) and height between 164 and 189 cm (176.3 ± 7.5 cm); BMI was between 22.04 and 40.68 (28.4 ± 5.2). Patients in the SA group were aged between 18 and 88 years (62 ± 15.3 years), with weight in the range 66 to 115 kg (81.2 ± 14.0 kg) and height between 150 and 190 cm (170 ± 9.2 cm); BMI was between 21.74 and 37.55 (27.6 ± 4.4).

4.2. Anesthetic Management

The group of patients taking general anesthesia (GA) underwent traumatological surgery in the lower limb or pelvis (open reduction with internal fixation, so-called ORIF). Patients received premedication of 1.5 mg bromazepam (Lexaurin, Kabu Pharma, Prague, Czech Republic) per os. GA was induced by propofol (Propofol, Fresenius Kabi, Bad Homburg, Germany) at a dose of 2 mg/kg. Analgesia was administered with sufentanil (Sufentanil Torrex, Chiesi Pharmaceuticals, Vienna, Austria) at an initial dose of 10 μg and further in accordance with monitored SPI (surgical plethysmography index). If muscle relaxation was required, atracurium (Tracrium, Aspen Pharma, Dublin, Ireland) was used at an initial dose of 0.5 mg/kg followed by monitoring of the depth of muscle relaxation (the monitor was an integral part of the Aisys anesthesia machine) to TOF 2 (train of four). GA was maintained by isoflurane (Forane, Abott Laboratories Ltd., Maidenhead, UK) in a carrier mixture of oxygen

(FiO_2 0.40) with nitrous oxide with a minimum alveolar concentration of 0.8–1.0. If decurarization was required at the end of the procedure, neostigmine (Syntostigmine, BB Pharma, Prague, Czech Republic) was administered 1.5 mg intravenously and atropine (Atropine Biotika, BB Pharma, Prague, Czech Republic) 0.5 mg intravenously. The demographic data of each patient was recorded and blood pressure, pulse, peripheral blood oxygen saturation, exhaled carbon dioxide concentration, and body temperature measured every 5 min during the surgical procedure. Fluid therapy was maintained with a balanced crystalloid solution (Ringerfundin, BBraun, Melsungen, Germany) via a peripheral venous catheter (18 G, BBraun, Melsungen, Germany) at a baseline rate of 100 mL/h and according to blood loss and circulatory stability. Body temperature was maintained in a physiological range between 36 and 37 °C with a heating pad (Astopad, Stihler electronic, Stuttgart, Germany) and an air heater (Warm Air, Polymed, Hradec Kralove, Czech Republic). Airways were secured by orotracheal intubation (SUMI, Sulejowek, Poland) or laryngeal mask (Teleflex, Dublin Road, Athlone, Westmeath, Ireland). Artificial lung ventilation was volume-controlled or pressure-controlled-volume-guaranteed to a tidal volume of 6 mL/kg with a respiratory rate of 12–16 breaths/minute (according to the exhaled concentration of carbon dioxide range between 35–45 mmHg) and a positive end expiratory pressure 4 cm H_2O. The Aisys anesthesiology machine (GE Healthcare. Prague, Czech Republic) was used.

The group of patients with neuraxial spinal anesthesia (SA) underwent orthopedic surgery—knee or hip replacement with total arthroplasty under subarachnoid anesthesia. Subarachnoid puncture was performed aseptically by Quinke needle (25G 88 mm, BBraun, Melsungen, Germany) in the L2/3 or L3/4 intervertebral space by a paramedial approach after a previous infiltration of the injection site with 1% mesocaine (Mesocaine, Zentiva, Prague). 2.2 mL of 0.5% levobupivacaine (Chirocaine, AbbVie, Prague, Czech Republic) and 2.5 μg of sufentanil (off-label administration; Sufentanil Torrex, Chiesi Pharmaceuticals, Vienna, Austria) were administered intrathecally. Circulatory instability following induction of subarachnoid anesthesia was resolved by titration with 10 mg doses of ephedrine (Ephedrin Biotika, BB Pharma, Prague, Czech Republic). During the procedure, patients received sedation with midazolam (Midazolam Accord, Accord Healthcare Polska, Warsaw, Poland). Patients were given oxygen by face mask at a supply of 5 L/min when the saturation dropped below 93%. Monitoring and data recording were similar to the GA group.

4.3. *Comet Assay*

4.3.1. The Lymphocyte Isolation

Four mL blood samples were obtained from patients before and immediately after the surgery into sodium citrate tubes and then were subjected to DNA damage analysis using the comet assay method previously described [39–41]. Briefly, blood was carefully laid over 4 mL LSM (Biotech, Austria) and tubes were centrifuged (30 min, 1500 rpm, 20 °C). The formed ring of lymphocytes (Figure 5) was transferred to centrifuge tubes (Sarstedt, Austria) and 10 mL of PBS (Sigma Aldrich, Saint Louis, MO, USA) was added. The concentration of cells in the sample was calculated using a Bürker chamber, and samples were centrifuged again (10 min, 1500 rpm, 8 °C). The sediment was re-suspended with PBS buffer (Sigma Aldrich, Saint Louis, MO, USA) and adjusted to a concentration of 1 million cells/mL.

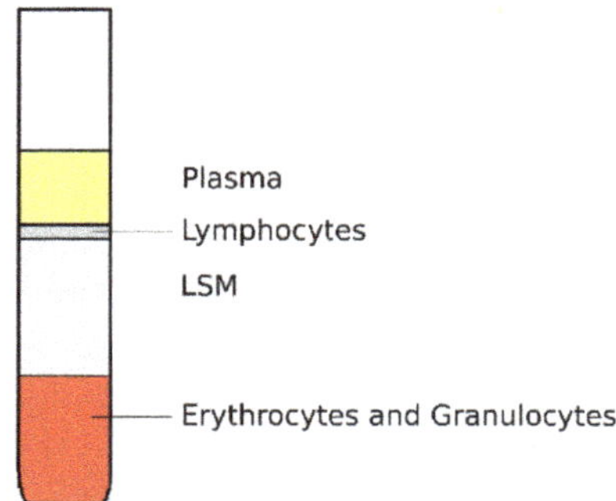

Figure 5. The separation of blood elements using PBS

4.3.2. Alkaline Version of Comet Assay

The alkaline modification of the comet assay was used for the determination of DNA damage [40]. For observation of individual types of damage (SSB breaks and oxidized pyrimidines and purines), lymphocytes were pipetted in 35 µL portions to the 1.5 mL microtubes (Eppendorf, Hamburg, Germany). To a microscope slide pre-coated with 1% aqueous agarose for electrophoresis (Serva, Heidelberg, Germany) was added 85 µL of a 1% solution of high melting point agarose in PBS (HMP agarose, Sigma Aldrich, Saint Louis, MO, USA), and the gel left at 4 °C to allow solidification. A mixture of the 35 µL cell suspension (approximately 35000 cells) and 85 µL of 1% low melting point agarose in PBS (LMP agarose, Sigma Aldrich, Saint Louis, MO, USA) was spread onto this high melting point agarose and again left at 4 °C to allow solidification.

4.3.3. Cell Lysis

Once the top layer had solidified, the slide was gently immersed in cold lysing solution (2.5 M NaCl, 100 mM EDTA, and 10 mM Tris-HCl pH 10 to which 1% Triton X-100 and 10% DMSO had been freshly added). The slides were left at 4 °C for at least 1 h.

4.3.4. Enzymatic Digestion

The slides which were used for enzymatic digestion were gently removed from the lysing solution, washed three times with ENDO buffer (0.1 M KCl, 40 mM HEPES, 0.5 mM EDTA, and 200 µL/mL BSA, pH 8, 37 °C), and 30 µL of the specific enzyme was added. We used the specific DNA endonuclease III (ENDO III) to detect oxidized pyrimidines and the FPG (formamidopyrimidine -DNA glycosylase) to identify altered purines. The slides were covered with coverslips and incubated in a thermobox at 37 °C for 1 h.

4.3.5. Unwinding, Electrophoresis, and Staining

The coverslips were removed from the incubated slides. All slides (incubated and unincubated) were placed in a horizontal electrophoresis tank (Model A5, Owl separation systems, Inc., Thermo Scientific, Waltham, MA, USA) filled with fresh electrophoresis buffer (300 mM NaOH and 1 mM EDTA, pH 13). The slides had to be covered with the buffer and exposed to alkali for 40 min to allow DNA unwinding and cleavage of alkali-labile sites. Electrophoresis was then conducted at 33 V, 300 mA for 30 min at 4 °C by using an electrophoresis power supply EPS 300 IIV (C.B.S. Scientific Company, Inc., CA, USA). The negatively charged DNA bases migrate to the positively charged anode and form the shape of a comet when subjected to an electric field. The size and the shape of the comet and the distribution of the DNA within the comet correlate with the extent of DNA damage. After electrophoresis, the slides were removed from the tank and washed three times for 5 min with neutralizing buffer (0.4 M Tris-HCl, pH 7.5) and finally with distilled water. Excess liquid was blotted from each slide and the DNA stained with 20 μL ethidium bromide. A clean coverslip was then placed over the slide.

4.3.6. DNA Damage Evaluation

Slides were evaluated under a fluorescent microscope with an optical system (NIKON INSTRUMENTS INC., Melville, NY, USA). One hundred cells per slide were scored according to % tail DNA by the software LUCIA Comet Assay (Laboratory Imaging, Prague, Czech Republic). This means a total of 300 cells were analyzed per subject. The DNA analysis is expressed as the ratio of the DNA intensity in the tail relative to the head of the comet. The percentage tail DNA was determined for single-strand breaks in the DNA (SSB), pyrimidine damage (ENDO III), and purine damage (FPG). After evaluation, the slides were washed and allowed to dry.

4.3.7. Statistical Evaluation

The acquired data were analyzed using the programs Graph-Pad Prism7 (GraphPad Software, La Jolla, CA, USA) and Excel 2016 (Microsoft, Redmont, WA, USA).

The normality distribution of the demographic data and the results of the laboratory tests was demonstrated by D'Agostino and Pearson test and the Shapiro–Wilk normality test. According to the normality of the distribution, the Mann–Whitney test was used for comparison of the demographic data and laboratory tests of the two experimental groups. It is used to evaluate unpaired experiments when comparing two sample files or two experimental reaches. It is used when due to the small number of the experimental data or their nature, there is uncertainty about the normality of the distribution. Results were expressed as the medians and first and third quartiles (25% percentile, 75% percentile).

For evaluation of the statistical significance of the DNA damage between the general anesthesia and spinal anesthesia groups, the Wilcoxon matched-pairs signed-rank test was used. It tests the match of two medians and does not assume normality of their distribution. The differences between patients in individual ASA grades between the GA group and SA group were tested by the Mann–Whitney test.

Author Contributions: M.K. (Monika Kucharova) – investigation. writing- original draft preparation. D.A. – investigation. data analysis. V.Z. – investigation. data analysis. M.K. (Maria Koscakova) – investigation. data analysis. R.S.—writing—review and editing. Z.Z.—funding acquisition. Supervision. M.H.—writing—review and editing. supervision. All authors have read and agreed to the published version of the manuscript.

Funding: This work was supported by the Development and Research of Drugs of Charles University [PROGRES Q42] and Ministry of Health Czech Republic—Development of Research Organization (University Hospital Hradec Kralove) [MH CZ—DRO UHHK 00179906] and by institutional support of University Hospital Hradec Kralove: MH CZ-DRO (UHHK, 00179906.

Acknowledgments: The authors are grateful to Ian McColl, (language reviewer, UK) for assistance with the manuscript.

Conflicts of Interest: The authors declare that they have no conflict of interest in relation to the topic of the work.

Abbreviations

GA	General anesthesia
NA	Neuraxial anesthesia
GABA	Gamma aminobutyric acid
NMDA	*N*-methyl-D-aspartic acid
ROS	Reactive oxygen species
ISF	Isoflurane
SVF	Sevoflurane
LSM	Lymphocyte separated medium
PBS	Phosphate buffered saline
EDTA	Ethylenediaminetetraacetic acid

Tris-HCl	Tris hydrochloride
Triton-X	Polyoxyethylene glycol *t*-octyl-phenyl ether
DMSO	Dimethylsulfoxide

References

1. Tavare, A.N.; Perry, A.J.S.; Benzonana, L.L.; Takata, M.; Ma, D. Cancer recurrence after surgery: Direct and indirect effecs of anesthetic agents. *Int. J. Cancer* **2012**, *130*, 1237–1250. [CrossRef]
2. Braz, M.G.; Karahalil, B. Genotoxicity of anesthetics evaluated in vitro (Animals). *Biomed. Res. Int.* **2015**, *2015*, 280802. [CrossRef]
3. Kaymak, C.; Kadioglu, E.; Coskun, E.; Basar, H.; Basar, M. Determination of DNA damage after exposure to inhalation anesthetics in human peripheral lymphocytes and sperm cells in vitro by comet assay. *Hum. Exp. Toxicol.* **2012**, *31*, 1207–1213. [CrossRef] [PubMed]
4. Braz, M.G.; Braz, L.G.; Barbosa, B.S.; Giacobino, J. DNA damage in patients who underwent minimally invasive surgery under inhalation or intravenous anesthesia. *Mutat. Res. Toxicol. Environ. Mutagenesis* **2011**, *726*, 251–254. [CrossRef] [PubMed]
5. Sardaş, S.; Karabiyik, L.; Aygün, N.; Karakaya, A.E. DNA damage evaluated by the alkaline comet assay in lymphocytes of humans anaesthetized with isoflurane. *Mutat. Res.* **1998**, *418*, 1–6. [CrossRef]
6. Da Costa Paes, E.R.; Braz, M.G.; de Lima, J.T.; da Silva, M.R.G.; de Sousa, L.B.; Lima, E.S.; de Vasconcellos, M.C.; Braz, J.R.C. DNA damage and antioxidant status in medical residents occupationally exposed to waste anesthetic gases. *Acta Cir. Bras.* **2014**, *29*, 280–286. [CrossRef] [PubMed]
7. Hoerauf, K.H.; Wiesner, G.; Schroegendorfer, K.F.; Jobst, B.P.; Spacek, A.; Hardt, M.; Sator-Katzenschalger, S.; Rüdiger, W. Waste anaesthetic gases induce sister chromatid exchanges in lymphocytes of operating room personnel. *Br. J. Anaesth.* **1999**, *82*, 764–766. [CrossRef]
8. Bienengraeber, M.W.; Weihrauch, D.; Kersten, J.R.; Pagel, P.S.; Warltier, D.C. Cardioprotection by volatile anesthetics. *Vasc. Pharmacol.* **2005**, *42*, 243–252. [CrossRef]
9. Corbett, T.H. Cancer and congenital anomalies associated with anesthetic. *Ann. N. Y. Acad. Sci.* **1976**, *271*, 58–66. [CrossRef]
10. Eger, E.I., 2nd; White, A.E.; Bown, C.L.; Biava, C.G.; Corbett, T.H.; Stevens, W.C. A test of the carcinogenicity of enflurane. isoflurane. halothane. methoxyflurane and nitrous oxide in mice. *Anesth. Analg.* **1978**, *57*, 678–694. [CrossRef]
11. Braz, M.G.; Mazoti, M.Á.; Giacobino, J.; Braz, L.G.; Golim Mde, A.; Ferrasi, A.C.; de Cavalho, L.R.; Braz, J.R.; Salvador, D.M. Genotoxicity cytotoxicity and gene expression in patients undergoing elective surgery under isoflurane anaesthesia. *Mutagenesis* **2011**, *26*, 415–420. [CrossRef] [PubMed]
12. Iwasaki, M.; Zhao, H.; Jaffer, T.; Unwith, S.; Benzonana, L.; Lian, Q.; Sakamoto, A.; Ma, D. Volatile anaesthetics enhance the metastasis related cellular signalling including CXCR2 of ovarian cancer cells. *Oncotarget* **2016**, *7*, 26042–26056. [CrossRef] [PubMed]
13. Yang, W.; Cai, J.; Zabkiewicz, C.; Zhang, H.; Ruge, F.; Jiang, W.G. The effects of anesthetics on recurrence and metastasis of cancer and clinical implications. *World J. Oncol.* **2017**, *8*, 63–70. [CrossRef] [PubMed]
14. Wigmore, T.J.; Mohammed, K.; Jhanji, S. Long-term survival for patients undergoing volatile versus IV anesthesia for cancer surgery. *Anesthesiology* **2016**, *124*, 69–79. [CrossRef] [PubMed]
15. Zhu, M.; Li, M.; Zhou, Y.; Dangelmajer, S.; Kahlert, U.D.; Xie, R.; Xi, Q.; Shahveranov, A.; Ye, D.; Lei, T. Isoflurane enhances the malignant potential of glioblastoma stem cells by promoting their viability. mobility in vitro and migratory capacity in vivo. *Br. J. Anaesth.* **2016**, *116*, 873–877. [CrossRef] [PubMed]
16. Benzonana, L.L.; Perry, N.J.; Watts, H.R.; Yang, B.; Perry, I.A.; Coombes, C.; Takata, M.; Ma, D. Isoflurane. a commonly used volatile anesthetic enhances renal cancer growth and malignant potential via the hypoxia-inducible factor cellular signaling pathway in vitro. *Anesthesiology* **2013**, *119*, 593–605. [CrossRef] [PubMed]
17. Arora, S.S.; Gooch, J.L.; Garsía, P.S. Postoperative cognitive dysfunction. Alzheimer's disease and anesthesia. *Int. J. Neurosci.* **2014**, *124*, 236–242. [CrossRef]
18. Inan, G.; Özköse Satirlar, Z. Alzheimer disease and anesthesia. *Turk. J. Med. Sci.* **2015**, *45*, 1026–1033. [CrossRef]

19. Bourne, E.; Wright, C.; Royse, C. A review of local anesthetic cardiotoxicity and treatment with lipid emulsion. *Local Reg. Anesth.* **2010**, *3*, 11–19.
20. Momoh, A.O.; Hilliard, P.E.; Chung, K.C. Regional and neuraxial analgesia for plastic surgery: Surgeon's and anesthesiologist's perspectives. *Plast. Reconstr. Surg.* **2014**, *134*, 58S–68S. [CrossRef]
21. Larsen, R. *Anestezie*, 7th ed.; Grada: Prague, Czech Republic, 2004.
22. Izdes, S.; Sardas, S.; Kadioglu, E.; Kaymak, C.; Ozcaglie, E. Assessment of genotoxic damage in nurses occupationally exposed to anaesthetic gases or antineoplastic drugs by the comet assay. *J. Occup. Health* **2009**, *51*, 283–286. [CrossRef] [PubMed]
23. Karabiyik, L.; Sardas, S.; Polat, U.; Kocabas, N.A.; Karakaya, A.E. Comparison of genotoxicity of sevoflurane and isoflurane in human lymphocytes studied in vivo using the comet assay. *Mutat. Res.* **2001**, *492*, 99–107. [CrossRef]
24. Eroglu, F.; Yavuz, L.; Ceylan, B.G.; Yilmaz, F.; Eroglu, E.; Delibas, N.; Naziroğlu, M. New volatile anesthetic. desflurane. reduces vitamin E level in blood of operative patients via oxidative stress. *Cell Biochem. Funct.* **2010**, *28*, 211–216. [CrossRef] [PubMed]
25. Alleva, R.; Tomasetti, M.; Solenghi, M.D.; Stagni, F.; Gamberini, F.; Bassi, A.; Fornasari, P.M.; Fanelli, G.; Borghi, B. Lymphocyte DNA damage precedes DNA repair or cell death after orthopaedic surgery under general anaesthesia. *Mutagenesis* **2003**, *18*, 423–428. [CrossRef]
26. Park, C.J.; Park, S.A.; Yoon, T.G.; Lee, S.J.; Yum, K.W.; Kim, H.J. Bupivacaine induces apoptosis via ROS in the Schwann cell line. *J. Dent. Res.* **2005**, *84*, 852–857. [CrossRef]
27. Lu, J.; Xu, S.Y.; Zhang, Q.G.; Le, H.Y. Bupivacaine induces reactive oxygen species production via activation of the AMP-activated protein kinase-dependent pathway. *Pharmacology* **2011**, *87*, 121–129. [CrossRef]
28. Werdehausen, R.; Braun, S.; Hermanns, H.; Kremer, D.; Küry, P.; Hollmann, M.W.; Bauer, I.; Stevens, M.F. The influence of adjuvants used in regional anesthesia on lidocaine-induced neurotoxicity in vitro. *Reg. Anesth. Pain Med.* **2011**, *36*, 436–443. [CrossRef]
29. Neri, M.; Milazzo, D.; Ugolini, D.; Milic, M.; Campolongo, A.; Pasqualetti, P.; Bonassi, S. Worldwide interest in the comet assay: A bibliometric study. *Mutagenesis* **2015**, *30*, 155–163. [CrossRef]
30. Kuchařová, M.; Hronek, M.; Rybáková, K.; Zadák, Z.; Štětina, R.; Josková, V.; Patková, A. Comet assay and its use for evaluating oxidative DNA damage in some pathological states. *Physiol. Res.* **2019**, *68*, 1–15. [CrossRef]
31. Apostolou, P.; Toloudi, M.; Kourtidou, E.; Mimikakou, G.; Vlachou, I.; Chatziioannou, M.; Papasotiriou, I. Use of the comet assay technique for quick and reliable prediction of in vitro response to chemotherapeutics in breast and colon cancer. *J. Biol. Res.* **2014**, *21*, 1–7. [CrossRef]
32. Alapetite, C.; Thirion, P.; la Rochefordiére, A.; Cosset, J.M.; Moustacchi, E. Analysis of alkaline comet assay of cancer patients with severe reactions to radiotherapy: Defective rejoining of radioinduced DNA strand breaks in lymphocytes of breast cancer patients. *Int. J. Cancer* **1999**, *83*, 83–90. [CrossRef]
33. Brozovic, G.; Orsolic, N.; Rozgaj, R.; Kasuba, V.; Knezevic, F.; Knezevic, A.H.; Benkovic, V.; Lisicic, D.; Borojevic, N.; Dikic, D. DNA damage and repair after exposure to sevoflurane in vivo evaluated in Swiss albino mice by the alkaline comet assay and micronucleus test. *J. Appl. Genet.* **2010**, *51*, 79–86. [CrossRef] [PubMed]
34. Kadioglu, E.; Sardas, S.; Erturk, S.; Ozatamer, O.; Karakaya, A.E. Determonation of DNA damage by alkaline halo and comet assay in patients under sevoflurane anesthesia. *Toxicol. Ind. Health* **2009**, *25*, 205–212. [CrossRef] [PubMed]
35. Nogueira, F.R.; Braz, L.G.; de Andrade, L.R.; de Carvalho, A.L.R.; Vane, L.A.; Módolo, N.S.P.; Aun, A.G.; Souza, K.M.; Reinaldo, J.; Braz CBraz, M.G. Evaluation of genotoxicity of general anesthesia maintained with desflurane in patients under minor surgery. *Environ. Mol. Mutagenesis* **2016**, *57*, 312–316. [CrossRef] [PubMed]
36. Sardaş, S.; Cuhruk, H.; Karakaya, A.E.; Atakurt, Z. Sister-chromatid exchanges in operating room personnel. *Mutat. Res.* **1992**, *279*, 117–120. [CrossRef]
37. Fikrova, P.; Stetina, R.; Hrciarik, M.; Hrnciarikova, D.; Hronek, M.; Zadak, Z. DNA crosslinks, DNA damage and repair in peripheral blood lymphocytes of non-small cell lung cancer patients treated with platinum derivatives. *Oncol. Rep.* **2014**, *31*, 391–396. [CrossRef]

38. Sardas, S.; Izdes, S.; Ozcagli, E.; Kanbak, O.; Kadioglu, E. The role of antioxidant supplementation in occupational exposure to waste anaesthetic gases. *Int. Arch. Occup. Environ. Health* **2006**, *80*, 154–159. [CrossRef]
39. Boyum, A. Separation of white blood cells. *Nature* **1964**, *204*, 793–794. [CrossRef]
40. Collins, A.R.; Dobson, V.L.; Dusinska, M.; Kennedy, G.; Stetina, R. The comet assay: What can it really tell us? *Mutat. Res.* **1997**, *375*, 183–193. [CrossRef]
41. Tice, R.; Vasques, M. Protocol for the Application of the pH>13 Alkaline Single Cell Gel (SCG) Assay to the Detection of DNA Damage in Mammalian Cells. Available online: http://cometassay.com/Tice%20and%20Vasques.pdf (accessed on 30 October 2019).

International Journal of
Molecular Sciences

Article

A Cross-Sectional Study on 3-(2-Deoxy-β-D-Erythro-Pentafuranosyl)Pyrimido [1,2-α]Purin-10(3H)-One Deoxyguanosine Adducts among Woodworkers in Tuscany, Italy

Filippo Cellai [1], Fabio Capacci [2], Carla Sgarrella [2], Carla Poli [3], Luciano Arena [3], Lorenzo Tofani [4], Roger W. Giese [5] and Marco Peluso [1,*]

1 Cancer Factor Risk Branch, Regional Cancer Prevention Laboratory, Institute for Cancer Research, Prevention and Clinical Network (ISPRO), 50139 Florence, Italy; f.cellai@ispro.toscana.it
2 Functional Unit for Prevention, Health and Safety in the Workplace, ASL10, 50139 Florence, Italy; fabio.capacci@uslcentro.toscana.it (F.C.); carla.sgarrella@uslcentro.toscana.it (C.S.)
3 Department of Prevention, Azienda USL Toscana Centro, 50139 Florence, Italy; carla.poli@uslcentro.toscana.it (C.P.); luciano.arena@uslcentro.toscana.it (L.A.)
4 Department of Neurosciences, Psychology, Drug Research and Child Health, University of Florence, 50139 Florence, Italy; lorenzo120787@gmail.com
5 Bouve College of Health Sciences, Barnett Institute, Northeastern University, Northeastern University, Boston, MA 02115, USA; r.giese@northeastern.edu
* Correspondence: m.peluso@ispro.toscana.it

Received: 14 May 2019; Accepted: 5 June 2019; Published: 5 June 2019

Abstract: Occupational exposure to wood dust has been estimated to affect 3.6 million workers within the European Union (EU). The most serious health effect caused by wood dust is the nasal and sinonasal cancer (SNC), which has been observed predominantly among woodworkers. Free radicals produced by inflammatory reactions as a consequence of wood dust could play a major role in SNC development. Therefore, we investigated the association between wood dust and oxidative DNA damage in the cells of nasal epithelia, the target site of SNC. We have analyzed oxidative DNA damage by determining the levels of 3-(2-deoxy-β-D-erythro-pentafuranosyl)pyrimido[1,2-α]purin-10(3H)-one deoxyguanosine (M_1dG), a major-peroxidation-derived DNA adduct and a biomarker of cancer risk in 136 woodworkers compared to 87 controls in Tuscany, Italy. We then examined the association of M_1dG with co-exposure to volatile organic compounds (VOCs), exposure length, and urinary 15-F_{2t} isoprostane (15-F_{2t}-IsoP), a biomarker of oxidant status. Wood dust at the workplace was estimated by the Information System for Recording Occupational Exposures to Carcinogens. M_1dG was measured using ^{32}P-postlabeling and mass spectrometry. 15-F_{2t}-IsoP was analyzed using ELISA. Results show a significant excess of M_1dG in the woodworkers exposed to average levels of 1.48 mg/m^3 relative to the controls. The overall mean ratio (MR) between the woodworkers and the controls was 1.28 (95% C.I. 1.03–1.58). After stratification for smoking habits and occupational status (exposure to wood dust alone and co-exposure to VOCs), the association of M_1dG with wood dust (alone) was even greater in non-smokers workers, MR of 1.43 (95% C.I. 1.09–1.87). Conversely, not consistent results were found in ex-smokers and current smokers. M_1dG was significantly associated with co-exposure to VOCs, MR of 1.95 (95% C.I. 1.46–2.61), and occupational history, MR of 2.47 (95% C.I. 1.67–3.62). Next, the frequency of M_1dG was significantly correlated to the urinary excretion of 15-F_{2t}-IsoP, regression coefficient (β) = 0.442 ± 0.172 (SE). Consistent with the hypothesis of a genotoxic mechanism, we observed an enhanced frequency of M_1dG adducts in woodworkers, even at the external levels below the regulatory limit. Our data implement the understanding of SNC and could be useful for the management of the adverse effects caused by this carcinogen.

Keywords: wood dust; nasal epithelia; VOCs; M_1dG

1. Introduction

Occupational exposure to wood dust has been estimated to affect 3.6 million workers within the European Union (EU) [1]. Among these, 1.5 million workers are exposed to low dust levels (<0.5 mg/m^3), whereas 0.2 million workers are exposed to higher dust values (>5 mg/m^3). The most serious health effect caused by wood dust is nasal and sinonasal cancer (SNC) [2], which has been observed predominantly among woodworkers. In 1960, the first link between SNC and wood dust was found in the British furniture industry [3]. In that study, a several hundred-fold higher risk of SNC was detected among woodworkers relative to unexposed controls. In 1995, based on epidemiological evidence, the International Agency for Research on Cancer (IARC) classified wood dust as a human carcinogen (Group 1) (IARC, 1995). SNC is a rare neoplasm, accounting for 3% of head and neck cancers, with squamous cell carcinoma and adenocarcinoma (AD) histotypes, where AD is more frequently correlated to wood dust. In a recent meta-analysis [4], the relative risk for SNC among woodworkers was estimated to be of 5.91 (95% Confidence Interval (C.I.) 4.31–8.11) in case-control studies and 1.61 (95% C.I. 1.10–2.37) in cohort studies. In that meta-analysis, the highest risk was observed for the AD subtype (29.43, 95% C.I. 16.46–52.61). The overall incidence of SNC was estimated at 0.5 cases per 100,000 between 1978–2002 in the EU [4]. Whereas the incidence of SNC was at 0.4–2.0 per 100,000 in men and at 0.1–0.5 per 100,000 in women in the period 1998–2002 in Italy. A significant risk of SNC was found in the wood and furniture industry in the Piedmont region, Italy, Odds Ratio (OR) of 4.4 (95% C.I. 1.41–13.4), and in Tuscany, Italy, OR of 5.4 (95% C.I. 1.7–17.2) [4]. The latter investigation showed the greatest correlation between wood dust and the AD subtype, OR of 89.7 (95% C.I. 19.8–407.3). In another case-control study in Piedmont, Italy [4], a significant risk of SNC was found to be associated with ever exposure to wood dust, OR of 11.4 with a risk for AD, OR of 58.6, ten-fold greater than for other histotypes. Moreover, the risk for AD subtype doubled every five-years of occupational history.

Free radicals released during inflammatory response to wood dust could play a major role in SNC development [5]. However, the genotoxic effects caused by inflammatory reactions subsequent to wood dust exposure are poorly known. Excessive radical oxygen species (ROS) induced by inflammation [6] can damage protein, membrane lipids, and genetic DNA [7]. In particular, the oxidative degradation of lipids by a free radical chain reaction mechanism can result in their non-enzymatic degradation to many compounds, including 4-hydroxynonenal and malondialdehyde, the latter a reactive aldehyde [8], capable of interacting with DNA to give 3-(2-deoxy-β-D-erythro-pentafuranosyl)pyrimido[1,2-α]purin-10(3H)-one deoxyguanosine (M_1dG) adducts [9]. M_1dG adducts, if not completely repaired, can block cellular replication and cause base pair and frameshift mutations [10]. M_1dG adducts have been associated to DNA methylation aberrations in the Long Interspersed Nuclear Element-1 repeated sequences and in the inflammatory cytokine interleukin(IL)-6 gene [11,12]. DNA aberrations and mutations are important signs of carcinogenic process and hospital-based studies showed that high M_1dG levels are linked to cancer development and tumor progression [13–15].

Significant associations between occupational air pollution exposure and increments of biomarker exposure and cancer risk are apparent [16–19]. Recently, we have examined the prevalence of urinary 15-F_{2t} isoprostane (15-F_{2t}-IsoP), a biomarker of oxidant status [20], in woodworkers compared to controls in Tuscany, Italy [21]. In that study, we found a significant excess of urinary 15-F_{2t}-IsoP excretion in woodworkers with an overall Means Ratio (MR) of 1.36, 95% C.I. 1.18–1.57. Subsequently, we wondered whether woodworkers experienced enhanced frequency of M_1dG adducts, a biomarker of oxidative stress and cancer risk [15,18,19,22–24], which could indicate potential health risks in later life. Therefore, we have examined the frequency of M_1dG in the cells of the nasal epithelia, the target site of SNC [5]. Our approach consisted in the conduction of a cross-sectional study to evaluate the prevalence of M_1dG, derived from peroxidation of DNA, in 136 woodworkers compared to 87 unexposed controls in Italy's Tuscany region. We then examined the association of M_1dG with co-exposure to volatile organic compounds (VOCs), exposure length, and the urinary levels of 15-F_{2t}-IsoP. Wood dust exposure was estimated by the Information System for Recording Occupational Exposures to Carcinogens (SIREP) [25]. SIREP is a system that provides the data that have been measured by employers in industry atmosphere according to Italian laws. M_1dG was examined by ^{32}P-postlabeling [26] and mass spectrometry [27]. Urinary 15-F_{2t}-IsoP data were obtained from a previous study [21]. Additional understanding of the link between M_1dG adducts and wood dust exposure can improve the knowledge of the mechanisms of actions of this carcinogen in the cells of the respiratory tract of woodworkers.

2. Results

2.1. Demographic Variables

Out of the 44 local wood companies contacted by medical doctors and local health services, 32 consented to participate in this study. There were 136 woodworkers—50 smokers and 25 ex-smokers (mean age 45.2 years ± 11 Standard Deviation (SD))—and 87 unexposed controls—26 smokers and 11 ex-smokers (mean age 45.6 years ± 9 SD). All participants were males, because most woodworkers are men. Controls were resident in areas of the Tuscany region with no proximity to major air pollutant sources. Demographic characteristics and lifestyle habits of workers and controls were comparable.

2.2. Wood Dust Exposure

The airborne levels of wood dust were quantified by an 8-h time-weighted average (TWA-8), according to Italian laws [25]. Wood dust values corresponded to a single value assessed from several consecutive samples obtained by fixed positions located at each workplace. The mean daily concentrations of airborne wood dust were of 1.48 mg/m^3 in the wood industries that participated in this study.

2.3. Reference M_1dG Adduct Standards by ^{32}P-Postlabeling and/or Mass Spectrometry

The levels of M_1dG, expressed as RAL, were 5.0 M_1dG adducts ± 0.6 (Standard Error (SE)) per 10^6 nn in the MDA-treated calf-thymus (DNA by ^{32}P-DNA post labeling) [26]. The presence of M_1dG in the MDA-treated calf-thymus DNA sample was confirmed by the matrix-assisted laser desorption/ionization time-of-flight mass spectrometry [27]. Relative to the accurate masses that were found in the spectrum from one spot, the exact mass of M_1dG was as follows, using the M nomenclature: M_1dG (581.166). A calibration curve was set up by diluting this reference DNA adduct standard sample with untreated calf-thymus DNA and measuring the decreasing levels of M_1dG, r-squared = 0.99.

2.4. M_1dG and Wood Dust

To investigate of the potential genotoxic of wood dust, we analyzed the levels of M_1dG using ^{32}P-postlabeling [26]. A characteristic profile of M_1dG adduct spot was detected in the chromatographic plates of the study population. The visual intensity of the M_1dG adduct spot was generally stronger in the plates of the woodworkers compared to the controls. When we analyzed the frequency of M_1dG in the woodworkers, our findings showed that the adduct frequency was significantly higher, up to 1.7-fold, within the workers exposed to wood dust as compared to the controls. Table 1 reports that there were 83.2 and 50.4 M_1dG adducts per 10^8 nn among the woodworkers and the controls, respectively. The overall mean ratio (MR) between the woodworkers and the unexposed controls was 1.28 (95% C.I. 1.03–1.58). Current smokers had also average amounts of adducts greater than ex-smokers and non-smokers, but the difference was not statistically significant. A significant relationship between adducts and occupational history was detected (147% excess). After stratification for both smoking habits and exposure status (exposure to wood dust alone and co-exposure to VOCs), results in Table 2 show that the relationship of M_1dG with wood dust (alone) was even greater in non-smokers workers, MR of 1.43 (95% C.I. 1.09–1.87). However, inconsistent results were found in ex-smokers and current smokers. Instead, M_1dG was significantly associated with a co-exposure to VOCs, MR of 1.95 (95% C.I. 1.46–2.61), and occupational history, MR of 2.47 (95% C.I. 1.67–3.62). The highest amount of M_1dG was measured in woodworkers who reported to be smokers as well as co-exposed to airborne organic solvents (158.4 M_1dG adducts per 10^8 nn) during woodworking.

Table 1. Mean level of three-(2-deoxy-β-D-erythro-pentafuranosyl)pyrimido[1,2-α]purin-10(3H)-one deoxyguanosine (M_1dG) adducts, expressed as a relative adduct level (RAL) per 10^8 normal nucleotides (nn), Mean Ratio (MR), and 95% Confidence Interval (C.I.), by exposure to wood dust and other variables.

M_1dG, Smoking Habits, Exposure to Wood Dust, and Occupational History				
	N	RAL Per 10^8 nn ± Standard Error	Mean Ratio and 95% C.I.	p-Value [a]
Smoking habits				
Non-smokers	111	63.7 ± 4.7	1	
Former smokers	36	68.7 ± 9.6	0.97 (95% C.I. 0.73–1.31)	0.873
Current smokers	76	81.0 ± 8.5	1.17 (95% C.I. 0.92–1.48)	0.190
Exposure to wood dust				
Controls	87	50.4 ± 2.7	1	
Woodworkers	136	83.2 ± 6.2	1.28 (95% C.I. 1.03–1.58)	0.021
Occupational history				
≤ 8 years	40	67.9 ± 15.1	1	
9–25 years	49	79.0 ± 8.8	1.67 (95% C.I. 1.16–2.38)	0.005
≥ 26 years	47	100.6 ± 8.2	2.47 (95% C.I. 1.67–3.62)	<0.001

[a] p-values (Test of Wald) were adjusted for age and smoking, as appropriate.

Table 2. Average level of three-(2-deoxy-β-D-erythro-pentafuranosyl)pyrimido[1,2-α]purin-10(3H)-one deoxyguanosine (M_1dG), expressed as relative adduct level (RAL) per 10^8 normal nucleotides, Mean Ratio (MR), and 95% Confidence Interval (C.I.), by smoking habits and occupational status, i.e., exposure to wood dust alone and co-exposure to volatile organic compounds (VOCs).

M_1dG, Exposure to Wood Dust (Alone), and Co-Exposure to VOCs				
	N	RAL per 10^8 nn ± Standard Error	Mean Ratio and 95% C.I.	p-Value [a]
Non-smokers				
Controls	50	45.5 ± 3.4	1	
Workers exposed to wood dust (alone)	48	78.8 ± 8.6	1.43 (95% C.I. 1.09–1.87)	0.009
Woodworkers exposed to VOCS	13	80.3 ± 17.5	1.59 (95% C.I. 1.05–2.43)	0.027
Former smokers				
Controls	11	50.9 ± 6.3	1	
Workers exposed to wood dust (alone)	17	67.2 ± 16.8	0.93 (95% C.I. 0.48–1.78)	0.829
Woodworkers exposed to VOCS	8	96.6 ± 21.1	1.52 (95% C.I. 0.71–3.26)	0.211
Current smokers				
Controls	26	59.8 ± 5.1	1	
Workers exposed to wood dust (alone)	34	60.8 ± 8.1	0.78 (95% C.I. 0.58–1.14)	0.211
Woodworkers exposed to VOCS	16	158.4 ± 29.3	2.33 (95% C.I. 1.45–3.70)	0.0003

[a] p-values (Test of Wald) were adjusted for age.

2.5. M_1dG and 15-F_{2t} Isoprostane

Next, we examined the correlation between the levels of M_1dG adducts with the urinary concentrations of 15-F_{2t}-IsoP, a biomarker of oxidant status [20], among woodworkers. Multivariate regression analysis showed that the M_1dG adducts were linearly correlated to the urinary excretion of 15-F_{2t}-IsoP, regression coefficient (β) = 0.442 ± 0.172 (SE), p-value = 0.011.

3. Discussion

In the current study, we examined the genotoxic effects associated with wood dust exposure using a comparative cross-sectional study, with larger numbers of subjects than prior studies [28–30]. The indoor concentrations of wood dust were used as a marker of external carcinogen exposure in the 32 wood companies located in Italy's Tuscany Region that participated in the study. The European Union Directive (1999/38) has classified wood dust as a carcinogenic agent and has set the occupational exposure limit (OEL) to 5.0 mg/m^3. The Scientific Committee for Occupational Exposure Limits (SCOEL) of the European Union has stated that exposure to wood dust above 0.5 mg/m^3 can cause pulmonary effects and should be avoided at workplace [1]. The air monitoring results, which were obtained from the SIREP system, showed that the indoor average levels of wood dust at the workplace was 1.48 mg/m^3.

Multivariate regression analysis showed that the frequency of M_1dG in exposed woodworkers was significantly enhanced as compared to the unexposed controls, MR of 1.28 (95% C.I. 1.03–1.58), even though the airborne exposure to wood dust was below the Italian regulatory limit of 5.0 mg/m^3 (Legislative Decree No 66/2000). The correlation of M_1dG with dust becomes even stronger in non-smokers who were exposed to wood dust (alone), MR of 1.43 (95% C.I. 1.43–1.87), after stratification for smoking habits and co-exposure status. Inconsistent results were found in ex-smokers and current smokers where the genotoxic effects of wood dust (alone) could be potentially merged with tobacco smoke carcinogens. Instead, strong associations were observed with co-exposure to VOCs, that reached the statistical significance among non-smokers and current smokers, MR of 1.59 (95% C.I. 1.05–2.43) and 2.33 (95% C.I. 1.45–3.70), respectively. In doing so, M_1dG generation was found to be significantly correlated to occupational history, MR of 2.47 (95% C.I. 1.67–3.62) and 15-F_{2t}-IsoP urinary excretion (p-value of 0.011).

The significant increment of M_1dG in woodworkers can be caused by ROS released during inflammatory reactions as a consequence of exposure to fine and abundant airborne dust created during woodworking. Significant evidence for inflammations as a consequence of dust exposure comes from an earlier study with experimental animals [31]. In that experiment, repeated airway exposure to wood dust induced inflammation, which was accompanied by several proinflammatory cytokines and chemokines, in the lungs of mice. During the inflammatory response, activated macrophages and neutrophils generate a variety of highly reactive oxidants, such as hydrogen peroxide and hypochlorite acid [32], which are capable of reacting with lipids of membrane leading to reactive aldehydes, which are further capable of interacting with DNA forming M_1dG adducts [7]. Our findings are in line with previous studies that report increased biomarkers in various tissues, such as peripheral blood and oral cells, of workers exposed to wood dust by micronucleus and comet (single-cell gel electrophoresis) techniques [28–30]. In those studies, high DNA strand breaks [28–30] and enhanced chromosomal instability values [2,28,33] were found in the woodworkers relative to the unexposed controls, but discrepant findings were also shown [34]. Results also provide evidence of a significant M_1dG increment (147% excess) in the long-term workers compared to those with shorter exposure, used as a reference level. Long-term adverse genotoxic effects due to inflammations and ROS production cannot be excluded in cells of respiratory epithelia. These findings are in line with previous studies [28,35]. For instance, the generation of a urinary biomarker of oxidative stress was found to be significantly correlated with the length of exposure in a study of workers occupationally exposed to asbestos. Rekhadevi et al. [28] found a significant association between the frequency of micronuclei and the length of occupational exposure to wood dust. Taken together, our results indicate that the production of aldehydes by oxidative degradation of lipids of cellular membranes can represent an obvious possible mechanism that could explain wood carcinogenicity.

Next, we observed a consistent increment of M_1dG (95% excess) with exposure to VOCs, another established risk factor in woodworking [36]. VOCs, such as benzene, xylene, and formaldehyde, are commonly used in wood companies, in particular by subjects involved in varnishing and cleaning of furniture elements [36]. This result is not surprising because exposure to benzene, toluene, and xylene is correlated to the production of ROS, heat shock proteins, and oxidative stress [37]. Benzene is a chemical that can be metabolized by CYP_2E_1 to various chemicals with the capability of redox-cycling, a reaction that generates ROS [38]. Formaldehyde is a substrate of CYP_2E_2 and can be oxidized by peroxidase, aldehyde oxidase, and xanthine oxidase with ROS formation [39]. Formaldehyde exposure has been associated with high levels of CYP_1A_1 and GSH and $GSTT_1$ downregulation [40].

15-F_{2t}-IsoP is a widely used biomarker of exposure [21,41]. Therefore, it is of primary importance to understand its biological relevance in terms of health risk. Recently, an early study has supported the hypothesis of a relationship between F_2-IsoP and 8-hydroxy-2′-deoxyguanosine, a sensitive marker of oxidative stress and cancer risk [24], in experimental animals [42]. Therefore, we have evaluated the association between 15-F_{2t}-IsoP and M_1dG in our study population. Results show that the levels of M_1dG were statistically significantly correlated with the urinary excretion of 15-F_{2t}-IsoPs, indicating that 15-F_{2t}-IsoP is significantly linked to the induction of early cancerogenic effects in the cells of nasal epithelia, the target site of SNC [5].

Although exposure registries are commonly used for the purposes of hazard control, exposure surveillance, and assessment of health risks [1], such the SIREP system [25], this approach has few limitations, including potential exposure misclassification rising from the heterogeneity of wood dust exposure levels within different wood industries. This does not completely reflect the exposure status of each woodworker. Indeed, there could be an underestimation of carcinogen exposure associated with some working operations [43]. Unreported variations in the use of the Personal Protective Equipment [44] could have influenced the individual levels of exposure to wood dust. In addition, a weakness of the study is that exposure levels to wood dust were measured with stationary stations and not personal samplers. Measurements from fixed samplers provide evidence of woodworkers' exposure via air but they are not well representative of individual exposures to dust due to spatial and temporal variations. Among the strengths of this study was the use of exfoliated nasal epithelia cells for adduct analysis, which consists of 89% of epithelial cells and 11% of neutrophils, with few eosinophils and lymphocytes. Unfortunately, we did not measure cell abnormalities in nasal epithelial cells in woodworkers to examine potential correlation between M_1dG levels and pre-cancerous lesions.

4. Material and Methods

4.1. Study Population

In this cross-sectional study, we randomly selected 44 wood industries among those which were under mandatory surveillance for assessing workplace health risks in Italy's Tuscany region. Companies were contacted by occupational physicians working in the local health service. Eligibility criteria were as follows: (1) to be employed in companies of the Florence province from at least one year (for the workers) and (2) to not have history of occupational or environmental carcinogen exposures (for the controls). All the participants were informed about the aims of the study and provided a written informed consent. Details about age, gender, professions, residence, lifestyle habits, occupational status and history, including exposures to carcinogens, such as VOCs and formaldehyde, were obtained by questionnaire. The study was approved by the Institutional Review Board of the General Hospital.

4.2. Wood Dust Exposure Measurement

Data contained in SIREP system [25] were used to compute the indoor concentrations of wood dust in the local wood companies. These data on carcinogen exposure were measured by fixed station air samplers at the workplace, collected by employers and regularly sent to the exposure registers, which contains quantitative measurements of volatile wood dust exposure.

4.3. Nasal Epithelia Brushing

Brushing is an easy and relatively noninvasive method to collect exfoliated cells from nasal epithelia [13]. Briefly, cells of the respiratory tract were collected from clean lower turbinate in each nostril with a cytobrush in the morning at workplace. Brushing samples were then treated with 10% acetylcysteine to break down mucus gel structure. After centrifugations, cellular pellets were stored at −80 °C.

4.4. Reference Adduct Standard

Calf-thymus DNA was exposed to 10 mM MDA (ICN Biomedicals, Irvine, CA, USA), as described [22]. MDA-treated calf-thymus DNA was diluted with untreated DNA to obtain lower amounts of the reference adduct standard to generate a calibration curve.

4.5. DNA Extraction and Purification

DNA was extracted and purified by digestion with ribonucleases A and T_1, proteinase K treatment, extraction with organic solvents and ethanol precipitation [45]. DNA concentration and purity of biological samples were determined by spectrophotometry. Coded DNA samples were stored at −80 °C until laboratory analysis.

4.6. Mass Spectrometry

The generation of M_1dG in MDA-treated calf-thymus DNA was analyzed by mass spectrometry (Voyager DE STR from Applied Biosystems, Framingham, MA), as previously reported [27].

4.7. ^{32}P-DNA Postlabeling

M_1dG formation was examined by ^{32}P-postlabeling [26]. In detail, DNA (2 μg) was incubated with micrococcal nuclease (21.4 mU/μL) and spleen phosphodiesterase (6.0 mU/μL) in hydrolysis buffer, pH 6.0 at 37 °C for 4.5 h. Digests were treated with nuclease P1 (0.1 U/μL) at 37 °C for 30′ [46]. Nuclease P1-resistant nucleotides were incubated with 25 μCi of carrier-free [γ-^{32}P]-ATP (3000 Ci/mM) and polynucleotide kinase T_4 (0.75 U/μL) to generate ^{32}P-labeled adducts in bicine buffer, pH 9.0, at 37 °C for 30 min. The generation of M_1dG was analyzed using the following chromatography system: $MgCl_2$ (0.35 M) for preparatory chromatography; while 2.1 M lithium formate, 3.75 M urea pH 3.75 and 0.24 M sodium phosphate, 2.4 M urea pH 6.4, were used for the bidimensional chromatography, respectively. M_1dG was detected by storage phosphor imaging employing intensifying screens, which were scanned with Typhoon 9210 (Amersham). The relative adduct labelling (RAL) of M_1dG was calculated by the following formula = (pixels in adducted nucleotides)/(pixels in nn). Adduct values were computed considering the recovery of the reference adduct standard.

4.8. Statistical Analysis

M_1dG levels were reported per 10^8 nn. RAL values were log transformed before performing the statistical analyses. Log-normal regression models, with age (years), smoking habits (never, ex, and current), and occupational status (woodworker vs. unexposed control) and history (years), such as independent variables, were employed to analyze the correlation of wood dust exposure with M_1dG. Workers were then sub-grouped according to VOC co-exposures in: (1) exposed to wood dust alone and (2) co-exposed to VOCs. MR estimates and its 95% C.I. were used as a measure of effect [47] for each level of the predictor variables relative to the control group. Statistical analysis was done using the software SAS9.3 and SPSS 20.0 (IBM SPSS Statistics, New York, NY, USA).

5. Conclusions

Our study adds relevant data to the body of literature, originating from population-based studies, showing significant induction of cellular anomalies indicative of genotoxicity in the target site of SNC, even at airborne levels of wood dust below the Italian regulatory limit. Consistent with a genotoxic mechanism, we observed an enhanced frequency of M_1dG adducts in the cells of the respiratory tract of workers occupationally exposed to wood dust. Certainly, other aldehyde component could contribute to the total adduct formation, which could result in a more complex pattern of genotoxicity. Worker surveillance using biomarkers of exposure and cancer risk, such as end points, could be relevant for managing health and safety risk of woodworkers.

Author Contributions: Conceptualization, F.C. (Filippo Cellai) and M.P.; methodology, M.P. and R.W.G.; formal analysis, F.C. (Fabio Capacci) and L.T.; investigation, C.S., C.P. and L.A.; writing—original draft preparation, M.P.; writing—review and editing, R.W.G. and F.C. (Filippo Cellai)".

Funding: This research was funded by Tuscany Region.

Acknowledgments: We are grateful to Dusca Bartoli, Giuseppe A. Farina, Tonina E. Iaia, Pierluigi Faina, and Luciano Monticelli for their useful assistance in the conduction of this project.

Conflicts of Interest: The authors declare no conflict of interest.

References

1. Kauppinen, T.; Vincent, R.; Liukkonen, T.; Grzebyk, M.; Kauppinen, A.; Welling, I.; Arezes, P.; Black, N.; Bochmann, F.; Campelo, F.; et al. Occupational exposure to inhalable wood dust in the member states of the european union. *Ann. Occup. Hyg.* **2006**, *50*, 549–561. [PubMed]
2. Çelik, A.; Kanık, A. Genotoxicity of occupational exposure to wood dust: Micronucleus frequency and nuclear changes in exfoliated buccal mucosa cells. *Environ. Mol. Mutagenesis* **2006**, *47*, 693–698. [CrossRef] [PubMed]
3. Acheson, E.D.; Cowdell, R.H.; Hadfield, E.; Macbeth, R.G. Nasal cancer in woodworkers in the furniture industry. *Br. Med. J.* **1968**, *2*, 587–596. [CrossRef] [PubMed]
4. Binazzi, A.; Ferrante, P.; Marinaccio, A. Occupational exposure and sinonasal cancer: A systematic review and meta-analysis. *BMC Cancer* **2015**, *15*, 49. [CrossRef] [PubMed]
5. IARC. Wood dust. A review of human carcinogens: Arsenic, metals, fibres, and dusts. In *IARC Monographs on the Evaluation of Carcinogenic Risks to Humans/World HEALTH Organization*; International Agency for Research on Cancer: Lyon, France, 2012.
6. Gungor, N.; Pennings, J.L.; Knaapen, A.M.; Chiu, R.K.; Peluso, M.; Godschalk, R.W.; Van Schooten, F.J. Transcriptional profiling of the acute pulmonary inflammatory response induced by lps: Role of neutrophils. *Respir. Res.* **2010**, *11*, 24. [CrossRef] [PubMed]
7. Marnett, L.J. Oxyradicals and DNA damage. *Carcinogenesis* **2000**, *21*, 361–370. [CrossRef] [PubMed]
8. Duong, A.; Steinmaus, C.; McHale, C.M.; Vaughan, C.P.; Zhang, L. Reproductive and developmental toxicity of formaldehyde: A systematic review. *Mutat. Res.* **2011**, *728*, 118–138. [CrossRef]
9. Jeong, Y.C.; Swenberg, J.A. Formation of m1g-dr from endogenous and exogenous ros-inducing chemicals. *Free Radic. Biol. Med.* **2005**, *39*, 1021–1029. [CrossRef]
10. Zhou, X.; Taghizadeh, K.; Dedon, P.C. Chemical and biological evidence for base propenals as the major source of the endogenous m1dg adduct in cellular DNA. *J. Biol. Chem.* **2005**, *280*, 25377–25382. [CrossRef]
11. Peluso, M.; Bollati, V.; Munnia, A.; Srivatanakul, P.; Jedpiyawongse, A.; Sangrajrang, S.; Piro, S.; Ceppi, M.; Bertazzi, P.A.; Boffetta, P.; et al. DNA methylation differences in exposed workers and nearby residents of the ma ta phut industrial estate, rayong, thailand. *Int. J. Epidemiol.* **2012**, *41*, 1753–1760. [CrossRef]
12. Peluso, M.E.; Munnia, A.; Bollati, V.; Srivatanakul, P.; Jedpiyawongse, A.; Sangrajrang, S.; Ceppi, M.; Giese, R.W.; Boffetta, P.; Baccarelli, A.A. Aberrant methylation of hypermethylated-in-cancer-1 and exocyclic DNA adducts in tobacco smokers. *Toxicol. Sci.* **2014**, *137*, 47–54. [CrossRef] [PubMed]
13. Munnia, A.; Bonassi, S.; Verna, A.; Quaglia, R.; Pelucco, D.; Ceppi, M.; Neri, M.; Buratti, M.; Taioli, E.; Garte, S.; et al. Bronchial malondialdehyde DNA adducts, tobacco smoking, and lung cancer. *Free Radic. Biol. Med.* **2006**, *41*, 1499–1505. [CrossRef]

14. Leuratti, C.; Watson, M.A.; Deag, E.J.; Welch, A.; Singh, R.; Gottschalg, E.; Marnett, L.J.; Atkin, W.; Day, N.E.; Shuker, D.E.; et al. Detection of malondialdehyde DNA adducts in human colorectal mucosa: Relationship with diet and the presence of adenomas. *Cancer Epidemiol. Biomarkers Prev.* **2002**, *11*, 267–273. [PubMed]
15. Peluso, M.; Munnia, A.; Risso, G.G.; Catarzi, S.; Piro, S.; Ceppi, M.; Giese, R.W.; Brancato, B. Breast fine-needle aspiration malondialdehyde deoxyguanosine adduct in breast cancer. *Free Radic. Res.* **2011**, *45*, 477–482. [CrossRef] [PubMed]
16. Peluso, M.E.; Munnia, A.; Giese, R.W.; Chellini, E.; Ceppi, M.; Capacci, F. Oxidatively damaged DNA in the nasal epithelium of workers occupationally exposed to silica dust in Tuscany region, Italy. *Mutagenesis* **2015**, *30*, 519–525. [CrossRef] [PubMed]
17. Bono, R.; Munnia, A.; Romanazzi, V.; Bellisario, V.; Cellai, F.; Peluso, M.E.M. Formaldehyde-induced toxicity in the nasal epithelia of workers of a plastic laminate plant. *Toxicol. Res.* **2016**, *5*, 752–760. [CrossRef] [PubMed]
18. Peluso, M.; Srivatanakul, P.; Munnia, A.; Jedpiyawongse, A.; Ceppi, M.; Sangrajrang, S.; Piro, S.; Boffetta, P. Malondialdehyde-deoxyguanosine adducts among workers of a Thai industrial estate and nearby residents. *Environ. Health Perspect.* **2010**, *118*, 55–59. [CrossRef] [PubMed]
19. Peluso, M.; Munnia, A.; Ceppi, M.; Giese, R.W.; Catelan, D.; Rusconi, F.; Godschalk, R.W.; Biggeri, A. Malondialdehyde-deoxyguanosine and bulky DNA adducts in schoolchildren resident in the proximity of the Sarroch industrial estate on Sardinia island, Italy. *Mutagenesis* **2013**, *28*, 315–321. [CrossRef]
20. Milne, L.G.; Dai, Q.; Roberts, J.L. The isoprostanes—25 years later. *Biochim. Biophys. Acta* **2015**, *1851*, 433–445. [CrossRef]
21. Bono, R.; Capacci, F.; Cellai, F.; Sgarrella, C.; Bellisario, V.; Trucco, G.; Tofani, L.; Peluso, A.; Poli, C.; Arena, L.; et al. Wood dust and urinary 15-f2t isoprostane in italian industry workers. *Environ. Res.* **2019**, *173*, 300–305. [CrossRef]
22. Bono, R.; Romanazzi, V.; Munnia, A.; Piro, S.; Allione, A.; Ricceri, F.; Guarrera, S.; Pignata, C.; Matullo, G.; Wang, P.; et al. Malondialdehyde-deoxyguanosine adduct formation in workers of pathology wards: The role of air formaldehyde exposure. *Chem. Res. Toxicol.* **2010**, *23*, 1342–1348. [CrossRef] [PubMed]
23. Phillips, D.H.; Venitt, S. DNA and protein adducts in human tissues resulting from exposure to tobacco smoke. *Int. J. Cancer* **2013**, *131*, 2733–2753. [CrossRef] [PubMed]
24. Brancato, B.; Munnia, A.; Cellai, F.; Ceni, E.; Mello, T.; Bianchi, S.; Catarzi, S.; Risso, G.G.; Galli, A.; Peluso, M.E. 8-oxo-7, 8-dihydro-2-deoxyguanosine and other lesions along the coding strand of the exon 5 of the tumour suppressor gene p53 in a breast cancer case-control study. *DNA Res.* **2016**, *23*, 395–402. [CrossRef] [PubMed]
25. Scarselli, A.; Binazzi, A.; Ferrante, P.; Marinaccio, A. Occupational exposure levels to wood dust in Italy, 1996–2006. *Occup. Environ. Med.* **2008**, *65*, 567–574. [CrossRef] [PubMed]
26. Bonassi, S.; Cellai, F.; Munnia, A.; Ugolini, D.; Cristaudo, A.; Neri, M.; Milic, M.; Bonotti, A.; Giese, R.W.; Peluso, M.E. 3-(2-deoxy-β-d-erythro-pentafuranosyl) pyrimido [1,2-α] purin-10 (3h)-one deoxyguanosine adducts of workers exposed to asbestos fibers. *Toxic. Lett.* **2017**, *270*, 1–7. [CrossRef] [PubMed]
27. Wang, P.; Gao, J.; Li, G.; Shimelis, O.; Giese, R.W. Nontargeted analysis of DNA adducts by mass-tag ms: Reaction of p-benzoquinone with DNA. *Chem. Res. Toxicol.* **2012**, *25*, 2737–2743. [CrossRef]
28. Rekhadevi, P.V.; Mahboob, M.; Rahman, M.F.; Grover, P. Genetic damage in wood dust-exposed workers. *Mutagenesis* **2009**, *24*, 59–65. [CrossRef] [PubMed]
29. Palus, J.; Dziubałtowska, E.; Rydzyński, K. DNA damage detected by the comet assay in the white blood cells of workers in a wooden furniture plant. *Mutat. Res.* **1999**, *444*, 61–74. [CrossRef]
30. Bruschweiler, D.E.; Wild, P.; Huynh, K.C.; Savova-Bianchi, D.; Danuser, B.; Hopf, B.N. DNA damage among woodworkers assessed with the comet assay. *Environ. Health Insights* **2016**, *10*, 105–112. [CrossRef]
31. Maatta, J.; Lehto, M.; Leino, M.; Tillander, S.; Haapakoski, R.; Majuri, M.-L.; Wolff, H.; Rautio, S.; Welling, I.; Husgafvel-Pursiainen, K.; et al. Mechanisms of particle-induced pulmonary inflammation in a mouse model: Exposure to wood dust. *Toxicol. Sci.* **2006**, *93*, 96–104. [CrossRef]
32. Vanhees, K.; van Schooten, F.J.; van Waalwijk van Doorn-Khosrovani, S.B.; van Helden, S.; Munnia, A.; Peluso, M.; Briede, J.J.; Haenen, G.R.; Godschalk, R.W. Intrauterine exposure to flavonoids modifies antioxidant status at adulthood and decreases oxidative stress-induced DNA damage. *Free Radic. Biol. Med.* **2013**, *57*, 154–161. [CrossRef] [PubMed]

33. Bruschweiler, E.D.; Hopf, N.B.; Wild, P.; Huynh, C.K.; Fenech, M.; Thomas, P.; Hor, M.; Charriere, N.; Savova-Bianchi, D.; Danuser, B. Workers exposed to wood dust have an increased micronucleus frequency in nasal and buccal cells: Results from a pilot study. *Mutagenesis* **2014**, *29*, 201–207. [CrossRef] [PubMed]
34. Wultsch, G.; Nersesyan, A.; Kundi, M.; Wagner, K.-H.; Ferk, F.; Jakse, R.; Knasmueller, S. Impact of exposure to wood dust on genotoxicity and cytotoxicity in exfoliated buccal and nasal cells. *Mutagenesis* **2015**, *30*, 701–709. [CrossRef] [PubMed]
35. Yoshida, R.; Ogawa, Y.; Shioji, I.; Yu, X.; Shibata, E.; Mori, I.; Kubota, H.; Kishida, A.; Hisanaga, N. Urinary 8-oxo-7, 8-dihydro-2′-deoxyguanosine and biopyrrins levels among construction workers with asbestos exposure history. *Ind. Health* **2001**, *39*, 186–188. [CrossRef] [PubMed]
36. Binazzi, A.; Corfiati, M.; Di Marzio, D.; Cacciatore, A.M.; Zajacovà, J.; Mensi, C.; Galli, P.; Miligi, L.; Calisti, R.; Romeo, E.; et al. Sinonasal cancer in the italian national surveillance system: Epidemiology, occupation, and public health implications. *Am. J. Ind. Med.* **2017**, *61*, 1–12. [CrossRef]
37. Singh, M.P.; Reddy, M.M.K.; Mathur, N.; Saxena, D.K.; Chowdhuri, D.K. Induction of hsp70, hsp60, hsp83 and hsp26 and oxidative stress markers in benzene, toluene and xylene exposed drosophila melanogaster: Role of ros generation. *Toxic. Appl. Pharm.* **2009**, *235*, 226–243. [CrossRef]
38. Chang, F.-K.; Mao, I.-F.; Chen, M.-L.; Cheng, S.-F. Urinary 8-hydroxydeoxyguanosine as a biomarker of oxidative DNA damage in workers exposed to ethylbenzene. *Ann. Work Expo. Health* **2011**, *55*, 519–525.
39. Gurel, A.; Coskun, O.; Armutcu, F.; Kanter, M.; Ozen, O.A. Vitamin e against oxidative damage caused by formaldehyde in frontal cortex and hippocampus: Biochemical and histological studies. *J. Chem. Neuroanat.* **2005**, *29*, 173–178. [CrossRef]
40. Zhang, Y.; Liu, X.; McHale, C.; Li, R.; Zhang, L.; Wu, Y.; Ye, X.; Yang, X.; Ding, S. Bone marrow injury induced via oxidative stress in mice by inhalation exposure to formaldehyde. *PLoS ONE* **2013**, *8*, e74974. [CrossRef]
41. Romanazzi, V.; Pirro, V.; Bellisario, V.; Mengozzi, G.; Peluso, M.; Pazzi, M.; Bugiani, M.; Verlato, G.; Bono, R. 15-f(2)t isoprostane as biomarker of oxidative stress induced by tobacco smoke and occupational exposure to formaldehyde in workers of plastic laminates. *Sci. Total Environ.* **2013**, *442*, 20–25. [CrossRef]
42. Belles, M.; Gonzalo, S.; Serra, N.; Esplugas, R.; Arenas, M.; Domingo, J.L.; Linares, V. Environmental exposure to low-doses of ionizing radiation. Effects on early nephrotoxicity in mice. *Environ. Res.* **2017**, *156*, 291–296. [CrossRef] [PubMed]
43. Pisaniello, D.L.; Connell, K.E.; Muriale, L. Wood dust exposure during furniture manufacture: Results from an australian survey and consideration for threshold limit value development. *Am. Ind. Hyg. Ass. J.* **1991**, *52*, 485–492. [CrossRef] [PubMed]
44. Alwis, U.; Mandryk, J.; Hocking, A.D.; Lee, J.; Mayhew, T.; Baker, W. Dust exposures in the wood processing industry. *Am. Ind. Hyg. Ass. J.* **1999**, *60*, 641–646. [CrossRef] [PubMed]
45. Peluso, M.; Castegnaro, M.; Malaveille, C.; Talaska, G.; Vineis, P.; Kadlubar, F.; Bartsch, H. 32p-postlabelling analysis of DNA adducted with urinary mutagens from smokers of black tobacco. *Carcinogenesis* **1990**, *11*, 1307–1311. [CrossRef] [PubMed]
46. Peluso, M.; Hainaut, P.; Airoldi, L.; Autrup, H.; Dunning, A.; Garte, S.; Gormally, E.; Malaveille, C.; Matullo, G.; Munnia, A.; et al. Methodology of laboratory measurements in prospective studies on gene-environment interactions: The experience of genair. *Mutat. Res.* **2005**, *574*, 92–104. [CrossRef]
47. Van Houwelingen, H.C.; Arends, L.R.; Stijnen, T. Advanced methods in meta-analysis: Multivariate approach and meta-regression. *Stat. Med.* **2002**, *21*, 589–624. [CrossRef] [PubMed]

International Journal of
Molecular Sciences

Review

Increased Thyroid Cancer Incidence in Volcanic Areas: A Role of Increased Heavy Metals in the Environment?

Pasqualino Malandrino [1,†], Marco Russo [1,†], Fiorenza Gianì [1], Gabriella Pellegriti [1], Paolo Vigneri [2], Antonino Belfiore [1], Enrico Rizzarelli [3,4,5] and Riccardo Vigneri [1,4,*]

1 Endocrinology, Department of Clinical and Experimental Medicine, University of Catania, Garibaldi-Nesima Medical Center, 95122 Catania, Italy; p.malandrino@unict.it (P.M.); mruss@hotmail.it (M.R.); fiorenza.giani@gmail.com (F.G.); g.pellegriti@unict.it (G.P.); antonino.belfiore@unict.it (A.B.)
2 Medical Oncology and the Center of Experimental Oncology and Hematology, Department of Clinical and Experimental Medicine, University of Catania, A.O.U. Policlinico Vittorio Emanuele, 95125 Catania, Italy; pvigneri@libero.it
3 Department of Chemical Sciences, University of Catania, 95125 Catania, Italy; erizzarelli@unict.it
4 Consiglio Nazionale delle Ricerche, Cristallography Institute (Catania Section), via P. Gaifami 18, 95126 Catania, Italy
5 Consorzio Interuniversitario di Ricerca in Chimica dei Metalli nei Sistemi Biologici (CIRCMSB), via Celso Ulpiani 27, 70126 Bari, Italy
* Correspondence: vigneri@unict.it; Tel.: +39-095-759-8747
† These authors contributed equally to this work.

Received: 28 April 2020; Accepted: 9 May 2020; Published: 12 May 2020

Abstract: Thyroid cancer incidence is significantly increased in volcanic areas, where relevant non-anthropogenic pollution with heavy metals is present in the environment. This review will discuss whether chronic lifelong exposure to slightly increased levels of metals can contribute to the increase in thyroid cancer in the residents of a volcanic area. The influence of metals on living cells depends on the physicochemical properties of the metals and their interaction with the target cell metallostasis network, which includes transporters, intracellular binding proteins, and metal-responsive elements. Very little is known about the carcinogenic potential of slightly increased metal levels on the thyroid, which might be more sensitive to mutagenic damage because of its unique biology related to iodine, which is a very reactive and strongly oxidizing agent. Different mechanisms could explain the specific carcinogenic effect of borderline/high environmental levels of metals on the thyroid, including (a) hormesis, the nonlinear response to chemicals causing important biological effects at low concentrations; (b) metal accumulation in the thyroid relative to other tissues; and (c) the specific effects of a mixture of different metals. Recent evidence related to all of these mechanisms is now available, and the data are compatible with a cause–effect relationship between increased metal levels in the environment and an increase in thyroid cancer incidence.

Keywords: thyroid; thyroid cancer; volcano; metals; metallome; carcinogens; hormesis; environment pollution; metal biocontamination

---

1. Introduction

Thyroid cancer is the most frequent endocrine cancer (1.0%–1.5% of all new cases in the US), and its incidence, which was stable until the 1980s, has constantly increased since that time [1,2]. It is now the fourth most frequent cancer in women [3], whereas it was the 14th in the early 1990s.

This increase has occurred worldwide, as documented by the annual percent change in all countries where this parameter has been calculated with very few exceptions [4,5].

The reasons underlying these changes are unknown. Many experts believe that the apparent increase is mostly due to the overdiagnosis of small papillary thyroid tumors without significant clinical relevance that were not detected in the past but have been identified in the last decades because of the increasing diffusion of sensitive imaging procedures such as ultrasound scans [6]. Although there is a general consensus that a higher detection rate contributes to the increasing thyroid cancer incidence, much evidence indicates that this cannot be the only explanation. In fact, large thyroid cancers have also increased [7], and these tumors are very unlikely to have gone undetected in the past. Moreover, thyroid cancer-related mortality, which should have decreased because of early detection and better treatment, is stable or increasing [8]. Finally, a temporal trend in the changes in the thyroid cancer molecular profile characterized by an increasing prevalence of BRAF and RAS mutations [2,9,10] supports the possibility that a true change in thyroid cancer biology is occurring.

The causes of these recent changes in both quantitative and qualitative thyroid cancer characteristics are most likely environmental, as suggested by the sharp increase in incidence during the last decades. Most malignancies have not increased within this period, indicating that the potential carcinogenic factors involved must in some way be thyroid specific. Therefore, general factors that are known to favor cancer, such as the obesity epidemic, are unlikely to play a major role specifically in thyroid cancer.

Other more plausible environmental risk factors have been suggested. Since the thyroid is very radiosensitive, especially at a young age, and exposure to radiation has doubled in the last 25 years in most industrialized countries (mainly because of medical diagnostic procedures), radiation is the most commonly referenced cause [11,12]. The higher radiation-related carcinogenicity in the thyroid might be a consequence of frequent dental X-rays in children/adolescents.

Another possible risk factor specific to the thyroid is the progressive increase in iodine intake due to prophylaxis programs for iodine deficiency and goiter carried out worldwide in recent decades. Iodine enrichment may favor chronic lymphocytic thyroiditis [13], which may in turn promote thyroid cancer by increasing thyroid-stimulating hormone (TSH) levels and inducing proinflammatory cytokine production and oxidative stress in the gland [14–16].

However, the large number of potential carcinogens associated with the westernized postindustrial lifestyle should also be considered. In recent decades, the population has been highly and progressively exposed to compounds and chemicals that may interfere with biological functions, including hormone homeostasis (endocrine disruptor chemicals) [17]. Many compounds used in agroindustrial activities (fertilizers, pesticides, repellents, and preservatives) may directly cause cancer or indirectly produce conditions that favor malignant transformation. For instance, the increased ingestion of nitrates, which are a frequent contaminant of drinking water in areas of intense agricultural industry and are present at high levels in processed meat, has been associated with an increased risk of thyroid cancer [18,19]. Other potential thyroid-specific carcinogens can originate from other industrial activities such as the organic compounds polybrominated diphenyl ethers and bisphenols [20–22]. However, many other environmental pollutants (solvents, plastic, heavy metals, diet preservatives, etc.) may be responsible for the increased thyroid cancer. Therefore, further investigations are warranted to identify these potential carcinogens and their mechanism of action on the thyroid to introduce preventive measures aimed at controlling the continuous thyroid cancer increase.

2. Volcanic Environment and Thyroid Cancer

A specific natural example of the thyroid cancer–environment relationship is the increased incidence of this cancer observed in residents of volcanic areas.

This association was first reported 40 years ago [23] and was then confirmed by observations made on islands with active volcanoes, such as Hawaii [24,25] and Iceland [26,27], in the late 1980s. An elevated thyroid cancer incidence has since been reported in numerous volcanic areas in the Pacific Ocean, such as Vanuatu [28], French Polynesia [29], and New Caledonia [30], leading to the hypothesis that several components of volcanic lava could be involved in the pathogenesis of thyroid cancer [31].

Many possible causative factors have been proposed, including the possibility that genetic characteristics present on isolated islands with a small population might be responsible. However, the observation that residents of Hawaii exhibit much higher rates of thyroid cancer relative to individuals with the same ethnic background living in other geographic areas suggests that environmental rather than genetic influences play a major role [24]. Among these environmental factors, geothermal causes such as high-temperature water containing hydrogen sulfide and radon [32] have been hypothesized to play a role, as has the possibility that specific dietary factors favoring thyroid cancer exist in these areas. The increased natural radioactivity found in the volcanic areas and mainly due to ^{222}Radon emission might also play a role, but recent studies found no association between Radon levels and thyroid cancer [33,34].

In the early 2000s, an epidemiological study was carried out in Sicily, which is a large Mediterranean island with over five million inhabitants on which a continuously active volcano (Mt. Etna, the highest volcano in Europe) is located in the northeastern area in the province of Catania. This volcanic area has a population of over 1,000,000 inhabitants, and it was therefore possible to compare two large populations with the same ethnic background, similar sex and age distributions, similar lifestyles, and similar access to medical assistance.

The incidence of thyroid cancer was more than doubled in residents of the volcanic area compared to the remaining population of Sicily [35]. The F/M ratio was 4.8:1, with no significant difference between the two areas. An increased incidence was also present in pediatric age [36]. Environmental factors such as iodine intake and industrial pollution did not differ in the two areas. Moreover, only thyroid cancer of the papillary histotype was increased in the volcanic area, reflecting the observation that this histotype is the main cause of the worldwide increase in thyroid cancer incidence (Table 1).

Table 1. Thyroid cancer in Sicily: age-standardized incidence rates for the world population (ASRw) in the volcanic and the control areas and the papillary/follicular histotypes ratio. Data from [35].

Environment	Inhabitants (millions)	Thyroid Cancer Incidence (ASRw)		Papillary/Follicular Ratio
		F	M	
Volcanic area (Catania province)	1116	31.7	6.4	25.9
Control area (all Sicily without Catania)	3853	14.1	3.0	9.8

Therefore, the findings in the Mt. Etna volcanic area strongly suggest an association between the volcanic environment and the increase in thyroid cancer with a cause–effect relationship but with no indication of the possible causative factors and their mechanisms of action.

Active volcanoes cause considerable non-anthropogenic pollution due to gas, ash, and lava emissions. This pollution may have different characteristics among different volcanoes depending on the chemical, physical, and geologic characteristics of each volcano and its effusive activity. In all cases, a variety of elements originating from the depths under the Earth's crust pollute the atmosphere, water, soil, and food and will cause biocontamination in the resident population via these routes [37,38]. Within these forms of pollution, heavy metals may play an important role. These natural components of the Earth's soil show complex interactions with organic elements (e.g., amino acids, carbohydrates, and nucleotides) and play an important role in biological events, including those related to cell growth and transformation [39].

3. Biological Bases of Metal Homeostasis

Once metals from the Earth's crust become available in the environment, they cannot be degraded or destroyed and play an important role in the biology of plants and animals. Some metals that are essential nutrients in trace amounts may become toxic compounds at higher concentrations. Other metals are toxic and carcinogenic even at very low levels. Therefore, the homeostatic regulation of metals is under strict control in cells, including the sensing, transport, and accumulation of metals.

However, the underlying regulatory mechanisms are poorly understood: metal-dependent processes may influence many aspects and functions of cell biology via mechanisms (and at concentrations) that are still unclear for many metals in many instances. Nature employs the unique chemical features of essential micronutrients belonging to the block d metals, such as Zn (Zinc), Cu (Copper), Ni (Nickel), and Co (Cobalt), because of their donor atom preferences, coordination structures, and redox capacity. These characteristics are important to obtain a reliable repertoire of structural and catalytic functions of many proteins.

Many of these proteins are present at high concentrations, and significant amounts of the associated metal ions are required to guarantee their functions. Tightly bound metal pools buried within proteins are found in the cytosol as well as all cell organelles, including the endoplasmic reticulum, Golgi, mitochondria, lysosomes/vacuoles, and nucleus. Metals in this form significantly contribute to the total metal ion content (0.01–1 mM) needed for optimal living cell survival [40,41].

In sharp contrast to the high level of metals bound to proteins (*static metallome*), the concentration of block d metal ions that are not tightly bound to proteins (*dynamic metallome*, i.e., labile or exchangeable metal ions) is very low. The analysis of metal-responsive sensor molecules [42] indicated that the level of labile cytosolic metal ions is in the pM–nM range, i.e., 5–9 orders of magnitude lower than that of the static metallome pool [43,44]. These findings demonstrate the need for an efficient network of players in metallostasis (metal homeostasis) that can provide a sufficient supply of metal ions for protein function while maintaining the dynamic metallome at an extremely low level [45,46]. This network can ensure metal homeostasis within cells even when extracellular levels of metal ions vary over a large range.

The main players in the metallostasis network are solute carriers (SLCs), which play a vital role in the healthy functioning of living cells, exerting strict control over the import and export of ions, metabolites, and nutrients across membranes [47]. It has been estimated that approximately 10% of the human genome is linked to the control of membrane transport [47]. The network also includes chaperones and storage molecules such as metallothionines (MTs) [48,49], transcription factors (TFs) [50,51] and small molecules involved in the detoxification system, such as glutathione (GSH) [52].

This fine control of the metallome is exemplified by the biology of the two best-characterized metal ions, Zn and Cu, both of which act as intracellular regulators of major signaling pathways [53] (Figure 1).

Zn transporters are classified into the two major families: SLC39A/ZIP (14 members) and SLC30A/ZnT (10 members), which are responsible for metal influx (both from outside the cell and from organelles to cytosol) and efflux (from cytosol to both outside the cell and to organelles), respectively [54,55]. The ZnT family includes eight cloned members, referred to as ZnT1-8 [54], and two others, ZnT9 and ZnT10, predicted from mouse and human genome resources [56].

A central role in metal homeostasis/detoxification is performed by metal-responsive transcription factor-1 (MTF-1), which binds DNA to modulate RNA transcription in response to altered cytosolic levels of Zn and/or Cu ions [57]; MTF-1 regulates the expression of both ZnT1 and MTs, which are proteins involved in metal release and storage, respectively [58].

For Cu, the high-affinity membrane copper transporter 1 (Ctr1) is the major cellular protein responsible for the uptake of this metal [59]. Minor SLCs for Cu include divalent metal ion transporter 1 (DMT1) and copper transporter 2 (Ctr2) [60].

Once the metal is inside the cell, intracellular chaperones transfer copper ions to specific targets. Such transfer is observed between copper chaperone for superoxide dismutase (CCS) and superoxide dismutase 1 (SOD1) in the cytosol; cytochrome c oxidase copper chaperone 17 (COX17) and cytochrome c oxidase (CCO) in mitochondria; and antioxidant 1 copper chaperone (Atox1) and ATPases such as copper-transporting P-type ATPase A and B (APT7A and APT7B) in the trans-Golgi network (TGN) [61–63].

When the copper concentration is excessive, Ctr1 is downregulated by specific protein 1 (Sp1), [64], which is a transcription factor that controls the homeostatic maintenance of Ctr1 mRNA. When there is a deficiency of these metal ions, Sp1 activity is reduced, and Ctr1 expression increases [55].

In addition, cytochrome c oxidase copper chaperone 11 (COX11) mRNA levels are related to Ctr1 expression and are upregulated in proliferating cells, in which there is the functional hyperactivity of copper trafficking pathways.

Owing to their chemical analogy to Zn^{2+}, Cd^{2+} and Hg^{2+} may exert a competitive effect in binding to transporters and other coplayers in zinc homeostasis. The same is true for Pd^{2+} with respect to Cu^{2+}, and similar competition may occur between Mo and toxic W (in the chemical forms of $MoO_4{}^{2-}$ and $WO_4{}^{2-}$)

For most other metals, similar complex machinery for extra- and intra-cellular metal trafficking might be present, but very little is known about their existence and function.

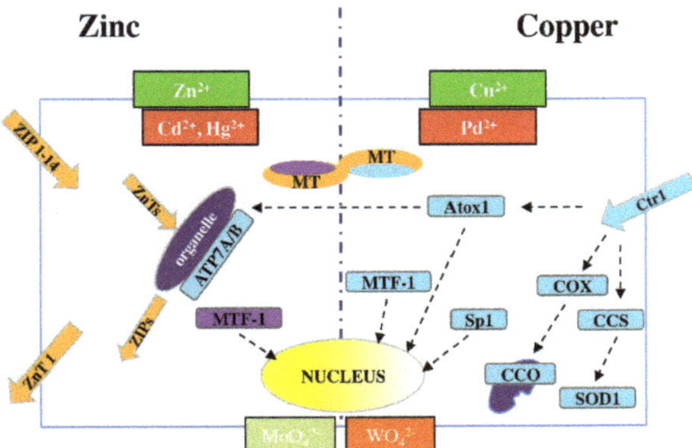

Figure 1. Schematic picture of dynamic metal homeostasis (metallostasis) players for Zn^{2+} (**left**) and Cu^{2+} (**right**). As a result of chemical similarities, Cd^{2+} and Hg^{2+} can compete with Zn^{2+}, while Pd^{2+} can compete with Cu^{+} for cellular transport and binding proteins. Similar competition can occur between molybdenum and tungsten. ZIP (14 members) = membrane protein transporters for Zn^{2+} influx; ZnT (10 members) = membrane protein transporters for Zn^{2+} efflux; MT = metallothioneins; MTF-1 = metal-responsive transcription factor-1; APT7A/B = metal-transporting P-type ATPase A and B; Ctr1 = membrane protein for copper uptake; Atox1 = antioxidant 1 copper chaperon; COX = cytochrome c oxidase copper chaperon; CCO = cytochrome c oxidase; CCS = copper chaperon for superoxide dismutase; SOD1 = superoxide dismutase 1; Sp1 = transcription factor specific protein 1.

4. Metal Carcinogenicity and the Thyroid

Metals include both essential metals that are required micronutrients for biological processes (Fe: Iron, Zn, Cu, Se: Selenium, etc.) and toxic chemicals that may damage cell biology and promote malignant transformation (As: Arsenic, Cd: Cadmium, Hg: Mercury, Ni, etc.).

As a result of the numerous complex factors involved in the interaction of metals with living cells and the possible combined action of different mechanisms, our present understanding of the carcinogenic potential of a single metal or a mixture of different metals in living cells in general and the thyroid in particular is very limited.

The carcinogenic effect of metals on target cells depends on several biological factors, such as bioavailability (entering cells through the cell membrane), the intracellular distribution, and interactions with cellular proteins and enzymes. These steps are tissue and cell-specific.

Once a metal enters cells, its genotoxicity is generally exerted by indirect rather than direct actions on DNA. The most common mechanisms include (a) the induction of oxidative stress, which may in turn activate intracellular signaling leading to oxidative DNA damage; (b) interference with DNA repair systems, causing the accumulation of mutations; (c) the deregulation of growth control by damaging the balance between proliferative and apoptotic pathways; and (d) modification of the DNA methylation pattern, affecting the expression of oncogenes and oncosuppressors.

In relation to these mechanisms, an important factor is metal speciation (both inorganic and metal–organic), which determines the physicochemical properties and bioavailability of metals and, therefore, their biological effects.

For instance, high levels of As in drinking water, soil, food, and the atmosphere are associated with several types of cancer (As is a recognized human carcinogen belonging to group 1 according to International Agency for Research on Cancer-IARC classification), but the underlying mechanisms are not fully understood and are probably different for different As compounds.

In humans, As is relatively nontoxic when it occurs as a metallo-organic species in seafood (arsenobetaine) [65]. Among the inorganic forms of As, arsenate poses higher toxicity to endocrine glands than arsenite [66]. Moreover, a methylated arsenic compound (dimethyl arsenic acid) promotes carcinogenesis in many rat organs, including the thyroid [67], but another As derivative (arsenic trioxide) reduces proliferation and increases apoptosis and iodine uptake in papillary and follicular thyroid cancer cells [68], acting as a differentiating anticancer agent.

Additional mechanisms of As carcinogenesis may involve microRNA dysregulation. In arsenic-transformed human lung epithelial cells, miR-222 is upregulated, and its inhibition decreases cell proliferation and migration and increases apoptosis [69]. Notably, miR-222 may be significantly upregulated in papillary thyroid cancer [70]. Finally, in French Polynesia, an area with a high incidence of thyroid cancer, the risk of this cancer is increased by 30% for each increase in As intake of 1 μg/d/kg body weight, despite being within the recommended daily intake indicated by the WHO. However, this increase in thyroid cancer particularly affected individuals with first-degree relatives with a history of cancer [71], suggesting that As may act as a cocarcinogen with a combined effect with genetic susceptibility.

Cadmium is another carcinogenic metal of the IARC group This metal has been identified as an endocrine-disrupting chemical for many endocrine glands, including the thyroid [72], but it is also a carcinogen with multifactorial mechanisms. The physiochemical properties of Cd^{2+} ions allow them to substitute for calcium ions in biological systems because of showing the same charge and a similar radius and to use zinc transporters and substitute for Zn^{2+} in many enzymes and transcription factors, competing for Zn finger motifs [73–75].

Cd can induce oxidative stress by inhibiting antioxidant enzymes, activating the PI3K (phosphoinositide 3-kinase) and ERK (extracellular signal-regulated kinase) signaling pathways, deregulating cell proliferation, and damaging DNA repair mechanisms. Through one or more of these mechanisms, Cd can induce cancer initiation and progression [76]. An additional cancer-promoting effect of Cd is its disrupting effect on E-cadherin: by displacing Ca^{2+} from this protein, Cd disrupts cadherin-mediated cell–cell adhesion and therefore favors tumor progression and invasiveness [77].

Finally, Cd may favor thyroid cancer with a unique mechanism due to its metalloestrogen characteristics. Cd can in fact mimic the effects of 17B-estradiol on the G protein-coupled estrogen receptor. This receptor is present in thyroid follicular cells, and via its stimulation, Cd can promote the proliferation, invasion, and migration of thyroid cancer cells [78].

Many other metals exhibit carcinogenic activity, but most of them have never been directly tested in the thyroid. In any case, the available data are fragmentary, inconclusive, and sometimes contradictory. Moreover, in most of the relevant studies, only high metal concentrations and short-term effects were investigated, representing quite different conditions from the chronic exposure and the low-level metal increases generally found in association with environmental pollution.

5. Heavy Metals in the Mt. Etna Volcanic Area and Resident Biocontamination

Based on the observation that an increased trace element concentration may be present in volcanic areas [79,80] and, more specifically, that increased levels of metals such as B (Boron), Fe, Mn (Manganese), and V (Vanadium) are found in the groundwater of the Mt. Etna volcanic area [35], a careful comparative study of heavy metal environmental pollution and the biocontamination of residents was carried out in volcanic and control areas of Sicily [81].

To investigate environmental pollution, metal concentrations were measured in water and lichens. Mt. Etna harbors a large aquifer that provides water to over 700,000 residents and is used for irrigation in most of Catania Province. Therefore, volcanic aquifer-originating water is an important vehicle for population biocontamination, both directly and indirectly via locally grown food. Lichens are composite organisms that bioaccumulate elements present in the atmosphere and are therefore used for the biomonitoring of atmospheric pollution [82,83].

To investigate human biocontamination, urine specimens were collected from two matched groups of individuals living in volcanic and control areas. In fact, under conditions of chronic exposure, urine is considered a reliable indicator of the chemicals absorbed by a subject through contact, inhalation, and ingestion.

By measuring 27 trace elements and heavy metals in the environment (water and lichens) and human biological samples (urine), considerable volcanically derived biocontamination was found. In the volcanic area, where the thyroid cancer incidence is doubled, many metals were significantly increased in both water and lichens, documenting metal pollution in the environment. The differences relative to the control areas were more marked in water, in which the concentrations of metals such as As, B, Cd, Hg, Mn, Mo (Molybdenum), Pd (Palladium), Se, U (Uranium), V, and W (Tungsten) were increased by three- to 50-fold, although their average concentrations never exceeded the reference values indicated by the World Health Organization [84].

When these elements were measured in the urine specimens, 18 elements were found to occur at a significantly increased level compared to values measured in the urine samples collected in the control areas.

In particular, the geometric mean value was two-fold or more than two-fold higher for eight metals: Cd, Hg, Mn, Pd, Tl, U, V, and W. Moreover, the values of B, Mo, Pd and W were higher than the 95th percentile of the Italian reference values in more than 20% of the urine specimens from the volcanic area [81] (Figure 2). This human biocontamination was confirmed by the increased concentration of metals in the scalp hair of children living in the Mt. Etna volcanic area [85].

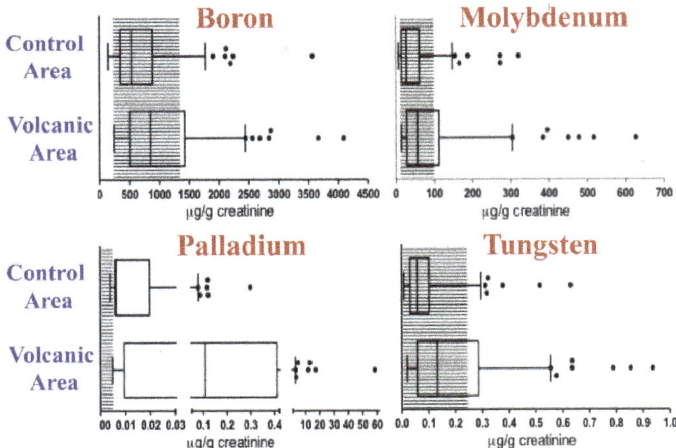

Figure 2. Metal biocontamination in the urine of residents of the volcanic area. The concentrations of boron, molybdenum, palladium, and tungsten were measured in the urine of 140 residents of the volcanic area and of 138 residents of the control area in Sicily. Data obtained from [81]. For B, Mo, Pd and W concentrations were significantly higher in the urine of residents of the volcanic area than in that of residents of the control nonvolcanic area and exceeded the normal reference values in over 20% of cases. The boxes indicate the 25th, 50th (median), and 75th percentiles. The whiskers indicate the 5th and 95th percentiles. The shaded area indicates the Italian reference values for urine. The dots indicate individuals with urinary concentrations higher than the 95th percentile.

These data document relevant metal pollution and consequent human biocontamination in subjects living in a volcanic area where the thyroid cancer incidence is greatly increased. The well-established carcinogenic effect of some metals and the observation that individuals living in volcanically active areas exhibit DNA damage more frequently than subjects living in nonvolcanic areas [86] may support (without proving) a cause–effect relationship between these findings.

6. Increases in Heavy Metals and Thyroid Cancer: A Cause–Effect Relationship?

A cause–effect relationship between chronic exposure to increased metal levels in the volcanic environment and thyroid cancer is difficult to demonstrate via clinical studies in residents of a volcanic area. Therefore, the problem has been approached through in vitro studies in human thyroid cells and in vivo studies in experimental animals.

A major unanswered question is how such small increases in environmental metals, not exceeding what is considered to be the normal range in most cases, can promote the malignant transformation of the human thyroid. A second important question is why the possible carcinogenic effect of increased metal levels in a volcanic area predominantly, though non-exclusively [87], affects the thyroid gland. The identification of the mechanisms involved in these processes is crucial for better understanding the possibility of a cause–effect relationship

6.1. Hormesis Effect

Chemical hormesis is a biological phenomenon characterized by a nonlinear response of biological activity (including cell growth and carcinogenesis) to a stimulator [88]. In hormesis, the biological response to increasing amounts of a chemical is biphasic, with biological effects increasing at low concentrations, followed by the inhibition of the effect at higher doses.

Many metals can cause hormetic responses in biological systems. Low levels of potentially toxic heavy metals such as Ag, As, Cd, Hg, and Y can cause stimulatory effects on cellular activities both in vitro [89–92] and in vivo [93–96]. The hormetic effects of these metals usually occur in the μM concentration range, and the stimulated increase is less than 100% above the basal level. The response depends on the concentration of the metal studied, the time of exposure, and the target cell examined. In vitro, malignant cells may be unresponsive to metal doses that cause a hormetic response in nontransformed cells [97], and in vivo, this response may be stage dependent, displaying toxicity only in a later developmental period [93,98].

The mechanism of the hormetic effect is unknown. In addition to possible cross-talk among metal ions with similar chemical features, it most likely involves the generation of reactive oxygen species (ROS), which could induce cell damage and activate reactive mechanisms [90,99,100]. Redox reactions can be modulated by the variable availability of transition metals that serve as donors of electrons. Through these mechanisms, low levels of ROS may stimulate the activation of the ERK/MAPK (mitogen-activated protein kinase) signaling pathway and, as a consequence, increase protein synthesis, cell proliferation and differentiation, and resistance to stress conditions [91,92,101]. A potential alternative mechanism is the binding to and inhibition of protein tyrosine phosphatases by metals [102], indirectly increasing the activity of tyrosine kinases.

Until very recently, no data were available on the hormetic effects of metals on thyroid cells. In 2019, in a study of cultured human thyrospheres (spheres containing thyroid stem and precursor cells), it was observed that chronic (days) exposure to low doses of the heavy metal W (applied as sodium tungstate dihydrate) caused a series of biological effects. In vitro, very low concentrations of W stimulated thyrosphere proliferation, as indicated by 5-bromo-2-deoxyuridine (BrdU) incorporation, increased DNA levels, and morphological changes observed under phase-contrast microscopy [103]. These effects were observed at very low W concentrations (nM) within the same range measured in the urine of the residents of the Mt. Etna volcanic area (where thyroid cancer incidence is markedly increased) and disappeared at higher (μM) concentrations. Similar effects on human thyrosphere proliferation were observed in preliminary experiments with Hg and Pd in the nanomolar range and with Zn in the micromolar range (Figure 3). In parallel experiments, no effect of W was observed in differentiated human thyrocytes in primary culture.

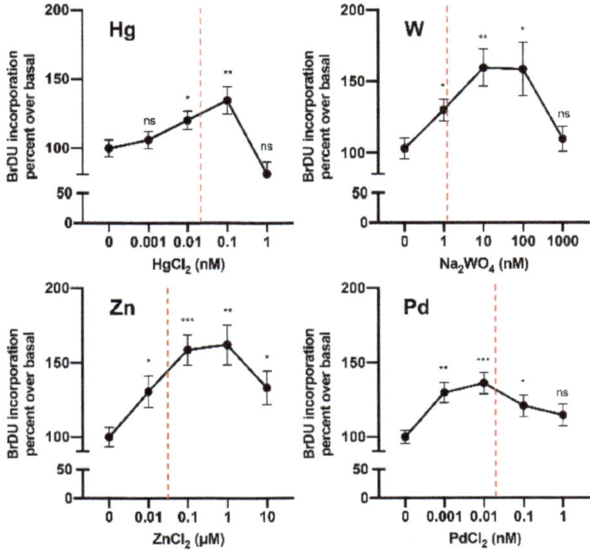

Figure 3. Metals at low concentrations stimulate human thyrosphere proliferation. Proliferation (measured by 5-bromo-2-deoxyuridine (BrDU) incorporation) of human thyrospheres (aggregates of thyroid stem/precursor cells) after chronic exposure to increasing concentrations of Hg, W, Zn, and Pd. Vertical dotted lines indicate the average concentration of each metal in the urine of residents in the volcanic area of Sicily [81].

A low-level metal-stimulated biological effect in thyrospheres was preceded by the activation of the ERK signaling pathway, while the inhibition of ERK phosphorylation with pertussis toxin inhibited thyrosphere growth [103]. These observations were recently confirmed in cultured thyroid nontransformed cells [96] and suggest a major role of the ERK intracellular pathway in the hormetic effect of tungstate on immature thyroid cell proliferation.

In the same model and at the same concentrations, W inhibited thyroid stem cell differentiation and reduced apoptosis. Moreover, mature thyrocytes derived from thyrospheres chronically exposed to tungstate presented some characteristics typical of transformed cells: they formed more and larger colonies in soft agar and in a clonogenic assay, and they showed a greater migration capacity in a scratch wound-healing assay [103].

Chronic exposure to low levels of W also altered the genetic profile of thyroid stem/precursor cells and affected DNA repair protein activity [103] (Figure 4).

These in vitro data indicate that chronic exposure to very low concentrations of W, while being harmless to mature thyrocytes, has relevant effects on undifferentiated or partially differentiated thyroid cells. Biphasic hormetic responses to metals have already been described for other types of undifferentiated cells, such as lung embryo fibroblasts and human embryonic kidney cells [100,104]. The novelty of this thyroid model is that exposed progenitor cell abnormalities produce a population of mature thyrocytes with biological characteristics compatible with a preneoplastic state.

The influence of thyroid progenitor cells' exposure to W on the characteristics of their progeny (mature thyrocytes) is reminiscent of the transgenerational transfer of the hormetic effects of metals reported in plants [105,106] and animals [95]. These findings may have important implications for the estimation of hazard assessments for carcinogenesis and cancer risk in later life. The question of safe metal levels in the environment during the prebirth period and neonatal life, when tissues (including the thyroid) exhibit a high prevalence of stem/precursor cells, is an important issue that deserves further study.

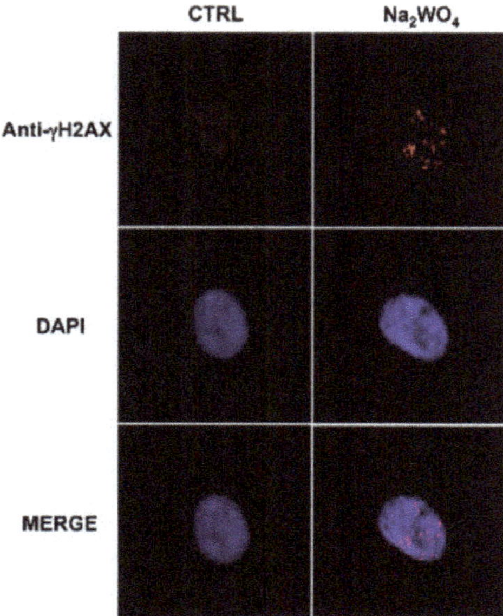

Figure 4. Exposure to a low dose of tungsten affects DNA repair proteins in human thyrospheres. The exposure of human thyrospheres to a low dose of W (Na_2WO_4, 30 nM for 90 min) increases the expression of the DNA repair protein γH2AX. γH2AX (red) was detected using a γH2AX antibody followed by an Alexa Fluor-594-conjugated secondary antibody. Nuclei were visualized with DAPI (4′,6-diamidino-2-phenylindole) (blu) [103].

6.2. Is Metal Accumulation in the Thyroid a Possible Mechanism Contributing to the Increase in Thyroid Cancer?

As mentioned above, many metals (B, Br, Cd, Co, Cu, Hg, Li, Mn, Mo, Pd, Se, Sn, Tl, U, V, W, and Zn) showed significantly increased levels ($p < 0.001$) in the urine of residents of the volcanic area in Sicily [81]. Therefore, individuals born and living in that area suffer lifelong biocontamination with metals, beginning very early in life.

However, high variability in the biocontamination levels of different metals is observed in individuals living in this area; as a consequence, the concentration ratio of each metal relative to those of the other trace elements is highly variable. Therefore, the effects of these metals on thyroid cells cannot be extrapolated from those observed in vitro with tungsten, as in vivo conditions are much more complex and heterogeneous. Each metal will most likely have different effects on the thyroid when acting separately than when multiple metals are simultaneously present in excess. Metal effects in fact depend not only on the dose but also on the synergistic or antagonistic influences of other metals. The ultimate effect of different, variable combinations of multiple excess heavy metals on the human thyroid is currently unknown.

In this regard, one factor that must be considered is the different capacities of the thyroid to specifically accumulate different metals, which depend not only on environmental exposure but also on metabolic processes such as cell uptake, retention, and clearance. According to the specificity and selectivity of these processes in thyroid cells, different metals may accumulate at higher levels in the thyroid than in other tissues. The selective accumulation of one or more trace elements with a carcinogenic effect could explain the predominant increase in thyroid cancer observed in the presence of environmental heavy metal biocontamination.

Recently, metals were comparatively measured (using inductively coupled plasma mass spectrometry) in normal human thyroid tissue and in sternothyroid muscle and neck subcutaneous fat tissues collected from the same euthyroid individuals.

As, Br, Cd, Hg, Mn, Se, and Sn showed significantly higher concentrations ($p < 0.01$) in the thyroid than in the other two tissues [107]. Among these elements, As, Cd, and Hg are recognized carcinogens (class 1, IARC). As and Hg (but not Cd, which was under the assay detection limit in all tissues) were also shown to be more concentrated in the thyroid of normal rats relative to the hindlimb muscle and abdominal visceral fat of the same rats.

Whether the relative accumulation of these carcinogenic elements contributes to the very frequent occurrence of thyroid cancer in the volcanic biocontaminated area is unclear. When the metal concentrations measured in the thyroid tissue of the residents of the volcanic biocontaminated area (n = 43) were compared to those found in the residents of the control area (n = 34), 11 out of the 18 examined elements were at slightly higher levels in the thyroid of subjects living in the volcanic area. However, a large overlap of metal levels was found between the two groups, and the differences were not statistically significant. Moreover, in residents of the volcanic area, most metals were increased also in muscle and adipose tissues, suggesting a generalized consequence of increased exposure, rather than a specific thyroid accumulation mechanism. However, this was not the case for As and Hg, which were slightly increased (+16.5% and +25%, respectively) in the thyroid but not in the other examined tissues of residents of the volcanic area. These small differences may be biologically relevant based on the hormetic mechanism.

In conclusion, studies on metal concentrations in the thyroid do not provide evidence of a clear role of the accumulation mechanism in metal-dependent thyroid carcinogenesis, but they are compatible with this possibility.

6.3. Metals and Thyroid-Specific Biology

In addition to the possibility that the selective accumulation of carcinogenic metals favors thyroid cancer more than cancers of other tissues, alternative mechanisms may explain why environmental metal pollution predominantly promotes thyroid cancer.

Follicular thyroid cells have peculiar biological properties: their main function is to produce thyroid hormones, which requires iodine. Iodine is a very reactive element and a strong oxidizing agent. Thyroid cells take up iodine via a specific sodium/iodine symporter (NIS) in the form of iodide anion (I^-), which is readily oxidized by the thyroid-specific enzyme thyroperoxidase (TPO). Therefore, thyroid cells are constantly exposed to free radicals produced by the continuous generation of hydrogen peroxide (H_2O_2) by the NADPH (nicotinamide adenine dinucleotide phosphate hydrogen) oxidase Duox. H_2O_2 is necessary for I^- oxidation to produce derivatives such as hypoiodite, hypoiodous acid, and iodate [108]. This intracellular chemistry produces high oxidative stress, which may in turn favor spontaneous mutagenesis. In fact, in an experimental model, the mutation rate in the thyroid is 8–10 times higher than that in the liver [109]; under these conditions, the additional free radicals produced by increased metal exposure can more easily cause DNA damage and cell transformation.

Other peculiar thyroid cell characteristics may be involved in increased thyroid sensitivity to metals present in excess, as indicated by the complete inhibition of the enzyme xanthine oxidase by tungsten in the rat thyroid [110].

The possible causal relationship between metals and thyroid cancer is supported by the effect of the heavy metal copper on $BRAF^{V600E}$ mutation-driven carcinogenesis. The papillary histotype is the most frequent (over 80%) thyroid cancer histotype and is due to the oncogenic mutation of BRAF in most cases (over 50%). Copper chelation inhibits MEK 1/2 kinase activity, and reduced MAPK signaling inhibits $BRAF^{V600E}$-driven melanoma growth [111]. The same effect was observed in $BRAF^{V600E}$-positive human PTC cells and in a genetically engineered mouse PTC model [112], indicating the possibility that increased copper may favor the occurrence of thyroid cancer with papillary histotype.

6.4. In Vivo Data in Experimental Animals

A well-accepted system for evaluating the cause–effect relationship between suspected carcinogens and the actual induction or promotion of cancer is to use experimental animal models. If the feeding of animals (mainly rodents) with a diet containing elevated concentrations of metals results in the appearance of signs of malignant transformation, it is assumed to indicate the carcinogenicity of the tested compound.

However, these models have important limitations. Animals show differences in sensitivity to metals relative to humans as a consequence of differences in absorbance, tissue accumulation, and clearance, metabolism, and excretion.

Moreover, species specificity has been observed; for example, As compounds may have carcinogenic effects in the mouse thyroid but not in the rat thyroid [113].

Many studies on metal carcinogenicity have been carried out in animals in a variety of tissues, but only a few of these studies focused on the thyroid, and most of them mainly concerned thyroid function, rather than carcinogenesis.

In addition, these animal experiments presented major problems regarding metal compound dosage and time of exposure, as in most cases, high doses (mM range) and a short exposure time (weeks) were used. When low exposure was examined, As compounds have been shown to disrupt T_4 homeostasis and influence related gene transcription after 8 weeks of exposure, but no morphological modifications were described [114]. Moreover, male rats exposed to bromine ($KBrO_3$) in quantities calculated to be within the high range expected in the environment exhibited morphological goiter-like changes in the thyroid after 66 days [115] and increases in the number of mitoses and vascularization after 133 days [116]. After longer exposure (animals exposed to similar $KBrO_3$ concentrations for up to two years), carcinogenic effects were observed in the thyroid and kidneys of rats but only in the kidneys of mice [117].

These data reflect an experimental condition very different from that observed in polluted volcanic areas, where not only is there a mixture of many metals present, but metal concentrations in the environment are also in the high–normal range and exposure occurs throughout an individual's life, including the prenatal stage [81].

To evaluate conditions better mimicking chronic exposure to low-dose metal pollution, a study was recently carried out in a well-established in vivo model of thyroid tumorigenesis [118]. Female rats were supplied with a goitrogenic diet, and B, Cd, and Mo were added to their drinking water at concentrations twice as high those measured in the urine of the residents of the volcanic area in Sicily. The thyroids of treated animals examined after 5 and 10 months of exposure exhibited progressive increases in follicular dyshomogeneity, nuclear pseudoinclusions, and papillary structures compared to the thyroids of the control rats (also hypothyroid under the goitrogenic diet) [119]. Papillary structures associated with nuclear aberrations are considered preneoplastic features of the thyroid. An increase in such structures indicates that chronic exposure to B, Cd, and Mo, even at low levels, accelerates the neoplastic characteristics of the thyroid induced by the goitrogenic diet. The study confirms that under conditions causing a predisposition to neoplastic transformation, chronic exposure to slightly increased B, Cd, and Mo levels may favor thyroid cancer initiation acting as tumor-promoting agents, rather than as true carcinogens. However, this model is quite different from the in vivo conditions of the residents of the volcanic area and is inadequate for documenting the carcinogenic potential of lifelong exposure to multiple metals.

7. Concluding Remarks

In recent decades, metal pollution of the environment has increased worldwide [120], highlighting the question of its possible deleterious effects on human health. The thyroid is a histologically and functionally complex gland. Its major functions of iodine uptake and incorporation into tyrosine residues to produce thyroid hormones require specific oxidoreduction processes that can make the thyroid more vulnerable to toxic heavy metals.

The health concern related to the anthropogenic pollution of the environment with metals is a serious issue and a worsening problem. As a result of its biochemistry and biology, the thyroid may act as a sensitive and precocious indicator of the possible health damage caused by environmental heavy metal pollution.

At present, we lack solid data on the carcinogenic effect of increased environmental metal levels on thyroid cancer initiation and progression. The available evidence is indirect, circumstantial, and incomplete.

More in general, our understanding of the metal pollution and its consequences on human health is totally inadequate considering the complexity and variability of the interactions of different cells with different metals, different metal doses and lengths of exposure, the different speciation of each metal, and the competing or potentiating effects metals on each others' activities.

Further rigorous and innovative studies on this issue aimed at identifying the mechanisms of action and the biological effects triggered by chronic exposure to slightly increased metal concentrations are warranted. More specifically, a better understanding of the additive, synergistic, or antagonistic effects of different metals with other organic and inorganic compounds in the environment is required. An improved understanding of these issues would greatly contribute to our comprehension of the relationship between environmental metal pollution and increased health damage, including thyroid cancer.

Author Contributions: Conceptualization and design of the review by R.V., M.R., P.M., F.G., and E.R. All authors collected and discussed the literature, with special participation of G.P., M.R., P.M., and A.B. for thyroid cancer, P.V. for oncology and E.R. for chemistry. F.G., M.R., P.M., and E.R. prepared the figures. R.V. and E.R. wrote the manuscript, which was substantially revised and approved in its final version by all authors. All authors have read and agreed to the published version of the manuscript.

Funding: This study was supported by a grant to R.V. by the AIRC Foundation (Milan, Italy, grant number 19897).

Acknowledgments: The authors thank Rita Pennisi for secretarial work.

Conflicts of Interest: The authors declare no conflict of interests.

References

1. Ito, Y.; Nikiforov, Y.E.; Schlumberger, M.; Vigneri, R. Increasing incidence of thyroid cancer: Controversies explored. *Nat. Rev. Endocrinol.* **2013**, *9*, 178–184. [CrossRef]
2. Vigneri, R.; Malandrino, P.; Vigneri, P. The changing epidemiology of thyroid cancer: Why is incidence increasing? *Curr. Opin. Oncol.* **2015**, *27*, 1–7. [CrossRef] [PubMed]
3. Ferlay, J.; Colombet, M.; Soerjomataram, I.; Mathers, C.; Parkin, D.M.; Piñeros, M.; Znaor, A.; Bray, F. Estimating the global cancer incidence and mortality in 2018: GLOBOCAN sources and methods. *Int. J. Cancer* **2019**, *144*, 1941–1953. [CrossRef]
4. Kilfoy, B.A.; Zheng, T.; Holford, T.R.; Han, X.; Ward, M.H.; Sjodin, A.; Zhang, Y.; Bai, Y.; Zhu, C.; Guo, G.L.; et al. International patterns and trends in thyroid cancer incidence, 1973–2002. *Cancer Causes Control* **2009**, *20*, 525–531. [CrossRef]
5. Pellegriti, G.; Frasca, F.; Regalbuto, C.; Squatrito, S.; Vigneri, R. Worldwide Increasing Incidence of Thyroid Cancer: Update on Epidemiology and Risk Factors. Available online: https://www.hindawi.com/journals/jce/2013/965212/?viewType=Print&%3BviewClass=Print&%3BB1 (accessed on 10 April 2020).
6. Davies, L.; Welch, H.G. Current Thyroid Cancer Trends in the United States. *JAMA Otolaryngol. Neck Surg.* **2014**, *140*, 317–322. [CrossRef] [PubMed]
7. Mao, Y.; Xing, M. Recent incidences and differential trends of thyroid cancer in the USA. *Endocr. Relat. Cancer* **2016**, *23*, 313–322. [CrossRef] [PubMed]
8. Lim, H.; Devesa, S.S.; Sosa, J.A.; Check, D.; Kitahara, C.M. Trends in Thyroid Cancer Incidence and Mortality in the United States, 1974–2013. *JAMA* **2017**, *317*, 1338–1348. [CrossRef] [PubMed]
9. Elisei, R. Molecular Profiles of Papillary Thyroid Tumors Have Been Changing in the Last Decades: How Could We Explain It? *J. Clin. Endocrinol. Metab.* **2014**, *99*, 412–414. [CrossRef] [PubMed]

10. Jung, C.K.; Little, M.P.; Lubin, J.H.; Brenner, A.V.; Wells, S.A.; Sigurdson, A.J.; Nikiforov, Y.E. The Increase in Thyroid Cancer Incidence During the Last Four Decades Is Accompanied by a High Frequency of BRAF Mutations and a Sharp Increase in RAS Mutations. *J. Clin. Endocrinol. Metab.* **2014**, *99*, E276–E285. [CrossRef] [PubMed]
11. Iglesias, M.L.; Schmidt, A.; Ghuzlan, A.A.; Lacroix, L.; De Vathaire, F.; Chevillard, S.; Schlumberger, M. Radiation exposure and thyroid cancer: A review. *Arch. Endocrinol. Metab.* **2017**, *61*, 180–187. [CrossRef]
12. Han, M.A.; Kim, J.H. Diagnostic X-Ray Exposure and Thyroid Cancer Risk: Systematic Review and Meta-Analysis. *Thyroid* **2017**, *28*, 220–228. [CrossRef]
13. Dong, W.; Zhang, H.; Zhang, P.; Li, X.; He, L.; Wang, Z.; Liu, Y. The changing incidence of thyroid carcinoma in Shenyang, China before and after universal salt iodization. *Med. Sci. Monit. Int. Med. J. Exp. Clin. Res.* **2013**, *19*, 49. [CrossRef] [PubMed]
14. Liu, C.-L.; Cheng, S.-P.; Lin, H.-W.; Lai, Y.-L. Risk of thyroid cancer in patients with thyroiditis: A population-based cohort study. *Ann. Surg. Oncol.* **2014**, *21*, 843–849. [CrossRef] [PubMed]
15. Oh, C.-M.; Park, S.; Lee, J.Y.; Won, Y.-J.; Shin, A.; Kong, H.-J.; Choi, K.-S.; Lee, Y.J.; Chung, K.-W.; Jung, K.-W. Increased prevalence of chronic lymphocytic thyroiditis in Korean patients with papillary thyroid cancer. *PLoS ONE* **2014**, *9*, e99054. [CrossRef] [PubMed]
16. Marcello, M.A.; Malandrino, P.; Almeida, J.F.M.; Martins, M.B.; Cunha, L.L.; Bufalo, N.E.; Pellegriti, G.; Ward, L.S. The influence of the environment on the development of thyroid tumors: A new appraisal. *Endocr. Relat. Cancer* **2014**, *21*, T235–T254. [CrossRef] [PubMed]
17. Benedetti, M.; Zona, A.; Beccaloni, E.; Carere, M.; Comba, P. Incidence of Breast, Prostate, Testicular, and Thyroid Cancer in Italian Contaminated Sites with Presence of Substances with Endocrine Disrupting Properties. *Int. J. Environ. Res. Public. Health* **2017**, *14*, 355. [CrossRef]
18. Kilfoy, B.A.; Zhang, Y.; Park, Y.; Holford, T.R.; Schatzkin, A.; Hollenbeck, A.; Ward, M.H. Dietary nitrate and nitrite and the risk of thyroid cancer in the NIH-AARP Diet and Health Study. *Int. J. Cancer* **2011**, *129*, 160–172. [CrossRef]
19. Ward, M.H.; Kilfoy, B.A.; Weyer, P.J.; Anderson, K.E.; Folsom, A.R.; Cerhan, J.R. Nitrate intake and the risk of thyroid cancer and thyroid disease. *Epidemiol. Camb. Mass* **2010**, *21*, 389. [CrossRef]
20. Fiore, M.; Oliveri Conti, G.; Caltabiano, R.; Buffone, A.; Zuccarello, P.; Cormaci, L.; Cannizzaro, M.A.; Ferrante, M. Role of Emerging Environmental Risk Factors in Thyroid Cancer: A Brief Review. *Int. J. Environ. Res. Public. Health* **2019**, *16*, 1185. [CrossRef]
21. Huang, H.; Sjodin, A.; Chen, Y.; Ni, X.; Ma, S.; Yu, H.; Ward, M.H.; Udelsman, R.; Rusiecki, J.; Zhang, Y. Polybrominated Diphenyl Ethers, Polybrominated Biphenyls, and Risk of Papillary Thyroid Cancer: A Nested Case-Control Study. *Am. J. Epidemiol.* **2020**, *189*, 120–132. [CrossRef]
22. Marotta, V.; Malandrino, P.; Russo, M.; Panariello, I.; Ionna, F.; Chiofalo, M.G.; Pezzullo, L. Fathoming the link between anthropogenic chemical contamination and thyroid cancer. *Crit. Rev. Oncol. Hematol.* **2020**, *150*, 102950. [CrossRef] [PubMed]
23. Kung, T.M.; Ng, W.L.; Gibson, J.B. Volcanoes and Carcinoma of the Thyroid: A Possible Association. *Arch. Environ. Health Int. J.* **1981**, *36*, 265–267. [CrossRef] [PubMed]
24. Goodman, M.T.; Yoshizawa, C.N.; Kolonel, L.N. Descriptive epidemiology of thyroid cancer in Hawaii. *Cancer* **1988**, *61*, 1272–1281. [CrossRef]
25. Kolonel, L.N.; Hankin, J.H.; Wilkens, L.R.; Fukunaga, F.H.; Hinds, M.W. An epidemiologic study of thyroid cancer in Hawaii. *Cancer Causes Control* **1990**, *1*, 223–234. [CrossRef]
26. Arnbjörnsson, E.; Arnbjörnsson, A.; Ólafsson, A. Thyroid Cancer Incidence in Relation to Volcanic Activity. *Arch. Environ. Health Int. J.* **1986**, *41*, 36–40. [CrossRef]
27. Hrafnkelsson, J.; Tulinius, H.; Jonasson, J.G.; Ólafsdottir, G.; Sigvaldason, H. Papillary Thyroid Carcinoma in Iceland: A study of the Occurrence in families and the coexistence of other primary tumours. *Acta Oncol.* **1989**, *28*, 785–788. [CrossRef]
28. Paksoy, N.; Montaville, B.; McCarthy, S. Cancer Occurrence in Vanuatu in the South Pacific, 1980–1986. *Asia Pac. J. Public Health* **1989**, *3*, 231–236. [CrossRef]
29. Curado, M.P.; Edwards, B.; Shin, H.R.; Storm, H.; Ferlay, J.; Heanue, M. *Cancer Incidence in Five Continents Vol. IX. IARC Scientific Publication No. 160. 2007*; IARC Press: Lyon, France, 2007.

30. Truong, T.; Rougier, Y.; Dubourdieu, D.; Guihenneuc-Jouyaux, C.; Orsi, L.; Hémon, D.; Guénel, P. Time trends and geographic variations for thyroid cancer in New Caledonia, a very high incidence area (1985–1999). *Eur. J. Cancer Prev.* **2007**, *16*, 62–70. [CrossRef]
31. Duntas, L.H.; Doumas, C. The 'rings of fire' and thyroid cancer. *Hormones* **2009**, *8*, 249–253. [CrossRef]
32. Kristbjornsdottir, A.; Rafnsson, V. Cancer incidence among population utilizing geothermal hot water: A census-based cohort study. *Int. J. Cancer* **2013**, *133*, 2944–2952. [CrossRef]
33. Goyal, N.; Camacho, F.; Mangano, J.; Goldenberg, D. Evaluating for a geospatial relationship between radon levels and thyroid cancer in Pennsylvania. *Laryngoscope* **2015**, *125*, E45–E49. [CrossRef] [PubMed]
34. Oakland, C.; Meliker, J.R. County-Level Radon and Incidence of Female Thyroid Cancer in Iowa, New Jersey, and Wisconsin, USA. *Toxics* **2018**, *6*, 17. [CrossRef] [PubMed]
35. Pellegriti, G.; De Vathaire, F.; Scollo, C.; Attard, M.; Giordano, C.; Arena, S.; Dardanoni, G.; Frasca, F.; Malandrino, P.; Vermiglio, F.; et al. Papillary Thyroid Cancer Incidence in the Volcanic Area of Sicily. *JNCI J. Natl. Cancer Inst.* **2009**, *101*, 1575–1583. [CrossRef] [PubMed]
36. Russo, M.; Malandrino, P.; Moleti, M.; D'Angelo, A.; Tavarelli, M.; Sapuppo, G.; Gianì, F.; Richiusa, P.; Squatrito, S.; Vigneri, R.; et al. Thyroid Cancer in the Pediatric Age in Sicily: Influence of the Volcanic Environment. *Anticancer Res.* **2017**, *37*, 1515–1522. [PubMed]
37. Dahal, B.M.; Fuerhacker, M.; Mentler, A.; Karki, K.B.; Shrestha, R.R.; Blum, W.E.H. Arsenic contamination of soils and agricultural plants through irrigation water in Nepal. *Environ. Pollut.* **2008**, *155*, 157–163. [CrossRef] [PubMed]
38. Abiye, T.A.; Sulaiman, H.; Hailu, A. Metal concentration in vegetables grown in the hydrothermally affected area in Ethiopia. *J. Geogr. Geol.* **2011**, *3*, 86.
39. World Health Organization; International Atomic Energy Agency; Food and Agriculture Organization of the United Nations. Trace Elements in Human Nutrition and Health. 1996. Available online: https://apps.who.int/iris/handle/10665/37931 (accessed on 24 April 2020).
40. Outten, C.E.; O'Halloran, T.V. Femtomolar Sensitivity of Metalloregulatory Proteins Controlling Zinc Homeostasis. *Science* **2001**, *292*, 2488–2492. [CrossRef]
41. Finney, L.A.; O'Halloran, T.V. Transition Metal Speciation in the Cell: Insights from the Chemistry of Metal Ion Receptors. *Science* **2003**, *300*, 931–936. [CrossRef]
42. Trusso Sfrazzetto, G.; Satriano, C.; Tomaselli, G.A.; Rizzarelli, E. Synthetic fluorescent probes to map metallostasis and intracellular fate of zinc and copper. *Coord. Chem. Rev.* **2016**, *311*, 125–167. [CrossRef]
43. Dittmer, P.J.; Miranda, J.G.; Gorski, J.A.; Palmer, A.E. Genetically Encoded Sensors to Elucidate Spatial Distribution of Cellular Zinc. *J. Biol. Chem.* **2009**, *284*, 16289–16297. [CrossRef]
44. Changela, A.; Chen, K.; Xue, Y.; Holschen, J.; Outten, C.E.; O'Halloran, T.V.; Mondragón, A. Molecular Basis of Metal-Ion Selectivity and Zeptomolar Sensitivity by CueR. *Science* **2003**, *301*, 1383–1387. [CrossRef] [PubMed]
45. Milardi, D.; Rizzarelli, E. *Neurodegeneration: Metallostasis and Proteostasis*; Royal Society of Chemistry: London, UK, 2011; ISBN 978-1-84973-301-4.
46. Grasso, G.; Giuffrida, M.L.; Rizzarelli, E. Metallostasis and amyloid β-degrading enzymes. *Metallomics* **2012**, *4*, 937–949. [CrossRef] [PubMed]
47. Nyquist, M.D.; Prasad, B.; Mostaghel, E.A. Harnessing Solute Carrier Transporters for Precision Oncology. *Molecules* **2017**, *22*, 539. [CrossRef] [PubMed]
48. Si, M.; Lang, J. The roles of metallothioneins in carcinogenesis. *J. Hematol. Oncol.* **2018**, *11*, 107. [CrossRef] [PubMed]
49. Calvo, J.; Jung, H.; Meloni, G. Copper metallothioneins. *IUBMB Life* **2017**, *69*, 236–245. [CrossRef] [PubMed]
50. Rutherford, J.C.; Bird, A.J. Metal-Responsive Transcription Factors That Regulate Iron, Zinc, and Copper Homeostasis in Eukaryotic Cells. *Eukaryot. Cell* **2004**, *3*, 1–13. [CrossRef]
51. Jung, J.; Lee, S.J. Biochemical and Biodiversity Insights into Heavy Metal Ion-Responsive Transcription Regulators for Synthetic Biological Heavy Metal Sensors. *J. Microbiol. Biotechnol.* **2019**, *29*, 1522–1542. [CrossRef]
52. Sies, H. Glutathione and its role in cellular functions. *Free Radic. Biol. Med.* **1999**, *27*, 916–921. [CrossRef]
53. Fukada, T.; Yamasaki, S.; Nishida, K.; Murakami, M.; Hirano, T. Zinc homeostasis and signaling in health and diseases. *JBIC J. Biol. Inorg. Chem.* **2011**, *16*, 1123–1134. [CrossRef]

54. Kambe, T.; Yamaguchi-Iwai, Y.; Sasaki, R.; Nagao, M. Overview of mammalian zinc transporters. *Cell. Mol. Life Sci. CMLS* **2004**, *61*, 49–68. [CrossRef]
55. La Mendola, D.; Giacomelli, C.; Rizzarelli, E. Intracellular Bioinorganic Chemistry and Cross Talk Among Different–Omics. Available online: https://www.ingentaconnect.com/content/ben/ctmc/2016/00000016/00000027/art00008 (accessed on 12 April 2020).
56. Seve, M.; Chimienti, F.; Devergnas, S.; Favier, A. In silico identification and expression of SLC30 family genes: An expressed sequence tag data mining strategy for the characterization of zinc transporters' tissue expression. *BMC Genom.* **2004**, *5*, 32. [CrossRef] [PubMed]
57. Dong, G.; Chen, H.; Qi, M.; Dou, Y.; Wang, Q. Balance between metallothionein and metal response element binding transcription factor 1 is mediated by zinc ions (Review). *Mol. Med. Rep.* **2015**, *11*, 1582–1586. [CrossRef] [PubMed]
58. Carpenter, M.C.; Palmer, A.E. Native and engineered sensors for Ca^{2+} and Zn^{2+}: Lessons from calmodulin and MTF1. *Essays Biochem.* **2017**, *61*, 237–243.
59. Nose, Y.; Kim, B.-E.; Thiele, D.J. Ctr1 drives intestinal copper absorption and is essential for growth, iron metabolism, and neonatal cardiac function. *Cell Metab.* **2006**, *4*, 235–244. [CrossRef] [PubMed]
60. Festa, R.A.; Thiele, D.J. Copper: An essential metal in biology. *Curr. Biol.* **2011**, *21*, R877–R883. [CrossRef] [PubMed]
61. Harrison, M.D.; Jones, C.E.; Dameron, C.T. Copper chaperones: Function, structure and copper-binding properties. *JBIC J. Biol. Inorg. Chem.* **1999**, *4*, 145–153. [CrossRef] [PubMed]
62. Palumaa, P. Copper chaperones. The concept of conformational control in the metabolism of copper. *FEBS Lett.* **2013**, *587*, 1902–1910. [CrossRef]
63. Devi, S.R.B.; Dhivya M, A.; Sulochana, K.N. Copper transporters and chaperones: Their function on angiogenesis and cellular signalling. *J. Biosci.* **2016**, *41*, 487–496. [CrossRef]
64. Yuan, S.; Chen, S.; Xi, Z.; Liu, Y. Copper-finger protein of Sp1: The molecular basis of copper sensing. *Metallomics* **2017**, *9*, 1169–1175. [CrossRef]
65. Hughes Michael, F. Biomarkers of Exposure: A Case Study with Inorganic Arsenic. *Environ. Health Perspect.* **2006**, *114*, 1790–1796. [CrossRef]
66. Sun, H.; Xiang, P.; Luo, J.; Hong, H.; Lin, H.; Li, H.-B.; Ma, L.Q. Mechanisms of arsenic disruption on gonadal, adrenal and thyroid endocrine systems in humans: A review. *Environ. Int.* **2016**, *95*, 61–68. [CrossRef] [PubMed]
67. Yamamoto, S.; Konishi, Y.; Matsuda, T.; Murai, T.; Shibata, M.-A.; Matsui-Yuasa, I.; Otani, S.; Kuroda, K.; Endo, G.; Fukushima, S. Cancer Induction by an Organic Arsenic Compound, Dimethylarsinic Acid (Cacodylic Acid), in F344/DuCrj Rats after Pretreatment with Five Carcinogens. *Cancer Res.* **1995**, *55*, 1271–1276. [PubMed]
68. Fröhlich, E.; Czarnocka, B.; Brossart, P.; Wahl, R. Antitumor Effects of Arsenic Trioxide in Transformed Human Thyroid Cells. *Thyroid* **2008**, *18*, 1183–1193. [CrossRef] [PubMed]
69. Wang, M.; Ge, X.; Zheng, J.; Li, D.; Liu, X.; Wang, L.; Jiang, C.; Shi, Z.; Qin, L.; Liu, J.; et al. Role and mechanism of miR-222 in arsenic-transformed cells for inducing tumor growth. *Oncotarget* **2016**, *7*, 17805–17814. [CrossRef] [PubMed]
70. He, H.; Jazdzewski, K.; Li, W.; Liyanarachchi, S.; Nagy, R.; Volinia, S.; Calin, G.A.; Liu, C.; Franssila, K.; Suster, S.; et al. The role of microRNA genes in papillary thyroid carcinoma. *Proc. Natl. Acad. Sci. USA* **2005**, *102*, 19075–19080. [CrossRef]
71. Zidane, M.; Ren, Y.; Xhaard, C.; Leufroy, A.; Côte, S.; Dewailly, E.; Noël, L.; Guérin, T.; Bouisset, P.; Bernagout, S.; et al. Non-Essential Trace Elements Dietary Exposure in French Polynesia: Intake Assessment, Nail Bio Monitoring and Thyroid Cancer Risk. *Asian Pac. J. Cancer Prev. APJCP* **2019**, *20*, 355–367. [CrossRef]
72. Iavicoli, I.; Fontana, L.; Bergamaschi, A. The Effects of Metals as Endocrine Disruptors. *J. Toxicol. Environ. Health Part B* **2009**, *12*, 206–223. [CrossRef]
73. Hartwig, A.; Asmuss, M.; Blessing, H.; Hoffmann, S.; Jahnke, G.; Khandelwal, S.; Pelzer, A.; Bürkle, A. Interference by toxic metal ions with zinc-dependent proteins involved in maintaining genomic stability. *Food Chem. Toxicol.* **2002**, *40*, 1179–1184. [CrossRef]
74. Buha, A.; Matovic, V.; Antonijevic, B.; Bulat, Z.; Curcic, M.; Renieri, E.A.; Tsatsakis, A.M.; Schweitzer, A.; Wallace, D. Overview of Cadmium Thyroid Disrupting Effects and Mechanisms. *Int. J. Mol. Sci.* **2018**, *19*, 1501. [CrossRef]

75. Beyersmann, D.; Hartwig, A. Carcinogenic metal compounds: Recent insight into molecular and cellular mechanisms. *Arch. Toxicol.* **2008**, *82*, 493. [CrossRef]
76. Huff, J.; Lunn, R.M.; Waalkes, M.P.; Tomatis, L.; Infante, P.F. Cadmium-induced Cancers in Animals and in Humans. *Int. J. Occup. Environ. Health* **2007**, *13*, 202–212. [CrossRef] [PubMed]
77. Prozialeck, W.C.; Lamar, P.C.; Lynch, S.M. Cadmium alters the localization of N-cadherin, E-cadherin, and β-catenin in the proximal tubule epithelium. *Toxicol. Appl. Pharmacol.* **2003**, *189*, 180–195. [CrossRef]
78. Zhu, P.; Liao, L.-Y.; Zhao, T.-T.; Mo, X.-M.; Chen, G.G.; Liu, Z.-M. GPER/ERK&AKT/NF-κB pathway is involved in cadmium-induced proliferation, invasion and migration of GPER-positive thyroid cancer cells. *Mol. Cell. Endocrinol.* **2017**, *442*, 68–80. [CrossRef] [PubMed]
79. Deheyn, D.D.; Gendreau, P.; Baldwin, R.J.; Latz, M.I. Evidence for enhanced bioavailability of trace elements in the marine ecosystem of Deception Island, a volcano in Antarctica. *Mar. Environ. Res.* **2005**, *60*, 1–33. [CrossRef]
80. Fiorentino, C.E.; Paoloni, J.D.; Sequeira, M.E.; Arosteguy, P. The presence of vanadium in groundwater of southeastern extreme the pampean region Argentina: Relationship with other chemical elements. *J. Contam. Hydrol.* **2007**, *93*, 122–129. [CrossRef]
81. Malandrino, P.; Russo, M.; Ronchi, A.; Minoia, C.; Cataldo, D.; Regalbuto, C.; Giordano, C.; Attard, M.; Squatrito, S.; Trimarchi, F.; et al. Increased thyroid cancer incidence in a basaltic volcanic area is associated with non-anthropogenic pollution and biocontamination. *Endocrine* **2016**, *53*, 471–479. [CrossRef]
82. Wolterbeek, B. Biomonitoring of trace element air pollution: Principles, possibilities and perspectives. *Environ. Pollut.* **2002**, *120*, 11–21. [CrossRef]
83. Augusto, S.; Máguas, C.; Branquinho, C. Guidelines for biomonitoring persistent organic pollutants (POPs), using lichens and aquatic mosses—A review. *Environ. Pollut.* **2013**, *180*, 330–338. [CrossRef]
84. World Health Organization. *Guidelines for Drinking-Water Quality*, 4th ed.; WHO Chronicle: Geneva, Switzerland, 2011; Volume 38.
85. Varrica, D.; Tamburo, E.; Dongarrà, G.; Sposito, F. Trace elements in scalp hair of children chronically exposed to volcanic activity (Mt. Etna, Italy). *Sci. Total Environ.* **2014**, *470–471*, 117–126. [CrossRef]
86. Rodrigues, A.S.; Arruda, M.S.C.; Garcia, P.V. Evidence of DNA damage in humans inhabiting a volcanically active environment: A useful tool for biomonitoring. *Environ. Int.* **2012**, *49*, 51–56. [CrossRef]
87. Russo, M.; Malandrino, P.; Addario, W.P.; Dardanoni, G.; Vigneri, P.; Pellegriti, G.; Squatrito, S.; Vigneri, R. Several Site-specific Cancers are Increased in the Volcanic Area in Sicily. *Anticancer Res.* **2015**, *35*, 3995–4001. [PubMed]
88. Calabrese, E.J. Hormesis: A revolution in toxicology, risk assessment and medicine. *EMBO Rep.* **2004**, *5*, S37–S40. [CrossRef]
89. Hao, C.; Hao, W.; Wei, X.; Xing, L.; Jiang, J.; Shang, L. The role of MAPK in the biphasic dose-response phenomenon induced by cadmium and mercury in HEK293 cells. *Toxicol. Vitr.* **2009**, *23*, 660–666. [CrossRef]
90. Llabjani, V.; Hoti, V.; Pouran, H.M.; Martin, F.L.; Zhang, H. Bimodal responses of cells to trace elements: Insights into their mechanism of action using a biospectroscopy approach. *Chemosphere* **2014**, *112*, 377–384. [CrossRef] [PubMed]
91. Mantha, M.; Jumarie, C. Cadmium-induced hormetic effect in differentiated Caco-2 cells: ERK and p38 activation without cell proliferation stimulation. *J. Cell. Physiol.* **2010**, *224*, 250–261. [CrossRef]
92. Damelin, L.H.; Vokes, S.; Whitcutt, J.M.; Damelin, S.B.; Alexander, J.J. Hormesis: A stress response in cells exposed to low levels of heavy metals. *Hum. Exp. Toxicol.* **2000**, *19*, 420–430. [CrossRef] [PubMed]
93. Nascarella, M.A.; Stoffolano, J.G.; Stanek, E.J.; Kostecki, P.T.; Calabrese, E.J. Hormesis and stage specific toxicity induced by cadmium in an insect model, the queen blowfly, Phormia regina Meig. *Environ. Pollut.* **2003**, *124*, 257–262. [CrossRef]
94. Renieri, E.A.; Sfakianakis, D.G.; Alegakis, A.A.; Safenkova, I.V.; Buha, A.; Matović, V.; Tzardi, M.; Dzantiev, B.B.; Divanach, P.; Kentouri, M.; et al. Nonlinear responses to waterborne cadmium exposure in zebrafish. An in vivo study. *Environ. Res.* **2017**, *157*, 173–181. [CrossRef]
95. Zhang, W.Z.; Sun, N.N.; Ma, S.Q.; Zhao, Z.C.; Cao, Y.; Zhang, C. Hormetic Effects of Yttrium on Male Sprague-Dawley Rats. *Biomed. Environ. Sci. BES* **2018**, *31*, 777–780. [CrossRef]
96. Maggisano, V.; Bulotta, S.; Celano, M.; Maiuolo, J.; Lepore, S.M.; Abballe, L.; Iannone, M.; Russo, D. Low Doses of Methylmercury Induce the Proliferation of Thyroid Cells In Vitro Through Modulation of ERK Pathway. *Int. J. Mol. Sci.* **2020**, *21*, 1556. [CrossRef]

97. Schmidt, C.M.; Cheng, C.N.; Marino, A.; Konsoula, R.; Barile, F.A. Hormesis effect of trace metals on cultured normal and immortal human mammary cells. *Toxicol. Ind. Health* **2004**, *20*, 57–68. [CrossRef] [PubMed]
98. Tyne, W.; Little, S.; Spurgeon, D.J.; Svendsen, C. Hormesis depends upon the life-stage and duration of exposure: Examples for a pesticide and a nanomaterial. *Ecotoxicol. Environ. Saf.* **2015**, *120*, 117–123. [CrossRef] [PubMed]
99. Hadacek, F.; Bachmann, G.; Engelmeier, D.; Chobot, V. Hormesis and a chemical raison d'ětre for secondary plant metabolites. *Dose Response* **2011**, *9*, 79–116. [CrossRef] [PubMed]
100. Yang, P.; He, X.-Q.; Peng, L.; Li, A.-P.; Wang, X.-R.; Zhou, J.-W.; Liu, Q.-Z. The Role of Oxidative Stress in Hormesis Induced by Sodium Arsenite in Human Embryo Lung Fibroblast (HELF) Cellular Proliferation Model. *J. Toxicol. Environ. Health A* **2007**, *70*, 976–983. [CrossRef]
101. Tan, Q.; Liu, Z.; Li, H.; Liu, Y.; Xia, Z.; Xiao, Y.; Usman, M.; Du, Y.; Bi, H.; Wei, L. Hormesis of mercuric chloride-human serum albumin adduct on N9 microglial cells via the ERK/MAPKs and JAK/STAT3 signaling pathways. *Toxicology* **2018**, *408*, 62–69. [CrossRef]
102. Singh, K.B.; Maret, W. The interactions of metal cations and oxyanions with protein tyrosine phosphatase 1B. *BioMetals* **2017**, *30*, 517–527. [CrossRef]
103. Gianì, F.; Pandini, G.; Scalisi, N.M.; Vigneri, P.; Fazzari, C.; Malandrino, P.; Russo, M.; Masucci, R.; Belfiore, A.; Pellegriti, G.; et al. Effect of low-dose tungsten on human thyroid stem/precursor cells and their progeny. *Endocr. Relat. Cancer* **2019**, *26*, 713–725. [CrossRef]
104. Jiang, G.; Duan, W.; Xu, L.; Song, S.; Zhu, C.; Wu, L. Biphasic effect of cadmium on cell proliferation in human embryo lung fibroblast cells and its molecular mechanism. *Toxicol. Vitr.* **2009**, *23*, 973–978. [CrossRef]
105. Tsui, M.T.K.; Wang, W.-X. Influences of maternal exposure on the tolerance and physiological performance of Daphnia magna under mercury stress. *Environ. Toxicol. Chem.* **2005**, *24*, 1228–1234. [CrossRef]
106. Carvalho, M.E.A.; Castro, P.R.C.; Azevedo, R.A. Hormesis in plants under Cd exposure: From toxic to beneficial element? *J. Hazard. Mater.* **2020**, *384*, 121434. [CrossRef]
107. Malandrino, P.; Russo, M.; Ronchi, A.; Moretti, F.; Gianì, F.; Vigneri, P.; Masucci, R.; Pellegriti, G.; Belfiore, A.; Vigneri, R. Concentration of Metals and Trace Elements in the Normal Human and Rat Thyroid: Comparison with Muscle and Adipose Tissue and Volcanic Versus Control Areas. *Thyroid* **2019**, *30*, 290–299. [CrossRef]
108. Kopp, P. Thyroid hormone synthesis. In *Werner and Ingbar's the Thyroid: A Fundamental and Clinical Text*, 10th ed.; Cooper, D., Ed.; Lippincott Williams Wilkins: Philadelphia, PA, USA, 2012; pp. 48–74.
109. Maier, J.; Van Steeg, H.; Van Oostrom, C.; Karger, S.; Paschke, R.; Krohn, K. Deoxyribonucleic Acid Damage and Spontaneous Mutagenesis in the Thyroid Gland of Rats and Mice. *Endocrinology* **2006**, *147*, 3391–3397. [CrossRef]
110. Kawada, J.; Shirakawa, Y.; Yoshimura, Y.; Nishida, M. Thyroid xanthine oxidase and its role in thyroid iodine metabolism in the rat: Difference between effects of allopurinol and tungstate. *J. Endocrinol.* **1982**, *95*, 117–124. [CrossRef] [PubMed]
111. Brady, D.C.; Crowe, M.S.; Turski, M.L.; Hobbs, G.A.; Yao, X.; Chaikuad, A.; Knapp, S.; Xiao, K.; Campbell, S.L.; Thiele, D.J. Copper is required for oncogenic BRAF signalling and tumorigenesis. *Nature* **2014**, *509*, 492–496. [CrossRef] [PubMed]
112. Xu, M.; Casio, M.; Range, D.E.; Sosa, J.A.; Counter, C.M. Copper chelation as targeted therapy in a mouse model of oncogenic BRAF-driven papillary thyroid cancer. *Clin. Cancer Res.* **2018**, *24*, 4271–4281. [CrossRef] [PubMed]
113. Assem, F.L.; Holmes, P.; Levy, L.S. The mutagenicity and carcinogenicity of inorganic manganese compounds: A synthesis of the evidence. *J. Toxicol. Environ. Health Part B* **2011**, *14*, 537–570. [CrossRef]
114. Sun, H.-J.; Li, S.-W.; Li, C.; Wang, W.-Q.; Li, H.-B.; Ma, L.Q. Thyrotoxicity of arsenate and arsenite on juvenile mice at organism, subcellular, and gene levels under low exposure. *Chemosphere* **2017**, *186*, 580–587. [CrossRef]
115. Velický, J.; TItlbach, M.; Dušková, J.; Vobecký, M.; Štrbák, V.; Raška, I. Potassium bromide and the thyroid gland of the rat: Morphology and immunohistochemistry, RIA and INAA analysis. *Ann. Anat. Anat. Anz.* **1997**, *179*, 421–431. [CrossRef]
116. Velický, J.; Titlbach, M.; Lojda, Z.; Dušková, J.; Vobecký, M.; Štrbák, V.; Raška, I. Long-term action of potassium bromide on the rat thyroid gland. *Acta Histochem.* **1998**, *100*, 11–23. [CrossRef]

117. DeAngelo, A.B.; George, M.H.; Kilburn, S.R.; Moore, T.M.; Wolf, D.C. Carcinogenicity of potassium bromate administered in the drinking water to male B6C3F1 mice and F344/N rats. *Toxicol. Pathol.* **1998**, *26*, 587–594. [CrossRef]
118. Kanno, J.; Matsuoka, C.; Furuta, K.; Onodera, H.; Miyajima, H.; Maekawa, A.; Hayashi, Y. Tumor promoting effect of goitrogens on the rat thyroid. *Toxicol. Pathol.* **1990**, *18*, 239–246. [CrossRef] [PubMed]
119. Luca, E.; Fici, L.; Ronchi, A.; Marandino, F.; Rossi, E.D.; Caristo, M.E.; Malandrino, P.; Russo, M.; Pontecorvi, A.; Vigneri, R. Intake of Boron, Cadmium, and Molybdenum enhances rat thyroid cell transformation. *J. Exp. Clin. Cancer Res.* **2017**, *36*, 73. [CrossRef] [PubMed]
120. Vareda, J.P.; Valente, A.J.; Durães, L. Assessment of heavy metal pollution from anthropogenic activities and remediation strategies: A review. *J. Environ. Manag.* **2019**, *246*, 101–118. [CrossRef] [PubMed]

International Journal of
Molecular Sciences

Review

Reactive Oxygen Species, Metabolic Plasticity, and Drug Resistance in Cancer

Vikas Bhardwaj [1] **and Jun He** [2,*]

1. College of Pharmacy, Thomas Jefferson University, Philadelphia, PA 19107, USA; vikas.bhardwaj@jefferson.edu
2. Department of Pathology, Anatomy & Cell Biology, Sidney Kimmel Medical College, Thomas Jefferson University, Philadelphia, PA 19107, USA

* Correspondence: Jun.he@jefferson.edu

Received: 29 April 2020; Accepted: 11 May 2020; Published: 12 May 2020

Abstract: The metabolic abnormality observed in tumors is characterized by the dependence of cancer cells on glycolysis for their energy requirements. Cancer cells also exhibit a high level of reactive oxygen species (ROS), largely due to the alteration of cellular bioenergetics. A highly coordinated interplay between tumor energetics and ROS generates a powerful phenotype that provides the tumor cells with proliferative, antiapoptotic, and overall aggressive characteristics. In this review article, we summarize the literature on how ROS impacts energy metabolism by regulating key metabolic enzymes and how metabolic pathways e.g., glycolysis, PPP, and the TCA cycle reciprocally affect the generation and maintenance of ROS homeostasis. Lastly, we discuss how metabolic adaptation in cancer influences the tumor's response to chemotherapeutic drugs. Though attempts of targeting tumor energetics have shown promising preclinical outcomes, the clinical benefits are yet to be fully achieved. A better understanding of the interaction between metabolic abnormalities and involvement of ROS under the chemo-induced stress will help develop new strategies and personalized approaches to improve the therapeutic efficiency in cancer patients.

Keywords: reactive oxygen species; metabolic adaptation; drug resistance; cancer

---

1. Introduction

Otto Warburg first observed alterations in cancer metabolism, wherein cancer cells produce most of their energy through glycolysis rather than mitochondrial oxidative phosphorylation in the presence of oxygen, in the 1920s [1]. Warburg observed that tumor cells convert majority of the glucose into lactate and not into CO_2 as observed in non-tumorous mammalian cells. The observation has since been witnessed by various other researchers and has been thoroughly reviewed [2,3]. Although glycolysis is an inefficient mechanism of energy production (glycolysis generates two ATP molecules, whereas the tricarboxylic acid (TCA) cycle produces 34 ATPs per glucose molecule), it provides cancer cells with ATP at a considerably faster rate than through mitochondria [4,5]. For quicker energy production, a cancer cell can enhance their uptake of glucose, a property commonly employed to visualize tumors using fluorodeoxyglucose, a radio-labeled glucose during positron emission tomography [6]. The upregulation of glycolysis also supports the proliferating cells by providing metabolites, such as serine, glycine, and alanine, for their anabolic processes [7–9]. Alternatively, the glycolytic metabolites can also be shunted into the pentose phosphate pathway (PPP) and provide cancer cells with ribose nucleotides and redox potential. Since the microenvironment provides tumor cells with ample nutrients (e.g., glucose, glutamine), ATP produced via glycolysis is sufficient to fulfill the tumor's requirements. The importance of glycolysis can be appreciated by studies demonstrating that its inhibition suppresses ATP production and leads to cancer cell apoptosis [10–12].

The Warburg effect shifts cancer cells from oxidative to reductive metabolism. The reductive metabolism is imperative for the biosynthesis of amino acids and metabolites to sustain cancer survival and growth. The shift in metabolism reduces the dependence on mitochondrial citrate and ATP as these are detrimental to the survival of cancer cells [13–16]. Genetic and epigenetic regulation of TCA cycle enzymes observed in cancer further support the metabolic shift away from mitochondrial respiration. The altered metabolism is widely considered as one of the hallmarks of cancer, and emerging evidence supports its key role in tumor development through its interaction with other drivers of tumor, namely oncogenes, tumor suppressors, and cellular redox balance [17].

Reactive oxygen species (ROS) are oxygen-containing and chemically reactive species formed by incomplete one-electron reduction of oxygen, which includes hydrogen peroxide (H_2O_2), superoxide anion (O_2^-), and hydroxyl radical (OH^-) [18]. ROS are naturally produced in cells through aerobic metabolism. Mitochondria respiratory chain, NADPH oxidase, and peroxisomes are the major endogenous sources of ROS. Under physiological conditions, normal cells maintain redox homeostasis with a low level of basal ROS by controlling the balance between ROS generation (pro-oxidants) and elimination (antioxidant capacity). A moderate increase in ROS favors cell proliferation and survival. However, when the amount of ROS reaches a certain level, it may overwhelm the antioxidant capacity of the cell and trigger cell death by oxidizing cellular macromolecules such as proteins, nuclear acids, and lipids. Growing evidence suggests that cancer cells exhibit increased intrinsic ROS stress due to metabolic abnormalities and oncogenic signaling. In order to maintain the redox dynamics of high ROS, cancer cells trigger an adaptation response by upregulating their antioxidant capacity.

The generation and maintenance of ROS homeostasis in cells largely rely on cellular metabolism while ROS also impacts energy metabolism by regulating key metabolic enzymes and critical oncogenic signaling pathways. In this review, we highlight studies that underline the interplay between the cellular redox balance and tumor metabolism, and explore how these mechanisms support tumor survival under drug-induced stresses (Figure 1).

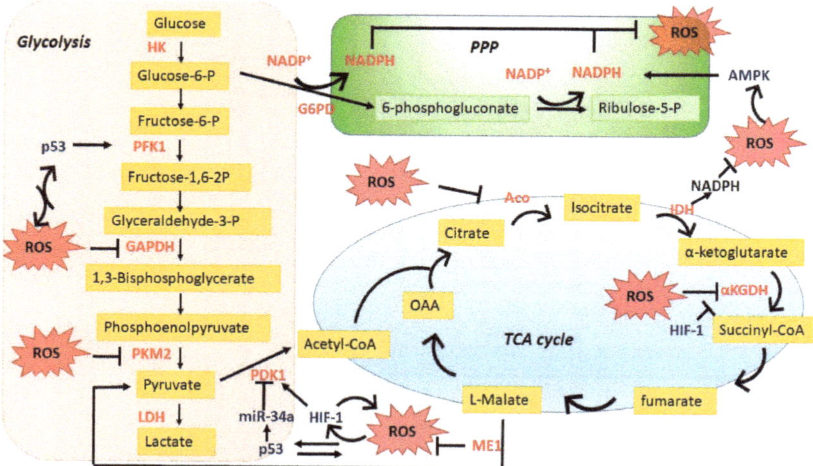

Figure 1. Interplay between reactive oxygen species (ROS) and the central carbon metabolism. ROS impact cellular metabolism through regulating key enzymes in glycolysis pathway/TCA cycle as well as redox signaling pathways; reciprocally, energy metabolic pathways especially PPP balance ROS homeostasis in cancer cells. PPP, pentose phosphate pathway; HK, hexokinase; PFK1, phosphofructokinase 1; PDK1, pyruvate dehydrogenase kinase; G6PD, glucose 6-phosphate dehydrogenase; Aco, aconitase; αKGDH, Alpha ketoglutarate dehydrogenase; IDH, isocitrate dehydrogenase; ME1, malic enzyme 1.

2. ROS Impacts Cancer Metabolic Reprogramming

2.1. Direct Regulation through Key Metabolic Enzymes

Although the Warburg hypothesis postulated elevated glycolysis in proliferating cells, the overall metabolic regulation in cancers is rather complex. In addition to glycolysis, cancer cells demonstrate elevated flux into the pentose phosphate pathway, enhanced glutamine consumption, enhanced rate of lipid biosynthesis, and utilization of protein as a fuel source [19–21]. These metabolic deregulations however do not occur in silo. It interacts with numerous signaling molecules to promote tumor phenotype. One such deregulation commonly observed in cancer cells is the elevated ROS level [22]. ROS plays a crucial role in maintaining and promoting the tumor phenotype via regulating oncogenic signaling and cellular metabolism [23–25]. The ROS levels however are to be kept within a certain range since high ROS levels can be detrimental to cancer survival [26]. Below, we have discussed how the cellular ROS and redox mechanism regulate tumor metabolism with a focus on central carbon metabolism, namely glycolysis, PPP, and the TCA cycle.

2.1.1. Glycolysis

Recent evidence suggests that upregulated NADPH oxidase (NOX) shifts the tumor's metabolism towards glycolysis in cells with mitochondrial dysfunction [27]. The uncovering of the novel role of NOX is notable since NOX catalyzes the conversion of molecular oxygen to the superoxide ion ($O_2 \rightarrow O^-$), and its upregulation is commonly observed in tumors [28].

Pyruvate kinase M2 (PKM2) is a rate-limiting glycolytic enzyme that converts phosphoenolpyruvate (PEP) and ADP to pyruvate and ATP [29]. The enzyme pyruvate kinase (PK) exists in two different isoforms: M1 and M2. Under physiological condition, the M1 isoform (PKM1) predominates, whereas cancer cells primarily express the M2 isoform (PKM2) [30]. PKM2 has significantly less pyruvate kinase activity, and it prevents the flow of glycolytic metabolites into the TCA cycle. The metabolites thus accumulated are utilized to meet the biosynthetic needs of the cancer cell [31]. PKM2 also promotes the Warburg effect by activation of the HIF-1α target genes *SLC2A1*, *LDHA*, and *PDK1* that facilitate the shift from oxidative phosphorylation to glycolytic metabolism to meet the nutrient demands of cancer cell proliferation [32]. PMK2 is highly expressed in various cancers, including lung, breast, and prostate, indicating a critical role in cancer progression beyond glycolysis [33]. PKM2 also represents one of the best examples of how ROS can directly regulate cellular metabolism. One study found that an increase in cellular ROS levels by hydrogen peroxide significantly reduced the pyruvate kinase activity of PKM2 through oxidation of Cys^{358}. However, this reduced pyruvate kinase activity recovered in the presence of a reducing agent, confirming that PK inhibition in the presence of hydrogen peroxide is ROS-dependent [25]. Further, the inhibition of pyruvate kinase activity promotes CO_2 production by the PPP and increases the production of reduced glutathione (GSH). The reduced pyruvate kinase activity thus promotes channeling of glycolytic metabolites into the PPP, which in turn increases GSH production to counter elevated ROS [25]. Mutation of Cys^{358} residue prevents ROS-induced inhibition of pyruvate kinase activity, leading to reduced GSH levels and sensitization of the cells to oxidative stress. Similarly, insulin-induced ROS inhibits pyruvate kinase activity in hepatocellular carcinoma [34,35]. Reduced pyruvate kinase activity was observed despite the induction of PKM2 protein levels in cells treated with insulin through suppression of miR-128 and miR-145 [34]. Although unclear, suppression of miR-128 and miR-145 may involve ROS induced DNA hypermethylation [36].

Glyceraldehyde 3-phosphate dehydrogenase (GAPDH) is another glycolytic enzyme readily regulated by the cellular redox system. It catalyzes the conversion of glyceraldehyde 3-phosphate to 1,3-diphosphoglycerate. GAPDH is often regarded as a housekeeping gene and used as reference control. However, its expression is upregulated in a wide variety of tumors and is associated with tumor proliferation, metastasis, and overall aggressive tumor behavior [37–40]. The elevated GAPDH levels are considered essential to maintaining the glycolytic phenotype present in tumors.

Early evidences show that accumulation of ROS is associated with reduced GAPDH activity [41,42]. In addition, treatment with oxidizing low-density lipoprotein reduced the expression of GAPDH in a ROS-dependent manner by increasing its proteosomal-mediated degradation [43]. Mechanistic analysis reveals that oxidizers, such as hydrogen peroxide, nitric oxide, and peroxides, cause oxidization of free cysteine thiols present on GAPDH [44–46]. The ROS induced GAPDH inhibition alters the function of GAPDH, leading to redirecting of glycolytic metabolites towards the PPP [37]. Since cancer cells usually express higher ROS levels than non-transformed cells, the alteration of expression and activity of glycolytic enzyme such as PKM2 and GAPDH may represent a necessary adaptation to enhance reducing power of the tumor cells by redirecting the metabolites into PPP for production of NADPH.

2.1.2. TCA Cycle

The tricarboxylic acid cycle plays an essential role in energy production, macromolecule synthesis, and maintenance of cellular redox balance. The TCA cycle through its series of biochemical reactions utilizes oxidized glycolytic product (acetyl CoA) to generate ATP, NADH, and FADH2. The electrons released from NADH and FADH2 enter the electron transport chain (ETC), where the electron is utilized to synthesize ATP in the presence of oxygen. The electron transport chain serves as the primary source of ROS in the cells (discussed below). Accumulating evidence uncovers the critical role of cellular redox status in direct or indirect regulation of TCA cycle activity. Mass spectrometry-based central carbon metabolic analysis reveal that induction of ROS by vitamin C inhibit the levels of various TCA cycle metabolites in breast cancer cells [47]. Few of the enzymes that are regulated by ROS are discussed below.

The enzyme aconitase (Aco) catalyzes the conversion of citrate to isocitrate. It has been demonstrated that the activity of enzyme Aco is often deregulated in cancers either due to mutation or reduced expression [48–50]. Aconitase is vulnerable to reactive oxygen and reactive nitrogen species [51–53]. The iron-sulfur cluster present in the aconitase enzyme is highly susceptible to oxidation, leading to the iron release and consequently, inactivation of the enzyme. Although the underlying benefit of ROS-induced Aco inactivation is still being uncovered, a recent study demonstrated that overexpression of Aco weakens Warburg-like features in breast cancer cells [54].

Alpha ketoglutarate dehydrogenase (αKGDH or 2KG) is a highly regulated TCA cycle enzyme that catalyzes conversion of α-ketoglutarate and coenzyme A to succinyl coA and in the process converts NAD^+ to NADH. Initial studies with cardiac mitochondria demonstrated that ROS inhibit NADH production and oxidative phosphorylation. Mechanistic analysis revealed that reduced mitochondrial activity observed was associated with reduced activity of αKGDH [55,56]. Although the studies assessing the effect of ROS on αKGDH in cancers are lacking, inhibition of αKGDH promotes utilization of glutamine derived αKG for fatty acid synthesis [57,58]. The citrate synthesized via reductive metabolism of αKG is important for viability and cancer biomass increase [59,60]. The activity of enzyme αKGDH is also inhibited by HIF-1, thus contributing to the reductive carboxylation of αKG [60].

2.2. Indirect Regulation through Oncogene or Tumor Suppressor Networks

Accumulating research evidence suggests that driver gene mutations found in cancers contribute to cell metabolic alterations, indicating that signaling pathways may influence the metabolic shift in cancer [61]. It has been revealed that ROS may control tumor cell metabolism by oxidation of oncogenes or tumor suppressors.

2.2.1. AMP-Activated Protein Kinase (AMPK)

AMPK is a key protein to control cellular energy homeostasis, which is generally a negative regulator of the Warburg effect. AMPK is activated by insufficient fuel supply and low oxygen to make nutrients for the anabolic/growth-promoting metabolic pathway [62]. As a stress-response molecule, AMPK acts as a tumor suppressor to prevent the carcinogenesis as a canonical downstream effector of LKB1. However, once tumors develop, AMPK becomes a tumor promoter by protecting against metabolic, oxidative, and genotoxic stresses, and is involved in cancer drug resistance [63]. AMPK is closely linked to redox homeostasis. Reduced nicotinamide adenine dinucleotide phosphate (NADPH) provides reducing power in many enzymatic reactions and also acts as an antioxidant to neutralize ROS [64]. AMPK regulates NADPH homeostasis by inhibition of the acetyl-CoA carboxylases ACC1 and ACC2, thus decreasing NADPH consumption and increasing NADPH generation where the pentose phosphate pathway is impaired [65]. AMPK itself is redox active in that it contains cysteine residues that can be oxidized by ROS. An earlier study showed that exposure to H_2O_2 can activate AMPK through the redox-sensitive cysteine residues (Cys-299/Cys-304) in the α1 catalytic subunit [66]. The mitochondrial ROS mediated AMPK activation is sufficient to mediate starvation-induced autophagy [67]. A recent study argued that mitochondria-derived ROS indirectly affects AMPK activity by decreasing the ATP/ADP ratio rather than the direct protein thiol oxidation [68].

2.2.2. Hypoxia-Inducible Factor 1 (HIF-1)

Hypoxia is a characteristic feature of solid tumors due to an imbalance between oxygen (O_2) supply and consumption, in which HIF-1 is a key regulator in response to low oxygen [69,70]. The activation of HIF-1 by hypoxia modulates erythropoiesis and angiogenesis, as well as glycolytic metabolism through multiple target genes. Many key glycolytic proteins are HIF-1 transcriptional target gene products, including glucose transporter 1 and 3 (GLUT1 and GLUT3), hexokinase (HK), 6-phosphofructo-2-kinase/fructose-2,6-bisphosphatases 3 (PFKEB3), and pyruvate kinase M2 (PKM2). Induction of these genes by HIF1 enhances glycolysis and the PPP pathway [71]. The overall metabolic outcome of HIF1 upregulation in cancer is to promote aerobic glycolysis. In addition to upregulation of glucose uptake, HIF-1 transcriptionally activates expression of pyruvate dehydrogenase kinase (PDK) [72]. The PDK inhibits the activity of pyruvate dehydrogenase, thus limiting the entry of glycolytic metabolites in the TCA cycle [72,73]. In addition, HIF-1 mediated metabolic reprogramming involves a reduction in cellular ROS levels via inhibition of electron transport chain complex 1 activity [74,75]. Studies show that HIF-1 is activated not only by hypoxia but also by growth factors and oncogenes. Our group previously found that ROS scavengers and antioxidant enzymes decreased HIF-1α expression levels in a dose-dependent manner under normoxic conditions in ovarian cancer cells [76]. Further studies revealed that heavy metals- or growth factors-induced HIF-1α activation in normoxia is mediated by cellular ROS production [77,78]. Jung et al. found that adenosine monophosphate-activated protein kinase (AMPK) mediates ROS-induced HIF-1α protein accumulation at the post-translational step by blocking its degradation resulting from the ubiquitination inhibition [29]. They also suggested that H_2O_2 might increase the transcriptional activity of HIF-1α through AMPK. Notably, the role of AMPK in hypoxia-induced HIF-1 activation is different than that in ROS-induced HIF-1 activation.

2.2.3. p53

p53 plays a key role in maintaining genome integrity in response to cellular stresses that lead to DNA damage. It is well-established that p53 acts as a negative regulator of glycolysis by inhibiting expression levels of glucose transporters, which limits activities of glycolytic enzymes phosphofructokinase 1 (PFK1) and phosphoglycerate mutase (PGM) [36]. There is a direct interplay between p53 and ROS leading to oxidative stress. On the one hand, p53 modulates cellular ROS levels through p53-inducible genes (PIGs) that encode a number of pro-oxidant enzymes to generate ROS or inhibit antioxidant genes, such as MnSOD, at the transcriptional level [79]. On the other hand, p53 contains cysteine (Cys) residues in its DNA binding domain, which can be oxidized by ROS. The oxidation of cysteines would impair the DNA-binding activity of p53 to specific genes. ROS can also activate protein kinases such as mitogen-activated protein kinase (MAPK), which in turn phosphorylate and thus activate p53 for apoptosis induction [80].

2.2.4. ROS-Responsive miRNAs

Deregulations of microRNA expression have been associated with tumor development, progression, metastasis, and therapeutic responses [81]. Notably, miRNAs are shown to mediate metabolic phenotypes through the regulation of glycolytic enzymes and mitochondria metabolism [82]. Growing evidence suggests a reciprocal connection between ROS signaling and the microRNA pathway, resulting in diverse biological effects in cancer cells [83]. Altered productions of ROS are associated with deregulated expression of miRNAs, suggesting that miRNAs play a role in regulating ROS production and vice versa. ROS control miRNAs expression levels at multiple layers. The proposed mechanisms include miRNA biogenesis, transcription, and epigenetic regulation [83]. For example, activation of miR-34a switches the glycolysis to mitochondria respiration in cancer cells by direct targeted-inhibition of glycolytic enzymes (e.g., hexokinase 1 (HK1), hexokinase 2 (HK2), glucose-6-phosphate isomerase, and pyruvate dehydrogenase kinase 1 (PDK1)) in cancer cells inp53-dependent manner [84]. As p53 is involved in microRNA processing pathways, such as Drosha-mediated pri-microRNA processing, ROS indirectly affect the miR-34a levels by promoting the transcription of the miR-34a gene through p53.

3. Glucose Metabolic Adaptation Alters the Redox Balance

As mentioned earlier, the mitochondrial electron chain and NOX are the main source of cellular ROS. In cancer cells, the high ROS levels are countered by enhanced antioxidant capacity of the cells [85]. The critical balance between ROS and antioxidant mechanisms is essential for cellular homeostasis as different ROS levels can initiate varied biological responses ranging from cellular signaling to oxidative damage of cellular proteins and genomic instability [86]. As expected, the ROS do not control varied cellular functions independently. ROS interact with cellular oncogenes and tumor suppressors and have complex interplay with tumor metabolism. As highlighted above, the mitochondrial citrate and ATP production is detrimental to cancer cells. In a similar vein, the mitochondrial ROS production is also significantly reduced to maintain ROS at a non-toxic level [87]. Below we highlight the accumulating evidence that emphasizes the role of metabolic pathways in regulating cellular ROS levels.

3.1. *Glycolysis*

ROS can regulate the expression of glycolytic enzymes, such as PKM2 and GAPDH. Recent studies have also elaborated the essential role of glycolysis in regulating cellular ROS levels. The enzyme lactate dehydrogenase A (LDHA) converts glycolytic pyruvate into lactate and is upregulated in various tumors [88]. siRNA mediated inhibition of enzyme LDHA diverting the pyruvate into TCA [89,90]. The inhibition of TCA flux by LDHA thus prevents the generation of mitochondrial ROS.

3.2. *Pentose Phosphate Pathway*

The most well understood tumor metabolic regulation that alters the cellular redox balance is the upregulation of the PPP. The PPP branches from glycolysis and provides the proliferating cancer cells with nucleotides (non-oxidative PPP) and NADPH (oxidative PPP). Cancer cells display elevated levels and activity of PPP enzymes involved in oxidative PPP, namely glucose 6-phosphate dehydrogenase (G6PD) [91–94]. The G6PD is the primary source of cellular NADPH and is upregulated in various tumors [95]. The importance of G6PD can be gauged through its essential role in cellular growth, neoplastic transformation, and tumorigenesis [96,97]. G6PD is the rate-limiting enzyme that controls the entry of glycolytic glucose-6-phosphate into the PPP. Since the NADPH plays a crucial role in reducing cellular oxidants, such as hydrogen peroxide (H_2O_2) and other ROS species [98], the cellular antioxidant system is directly or indirectly dependent on NADPH for its functioning. The enzyme glutathione reductase reduces glutathione (GSH) in the presence of NADPH [99]. The GSH is an essential antioxidant defense mechanism in the cell that converts H_2O_2 into water. Other than glutathione, the antioxidant activity of nitric oxide synthetase is also dependent on NADPH derived from the oxidative PPP, namely the enzyme G6PD [100,101].

3.3. Tri-Carboxylic Acid (TCA) Cycle

As outlined above, the TCA cycle is a major source of cellular reducing equivalent that transfers electrons to the electron transport chain (ETC). However, up to 2% of electrons leak out of ETC and interact with mitochondrial oxygen, leading to the formation of superoxide ions, which can ultimately result in the formation of hydrogen peroxide and hydroxyl and peroxynitrite radical [102–106]. Hydroxyl radical and peroxynitrite are strong oxidants that can interact with cellular components leading to signaling, or oxidation of lipids, proteins, and DNA in a concentration-dependent manner [107–109].

The reduced oxidative phosphorylation in cancer does mean that the TCA is non-functional [110]. Recent studies have demonstrated that the TCA cycle continues to generate essential macromolecules and energy by glutamine-induced anaplerosis [111–113]. The intracellular glutamine is converted into α-ketoglutarate, which is then converted into citrate by the activity of enzyme isocitrate dehydrogenase (IDH). The synthesized citrate serves as a precursor for the synthesis of fatty acids and other macromolecules for tumor anabolic processes [58,114].

The isocitrate dehydrogenase enzyme (IDH) is expressed in three isoforms where IDH1 is found in cytoplasm and IDH2 and IDH3 are mitochondrial bound [115]. Of these three, the IDH3 predominates at the TCA cycle, is NAD^+ dependent, and catalyzes irreversible oxidative decarboxylation of isocitrate (ICT) in the presence of NAD^+ to produce 2KG and NADH. IDH1 and IDH2, on the other hand, are NADP-dependent and can catalyze in both reactions (ICT to 2KG and vice versa). The two NADP-bound IDH enzymes are essential for synthesis of NADPH and play a crucial role in cellular defense against oxidative injury [116–118]. Antisense mediated inhibition of NADP-dependent IDH (IDH1 and IDH2) significantly enhanced γ-radiation-induced ROS production, lipid peroxidation, and protein oxidation [119]. Analysis of TCGA data demonstrates that IDH1 is upregulated in several malignancies, including anaplastic large cell lymphoma, glioblastoma, and pancreatic ductal adenocarcinomas, and it is associated with poor prognosis among leukemia patients [120–123]. Knockdown of IDH1 significantly reduces NADPH content, leading to reduction in GSH levels and induction of ROS in glioblastoma cells. Recently, mutant forms of IDH have been observed in a variety of malignancies, including glioblastomas, leukemias, osteosarcomas, and thyroid tumors [124–127]. The enzymatic activity of mutated IDH1/2 displays conversion of 2KG into an oncometabolite D-2-hydroxyglutarate and, in the process, IDH1/2 consumes NADPH. Overexpression of mutant IDH1 (R132H) in glioblastoma cells reduces NADPH and GSH levels leading to elevated ROS levels [128]. The R132H mutant IDH1 also has a dominant negative affect on IDH1 activity that may contribute to reduced NADPH and elevated ROS levels [129]. Though the production of oncometabolites by R132H mutants enhances HIF1α levels and induces tumor formation, the reason for the R132H mutation being associated with improved glioma patient survival remains elusive [128–131].

Malic enzymes (ME) catalyze the conversion of malate into pyruvate and, in the process, synthesize reducing equivalents [132]. ME is encoded by three homologous genes, of which ME1 is located in the cytoplasm, whereas ME2 and ME3 are present in mitochondria. The ME1 is an NADP bound enzyme and, along with G6PD, is the major source of cellular NADPH [133,134]. Recent reports have suggested that ME1 behaves as an oncogene and is associated with tumor growth and invasion [135,136]. Investigation of the role of ME1 in gastric cancer reveals that knockdown of ME1 is associated with elevated ROS and decreased NADPH levels in glucose limiting conditions [137].

4. Metabolic Deregulations Lead to Drug Resistance

Overcoming therapy resistance remains one of the most important unmet needs for cancer treatment. Various mechanisms contribute to the development of drug resistance in cancers, such as an increase in drug efflux, alteration of target genes, intracellular inactivation of drugs, and intracellular signaling leading to epithelial–mesenchymal transition and DNA repair. Although the role of altered metabolism in tumor cell survival and proliferation has been known for decades, the importance of metabolism in regulating the therapy response has been realized only recently. Below we highlight key studies pertaining to the role of metabolic deregulations that underline tumor resistance to therapeutic agents (Figure 2).

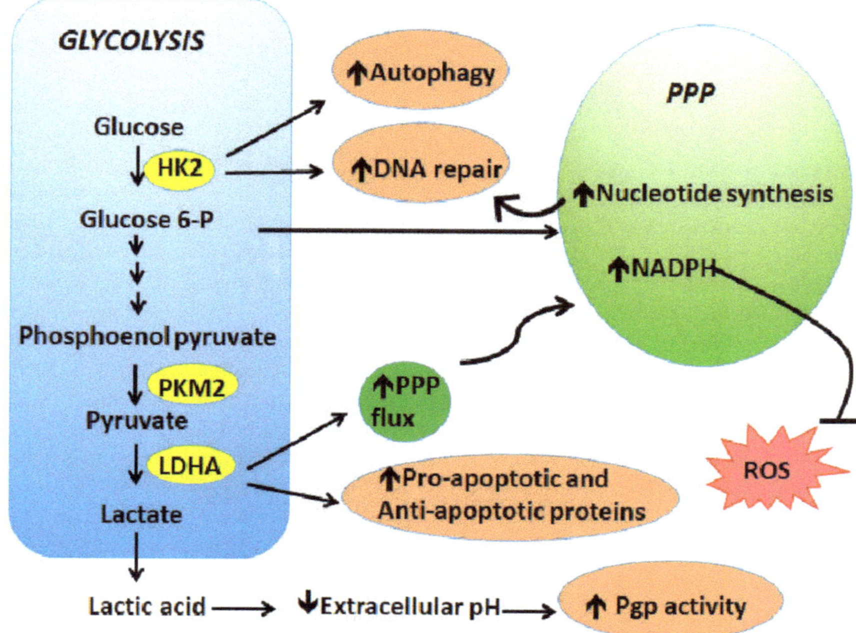

Figure 2. Metabolic deregulations lead to drug resistance. The metabolic shift favors cancer cell proliferation and survival in response to therapy via upregulation of DNA repair, increase of prosurvival signaling and autophagy, activation of drug efflux pumps, and neutralization of ROS.

4.1. Glycolysis

One of the most well characterized mechanisms of drug resistance is related to the enhanced level and/or activity of the efflux pumps that expel the drug out of the cells. Studies show that ROS or oxidative stress regulates ATP-binding cassette (ABC) transporters, which are associated with chemoresistance [138]. The role of ABC transporters in therapy resistance has been well-reviewed [138,139]. Mechanistic analyses revealed that the activity of p-glycoprotein (Pgp, an ABC transporter), a crucial drug-efflux transporter involved in multidrug resistance (MDR), is more than doubled in prostate cancer cells exposed to acidic media (pH 6.6) [140]. The enhanced Pgp activity reduces cellular sensitivity to the cytotoxic agent daunorubicin. Similarly, the acidification of extracellular milieu reduces the cytotoxicity of weak base therapeutic agents such as doxorubicin, thus contributing drug inaction [141]. Incidentally, neutralization of tumor pH enhanced the cytotoxicity of doxorubicin, confirming the direct role of pH in regulating the tumor response to therapy. Similarly, the toxicity of paclitaxel, mitoxantrone, and topotecan is reduced in slightly acidic (pH 6.5) compared to neutral conditions (pH 7.4) [142]. In another study, the authors showed that changes in glucose levels markedly enhanced cellular ROS via NADPH oxidase 4 and thus activated HIF-1/Pgp leading to resistance to doxorubicin [143]. ROS-induced Pgp activity can be reversed by ROS scavenger NAC treatment.

Another important mechanism associated with drug resistance is enhanced DNA repair capacity of the cells. Upregulation of glycolysis is associated with enhanced repair of damaged DNA, ultimately leading to reduced sensitivity of cells to chemotherapeutic agents. A recent study showed that upregulation of glycolysis using mitochondrial respiratory modifiers protected the cancer cells from radiation-induced cytotoxicity [144]. The modifiers enhance glycolysis via induction of GLUT-1 and hexokinase, allowing uptake and utilization of glucose. The enhanced glycolysis promoted repair of radiation induced damaged DNA by activating both homologous recombination and non-homologous end joining [144]. Wagner et al. demonstrated that lactate enhances cellular DNA repair capacity by increasing the activity of DNA-PKc, leading to the protection of cells from doxorubicin (DOX) and cisplatin (CDDP) induced cytotoxicity [145]. In other studies, inhibition of glycolysis effectively overcame resistance to DNA damaging agents such as 5-fluorouracil and doxorubicin [146,147].

Hexokinase 2: The first enzyme of glycolysis is upregulated in various tumors, and its elevated level is associated with cisplatin resistance [148]. Inhibition of HK2 sensitizes the resistant cells to cisplatin-induced cell death and apoptosis [148,149]. Activation of ERK has been linked to the protective effect of HK2 against cisplatin. Cisplatin induces an ERK-mediated autophagic response that protects the cells from the drug-induced toxicity. In cells that overexpress HK2, enhanced autophagic response is observed, whereas, inhibition of HK2 causes suppression of autophagy, thus sensitizing resistant ovarian cancer xenografts to cisplatin [150]. In another study, Vartanian et al. demonstrated that activation of Erk by HK2 as a potential mechanism for radiation resistance in glioblastoma cells [151]. Expression of hexokinase 2 has also been implicated in tumor resistance to antimetabolites, such as 5-fluorouracil (5FU) and gemcitabine [152–154]. Although two inhibitors of hexokinase-2, namely 2-deoxyglucose and 3-bromopyruvate, have shown an excellent preclinical response, their clinical benefits are yet to be proved as efficient chemotherapeutic agents.

In addition to its glycolytic function, PKM2 has been shown to provide cancer cells with prosurvival and antiapoptotic properties by transcriptionally upregulating BcL-xl expression [155]. Earlier studies suggested that reduced expression of PKM2 is associated with reduced responsiveness of ovarian and colorectal cancers to platinum compounds [156,157]. However, subsequent studies demonstrated that elevated PKM2 expression is associated with reduced cisplatin sensitivity [158–162]. Understandably, inhibition of PKM2 in the later studies resulted in increased responsiveness of resistant cells to the drug. Similarly, contradictory observations on PKM2's protective effect against 5FU-induced cytotoxicity has potentially impacted its use as a drug target. In colon cancer patients, high expression of PKM2 is associated with a poor response to 5FU-based therapy; however, no association was observed in gastric cancer cells [157,163]. The contradictory role in platinum-resistance suggests a cell-type dependent protective effect of PKM2 and has raised questions on the validity of PKM2 as a target for cancer therapy.

GAPDH is a pleiotropic enzyme whose function is dictated by its subcellular localization [164]. Localization in the nucleus regulated various non-metabolic functions of GAPDH, such as telomere protection, DNA repair, and regulation of autophagy and cell death [165]. Incidentally, two studies have reported that the presence of GAPDH enhances the sensitivity of cancer cells to therapeutic drugs. These studies demonstrated that depletion of GAPDH sensitizes tumor cells to antimetabolite agents, however, this depletion did not alter the cell's response to other chemotherapeutic drugs, such as doxorubicin and fludarabine [166,167]. In pancreatic cancer, GAPDH translocation into the nucleus due to drug-induced oxidative stress is associated with cell death in vitro and in vivo [168]. Other studies have also demonstrated proapoptotic effect of GAPDH due to its regulation of autophagy and its interaction with one or more apoptotic cascades [165]. Although the mechanism remains unclear, the proapoptotic role of GAPDH is believed to be due to its interaction with p53. Hara et al. showed that translocation of GAPDH to the nucleus is initiated upon binding with SIAH1. The formation of complex stabilizes SIAH1 expression in the nucleus, leading to apoptosis [169,170]. p53 plays a pivotal role in stimulating the interaction between GAPDH and SIAH1 by transcriptionally upregulating the expression of both [171,172]. Evidence also suggests a direct interaction between p53 and GAPDH, wherein GAPDH enhances the proapoptotic functions of p53 through its acetylation and serine 46 phosphorylation [173]. The post-translational modification of p53 is essential for its translocation into the mitochondria to initiate Bax mediated apoptosis. The inter-regulation between GAPDH and p53 is further validated by a report highlighting that mutated p53 prevents the translocation of

GAPDH into the nucleus [174]. The stabilization of GAPDH in the cytoplasm is critical for mutant p53 induced antiapoptosis. A recent study by Li et al. showed that GAPDH's translocation into the nucleus increases transcription of p53 gene, providing additional evidence on interplay between p53 and GAPDH in regulating cellular apoptosis [175].

4.2. Pentose Phosphate Pathway

Many of the chemotherapeutic and targeted agents are dependent on ROS for their cytotoxicity. Due to their role in the synthesis of NADPH leading to ROS detoxification, two of the enzymes involved in oxidative PPP—glucose 6-phosphate dehydrogenase (G6PD) and 6-phosphogluconate dehydrogenase (6PGD)—have been implicated in imparting therapeutic resistance [176,177]. Analysis of doxorubicin-resistant model of colon cancer cells reveals an enhanced PPP activity and elevated levels of G6PD and glutathione [178]. The resistant cells also exhibited elevated levels of multi drug-resistant associated proteins (MRP1 and MRP2). Since the inhibition of G6PD using chemical inhibitors overcame multi drug resistance, the authors concluded that elevated glutathione levels are necessary for extruding drugs out of cells. Catanzaro et al. demonstrated that cisplatin resistant cells express elevated levels of enzyme G6PD, and the resistant cells are sensitive to G6PD inhibition [179]. Using a lung cancer model of cisplatin resistance, Hong et al. further verified that inhibition of G6PD sensitized resistant cells to cisplatin [180]. Furthermore, Zhang et al. established that the TGFβ1-FOXM1-HMGA1-TGFβ1 positive feedback loop maintain G6PD expression in cisplatin resistant cells, and that disruption of this axis sensitizes the cells to the drug [181]. Aberrant expression of 6-phosphogluconate has also been shown to be involved in chemotherapeutic and radiation resistance [182–184]. Although preliminary, recent evidence has shown that epidermal growth factor receptor (EGFR) phosphorylates 6PGD at tyrosine (Y481), which enhances 6PGD activity by increasing its affinity to NAD^+. The phosphorylation also appears essential to EGFR induced radiation resistance in glioma cells [182].

The non-oxidative branch of PPP has also been implicated in chemotherapeutic resistance. Li et al. found that overexpression of Rac1 is associated with multi-drug resistance, and Rac1 mediated non-oxidative PPP is a key driver of cisplatin resistance in breast cancer cells [185]. The non-oxidative branch of PPP assists in the DNA repair process by providing the damaged cells with nucleosides. Shukla et al. identified that pancreatic cancer cells that are resistant to gemcitabine have enhanced flux of glucose carbon into the non-oxidative branch of PPP [186]. The enhanced flux is aided by increased expression of non-oxidative PPP enzymes transketolase (TKT). The HIF1α induced TKT expression in the resistant cells assist in increased pyrimidine synthesis that protects the cancer cells from gemcitabine-induced cytotoxicity [186]. Although knockdown of TKT enhances the sensitivity of cancer cells to cisplatin, in cervical cancer the underlying mechanism for that is yet to be realized [187].

5. Conclusions

Otto Warburg first observed the aberrant features in tumor cells characterized by a shift in their energy metabolism towards glycolysis even in conditions with ample oxygen [1]. The reprograming of tumor energetics has gained interest in the last two decades due to its association with oncogenes, tumor suppressors, and the cellular redox system [3]. A highly coordinated interplay between tumor energetics and reactive oxygen species (ROS) generates a powerful phenotype that provides the tumor cells with proliferative, antiapoptotic, and overall aggressive characteristics. Through this review, we summarized the current literature on (1) how ROS regulates tumor metabolism and (2) how metabolic adaptations in tumors regulate ROS. Albeit high levels of intracellular ROS being frequently observed in cancer cells, ROS induction is applied as a principle for most non-surgical treatments including chemotherapy and radiotherapy. By modulating the metabolic flux from oxidative phosphorylation to glycolysis and PPP, tumor cells greatly enhance its antioxidant system to maintain ROS homeostasis and prevent ROS-mediated cell death. It is no surprise that the interaction between tumor energetics and ROS plays a fundamental role in regulating tumor's response to chemotherapeutic drugs. Though attempts of targeting tumor energetics have shown promising preclinical outcomes, the clinical benefits are yet to be fully achieved [188]. A key missing link in realizing metabolic abnormalities as a druggable target would be to understand if outcomes of altered energetics involve ROS modulation. For example, the hexokinase 2 induced autophagic response is known to involve ROS generation [189]. Since hexokinase 2 induced autophagy is associated with drug-resistance, a potential approach would include targeting metabolism with inhibitors while disabling antioxidant systems induced by radio- or chemotherapy to improve patient outcome [150]. Another important avenue to explore is to delineate the personalization of metabolic inhibitors. Since the metabolic enzymes may preferentially benefit certain cancers, a personalized therapeutic approach that involves validating the importance of the enzymes in a specific context, e.g., cancer types, may be a viable area to discover.

Author Contributions: Both authors participated in the work of drafting and revising the article. All authors have read and agreed to the published version of the manuscript.

Funding: This work is supported by National Cancer Institute (R00CA215316).

Acknowledgments: We appreciate Jennifer Fisher Wilson (Science and Medical Writer, Thomas Jefferson University) for her help with English editing.

Conflicts of Interest: The authors declare that they have no competing financial interests.

References

1. Warburg, O.; Wind, F.; Negelein, E. The Metabolism of Tumors in the Body. *J. Gen. Physiol.* **1927**, *8*, 519–530. [CrossRef] [PubMed]
2. Gillies, R.J.; Gatenby, R.A. Adaptive landscapes and emergent phenotypes: why do cancers have high glycolysis? *J. Bioenerg. Biomembr.* **2007**, *39*, 251–257. [CrossRef] [PubMed]
3. Pavlova, N.N.; Thompson, C.B. The Emerging Hallmarks of Cancer Metabolism. *Cell Metab.* **2016**, *23*, 27–47. [CrossRef] [PubMed]
4. Nakashima, R.A.; Paggi, M.G.; Pedersen, P.L. Contributions of glycolysis and oxidative phosphorylation to adenosine 5′-triphosphate production in AS-30D hepatoma cells. *Cancer Res.* **1984**, *44*, 5702–5706.
5. Pfeiffer, T.; Schuster, S.; Bonhoeffer, S. Cooperation and competition in the evolution of ATP-producing pathways. *Science* **2001**, *292*, 504–507. [CrossRef]
6. Papathanassiou, D.; Bruna-Muraille, C.; Jouannaud, C.; Gagneux-Lemoussu, L.; Eschard, J.P.; Liehn, J.C. Single-photon emission computed tomography combined with computed tomography (SPECT/CT) in bone diseases. *Joint Bone Spine* **2009**, *76*, 474–480. [CrossRef]
7. Sowers, M.L.; Herring, J.; Zhang, W.; Tang, H.; Ou, Y.; Gu, W.; Zhang, K. Analysis of glucose-derived amino acids involved in one-carbon and cancer metabolism by stable-isotope tracing gas chromatography mass spectrometry. *Anal. Biochem.* **2019**, *566*, 1–9. [CrossRef]
8. Madhu, B.; Jauhiainen, A.; McGuire, S.; Griffiths, J.R. Exploration of human brain tumour metabolism using pairwise metabolite-metabolite correlation analysis (MMCA) of HR-MAS 1H NMR spectra. *PLoS ONE* **2017**, *12*, e0185980. [CrossRef]
9. Cortes-Cros, M.; Hemmerlin, C.; Ferretti, S.; Zhang, J.; Gounarides, J.S.; Yin, H.; Muller, A.; Haberkorn, A.; Chene, P.; Sellers, W.R.; et al. M2 isoform of pyruvate kinase is dispensable for tumor maintenance and growth. *Proc. Natl. Acad. Sci. USA* **2013**, *110*, 489–494. [CrossRef] [PubMed]
10. Shaw, R.J.; Kosmatka, M.; Bardeesy, N.; Hurley, R.L.; Witters, L.A.; DePinho, R.A.; Cantley, L.C. The tumor suppressor LKB1 kinase directly activates AMP-activated kinase and regulates apoptosis in response to energy stress. *Proc. Natl. Acad. Sci. USA* **2004**, *101*, 3329–3335. [CrossRef] [PubMed]
11. Lum, J.J.; Bauer, D.E.; Kong, M.; Harris, M.H.; Li, C.; Lindsten, T.; Thompson, C.B. Growth factor regulation of autophagy and cell survival in the absence of apoptosis. *Cell* **2005**, *120*, 237–248. [CrossRef] [PubMed]
12. Ganapathy-Kanniappan, S.; Kunjithapatham, R.; Geschwind, J.F. Anticancer efficacy of the metabolic blocker 3-bromopyruvate: specific molecular targeting. *Anticancer Res.* **2013**, *33*, 13–20. [PubMed]
13. Samudio, I.; Fiegl, M.; Andreeff, M. Mitochondrial uncoupling and the Warburg effect: Molecular basis for the reprogramming of cancer cell metabolism. *Cancer Res.* **2009**, *69*, 2163–2166. [CrossRef] [PubMed]
14. Lu, Y.; Zhang, X.; Zhang, H.; Lan, J.; Huang, G.; Varin, E.; Lincet, H.; Poulain, L.; Icard, P. Citrate induces apoptotic cell death: A promising way to treat gastric carcinoma? *Anticancer Res.* **2011**, *31*, 797–805.

15. Lincet, H.; Kafara, P.; Giffard, F.; Abeilard-Lemoisson, E.; Duval, M.; Louis, M.H.; Poulain, L.; Icard, P. Inhibition of Mcl-1 expression by citrate enhances the effect of Bcl-xL inhibitors on human ovarian carcinoma cells. *J. Ovarian Res.* **2013**, *6*, 72. [CrossRef]
16. Kruspig, B.; Nilchian, A.; Orrenius, S.; Zhivotovsky, B.; Gogvadze, V. Citrate kills tumor cells through activation of apical caspases. *Cell Mol. Life Sci.* **2012**, *69*, 4229–4237. [CrossRef]
17. Hanahan, D.; Weinberg, R.A. Hallmarks of cancer: the next generation. *Cell* **2011**, *144*, 646–674. [CrossRef]
18. D'Autreaux, B.; Toledano, M.B. ROS as signalling molecules: mechanisms that generate specificity in ROS homeostasis. *Nat. Rev. Mol. Cell Biol.* **2007**, *8*, 813–824. [CrossRef]
19. Maraini, G.; Gozzoli, F. Binding of retinol to isolated retinal pigment epithelium in the presence and absence of retinol-binding protein. *Investig. Ophthalmol.* **1975**, *14*, 785–787. [CrossRef]
20. Keenan, M.M.; Chi, J.T. Alternative fuels for cancer cells. *Cancer J.* **2015**, *21*, 49–55. [CrossRef]
21. Galluzzi, L.; Kepp, O.; Vander Heiden, M.G.; Kroemer, G. Metabolic targets for cancer therapy. *Nat. Rev. Drug Discov.* **2013**, *12*, 829–846. [CrossRef] [PubMed]
22. Kim, J.; Kim, J.; Bae, J.S. ROS homeostasis and metabolism: a critical liaison for cancer therapy. *Exp. Mol. Med.* **2016**, *48*, e269. [CrossRef] [PubMed]
23. Ma, L.; Fu, Q.; Xu, B.; Zhou, H.; Gao, J.; Shao, X.; Xiong, J.; Gu, Q.; Wen, S.; Li, F.; et al. Breast cancer-associated mitochondrial DNA haplogroup promotes neoplastic growth via ROS-mediated AKT activation. *Int. J. Cancer* **2018**, *142*, 1786–1796. [CrossRef] [PubMed]
24. Juarez, J.C.; Manuia, M.; Burnett, M.E.; Betancourt, O.; Boivin, B.; Shaw, D.E.; Tonks, N.K.; Mazar, A.P.; Donate, F. Superoxide dismutase 1 (SOD1) is essential for H_2O_2-mediated oxidation and inactivation of phosphatases in growth factor signaling. *Proc. Natl. Acad. Sci. USA* **2008**, *105*, 7147–7152. [CrossRef]
25. Anastasiou, D.; Poulogiannis, G.; Asara, J.M.; Boxer, M.B.; Jiang, J.K.; Shen, M.; Bellinger, G.; Sasaki, A.T.; Locasale, J.W.; Auld, D.S.; et al. Inhibition of pyruvate kinase M2 by reactive oxygen species contributes to cellular antioxidant responses. *Science* **2011**, *334*, 1278–1283. [CrossRef]
26. Wellen, K.E.; Thompson, C.B. Cellular metabolic stress: considering how cells respond to nutrient excess. *Mol. Cell* **2010**, *40*, 323–332. [CrossRef]
27. Lu, W.; Hu, Y.; Chen, G.; Chen, Z.; Zhang, H.; Wang, F.; Feng, L.; Pelicano, H.; Wang, H.; Keating, M.J.; et al. Correction: Novel Role of NOX in Supporting Aerobic Glycolysis in Cancer Cells with Mitochondrial Dysfunction and as a Potential Target for Cancer Therapy. *PLoS Biol.* **2017**, *15*, e1002616. [CrossRef]
28. Bedard, K.; Krause, K.H. The NOX family of ROS-generating NADPH oxidases: physiology and pathophysiology. *Physiol. Rev.* **2007**, *87*, 245–313. [CrossRef]
29. Jurica, M.S.; Mesecar, A.; Heath, P.J.; Shi, W.; Nowak, T.; Stoddard, B.L. The allosteric regulation of pyruvate kinase by fructose-1,6-bisphosphate. *Structure* **1998**, *6*, 195–210. [CrossRef]
30. Anastasiou, D.; Yu, Y.; Israelsen, W.J.; Jiang, J.K.; Boxer, M.B.; Hong, B.S.; Tempel, W.; Dimov, S.; Shen, M.; Jha, A.; et al. Pyruvate kinase M2 activators promote tetramer formation and suppress tumorigenesis. *Nat. Chem. Biol.* **2012**, *8*, 839–847. [CrossRef]
31. Christofk, H.R.; Vander Heiden, M.G.; Harris, M.H.; Ramanathan, A.; Gerszten, R.E.; Wei, R.; Fleming, M.D.; Schreiber, S.L.; Cantley, L.C. The M2 splice isoform of pyruvate kinase is important for cancer metabolism and tumour growth. *Nature* **2008**, *452*, 230–233. [CrossRef] [PubMed]
32. Luo, W.; Semenza, G.L. Emerging roles of PKM2 in cell metabolism and cancer progression. *Trends Endocrinol. Metab.* **2012**, *23*, 560–566. [CrossRef] [PubMed]
33. Bluemlein, K.; Gruning, N.M.; Feichtinger, R.G.; Lehrach, H.; Kofler, B.; Ralser, M. No evidence for a shift in pyruvate kinase PKM1 to PKM2 expression during tumorigenesis. *Oncotarget* **2011**, *2*, 393–400. [CrossRef] [PubMed]
34. Li, Q.; Liu, X.; Yin, Y.; Zheng, J.T.; Jiang, C.F.; Wang, J.; Shen, H.; Li, C.Y.; Wang, M.; Liu, L.Z.; et al. Insulin regulates glucose consumption and lactate production through reactive oxygen species and pyruvate kinase M2. *Oxid. Med. Cell. Longev.* **2014**, *2014*, 504953. [CrossRef] [PubMed]
35. Iqbal, M.A.; Siddiqui, F.A.; Gupta, V.; Chattopadhyay, S.; Gopinath, P.; Kumar, B.; Manvati, S.; Chaman, N.; Bamezai, R.N. Insulin enhances metabolic capacities of cancer cells by dual regulation of glycolytic enzyme pyruvate kinase M2. *Mol. Cancer* **2013**, *12*, 72. [CrossRef]
36. Maddocks, O.D.; Vousden, K.H. Metabolic regulation by p53. *J. Mol. Med. (Berl.)* **2011**, *89*, 237–245. [CrossRef]

37. Ralser, M.; Wamelink, M.M.; Kowald, A.; Gerisch, B.; Heeren, G.; Struys, E.A.; Klipp, E.; Jakobs, C.; Breitenbach, M.; Lehrach, H.; et al. Dynamic rerouting of the carbohydrate flux is key to counteracting oxidative stress. *J. Biol.* **2007**, *6*, 10. [CrossRef]
38. Nguewa, P.A.; Agorreta, J.; Blanco, D.; Lozano, M.D.; Gomez-Roman, J.; Sanchez, B.A.; Valles, I.; Pajares, M.J.; Pio, R.; Rodriguez, M.J.; et al. Identification of importin 8 (IPO8) as the most accurate reference gene for the clinicopathological analysis of lung specimens. *BMC Mol. Biol.* **2008**, *9*, 103. [CrossRef]
39. Jung, M.; Ramankulov, A.; Roigas, J.; Johannsen, M.; Ringsdorf, M.; Kristiansen, G.; Jung, K. In search of suitable reference genes for gene expression studies of human renal cell carcinoma by real-time PCR. *BMC Mol. Biol.* **2007**, *8*, 47. [CrossRef]
40. Rubie, C.; Kempf, K.; Hans, J.; Su, T.; Tilton, B.; Georg, T.; Brittner, B.; Ludwig, B.; Schilling, M. Housekeeping gene variability in normal and cancerous colorectal, pancreatic, esophageal, gastric and hepatic tissues. *Mol. Cell. Probes* **2005**, *19*, 101–109. [CrossRef]
41. McKenzie, S.J.; Baker, M.S.; Buffinton, G.D.; Doe, W.F. Evidence of oxidant-induced injury to epithelial cells during inflammatory bowel disease. *J. Clin. Investig.* **1996**, *98*, 136–141. [CrossRef] [PubMed]
42. McKenzie, S.M.; Doe, W.F.; Buffinton, G.D. 5-aminosalicylic acid prevents oxidant mediated damage of glyceraldehyde-3-phosphate dehydrogenase in colon epithelial cells. *Gut* **1999**, *44*, 180–185. [CrossRef] [PubMed]
43. Sukhanov, S.; Higashi, Y.; Shai, S.Y.; Itabe, H.; Ono, K.; Parthasarathy, S.; Delafontaine, P. Novel effect of oxidized low-density lipoprotein: cellular ATP depletion via downregulation of glyceraldehyde-3-phosphate dehydrogenase. *Circ. Res.* **2006**, *99*, 191–200. [CrossRef] [PubMed]
44. Hwang, N.R.; Yim, S.H.; Kim, Y.M.; Jeong, J.; Song, E.J.; Lee, Y.; Lee, J.H.; Choi, S.; Lee, K.J. Oxidative modifications of glyceraldehyde-3-phosphate dehydrogenase play a key role in its multiple cellular functions. *Biochem. J.* **2009**, *423*, 253–264. [CrossRef] [PubMed]
45. Rodacka, A.; Strumillo, J.; Serafin, E.; Puchala, M. Effect of resveratrol and tiron on the inactivation of glyceraldehyde-3- phosphate dehydrogenase induced by superoxide anion radical. *Curr. Med. Chem.* **2014**, *21*, 1061–1069. [CrossRef] [PubMed]
46. Ishii, T.; Sunami, O.; Nakajima, H.; Nishio, H.; Takeuchi, T.; Hata, F. Critical role of sulfenic acid formation of thiols in the inactivation of glyceraldehyde-3-phosphate dehydrogenase by nitric oxide. *Biochem. Pharmacol.* **1999**, *58*, 133–143. [CrossRef]
47. Uetaki, M.; Tabata, S.; Nakasuka, F.; Soga, T.; Tomita, M. Metabolomic alterations in human cancer cells by vitamin C-induced oxidative stress. *Sci. Rep.* **2015**, *5*, 13896. [CrossRef]
48. Wang, P.; Mai, C.; Wei, Y.L.; Zhao, J.J.; Hu, Y.M.; Zeng, Z.L.; Yang, J.; Lu, W.H.; Xu, R.H.; Huang, P. Decreased expression of the mitochondrial metabolic enzyme aconitase (ACO2) is associated with poor prognosis in gastric cancer. *Med. Oncol.* **2013**, *30*, 552. [CrossRef]
49. Ternette, N.; Yang, M.; Laroyia, M.; Kitagawa, M.; O'Flaherty, L.; Wolhulter, K.; Igarashi, K.; Saito, K.; Kato, K.; Fischer, R.; et al. Inhibition of mitochondrial aconitase by succination in fumarate hydratase deficiency. *Cell Rep.* **2013**, *3*, 689–700. [CrossRef]
50. Singh, K.K.; Desouki, M.M.; Franklin, R.B.; Costello, L.C. Mitochondrial aconitase and citrate metabolism in malignant and nonmalignant human prostate tissues. *Mol. Cancer* **2006**, *5*, 14. [CrossRef]
51. Han, D.; Canali, R.; Garcia, J.; Aguilera, R.; Gallaher, T.K.; Cadenas, E. Sites and mechanisms of aconitase inactivation by peroxynitrite: modulation by citrate and glutathione. *Biochemistry* **2005**, *44*, 11986–11996. [CrossRef] [PubMed]
52. Castro, L.; Rodriguez, M.; Radi, R. Aconitase is readily inactivated by peroxynitrite, but not by its precursor, nitric oxide. *J. Biol. Chem.* **1994**, *269*, 29409–29415. [PubMed]
53. Cantu, D.; Schaack, J.; Patel, M. Oxidative inactivation of mitochondrial aconitase results in iron and H_2O_2-mediated neurotoxicity in rat primary mesencephalic cultures. *PLoS ONE* **2009**, *4*, e7095. [CrossRef] [PubMed]
54. Ciccarone, F.; Di Leo, L.; Lazzarino, G.; Maulucci, G.; Di Giacinto, F.; Tavazzi, B.; Ciriolo, M.R. Aconitase 2 inhibits the proliferation of MCF-7 cells promoting mitochondrial oxidative metabolism and ROS/FoxO1-mediated autophagic response. *Br. J. Cancer* **2020**, *122*, 182–193. [CrossRef]
55. Applegate, M.A.; Humphries, K.M.; Szweda, L.I. Reversible inhibition of alpha-ketoglutarate dehydrogenase by hydrogen peroxide: Glutathionylation and protection of lipoic acid. *Biochemistry* **2008**, *47*, 473–478. [CrossRef]

56. Humphries, K.M.; Yoo, Y.; Szweda, L.I. Inhibition of NADH-linked mitochondrial respiration by 4-hydroxy-2-nonenal. *Biochemistry* **1998**, *37*, 552–557. [CrossRef]
57. Metallo, C.M.; Gameiro, P.A.; Bell, E.L.; Mattaini, K.R.; Yang, J.; Hiller, K.; Jewell, C.M.; Johnson, Z.R.; Irvine, D.J.; Guarente, L.; et al. Reductive glutamine metabolism by IDH1 mediates lipogenesis under hypoxia. *Nature* **2011**, *481*, 380–384. [CrossRef]
58. Mullen, A.R.; Wheaton, W.W.; Jin, E.S.; Chen, P.H.; Sullivan, L.B.; Cheng, T.; Yang, Y.; Linehan, W.M.; Chandel, N.S.; DeBerardinis, R.J. Reductive carboxylation supports growth in tumour cells with defective mitochondria. *Nature* **2011**, *481*, 385–388. [CrossRef]
59. Nitsch, L.; Garbi, C.; Gentile, R.; Mascia, A.; Negri, R.; Polistina, C.; Vergani, G.; Zurzolo, C. Morphological changes induced by prolonged TSH stimulation or starvation in the rat thyroid cell line FRTL. *Horm. Metab. Res. Suppl.* **1990**, *23*, 32–37.
60. Sun, R.C.; Denko, N.C. Hypoxic regulation of glutamine metabolism through HIF1 and SIAH2 supports lipid synthesis that is necessary for tumor growth. *Cell Metab.* **2014**, *19*, 285–292. [CrossRef]
61. Levine, A.J.; Puzio-Kuter, A.M. The control of the metabolic switch in cancers by oncogenes and tumor suppressor genes. *Science* **2010**, *330*, 1340–1344. [CrossRef] [PubMed]
62. Faubert, B.; Boily, G.; Izreig, S.; Griss, T.; Samborska, B.; Dong, Z.; Dupuy, F.; Chambers, C.; Fuerth, B.J.; Viollet, B.; et al. AMPK is a negative regulator of the Warburg effect and suppresses tumor growth in vivo. *Cell Metab.* **2013**, *17*, 113–124. [CrossRef] [PubMed]
63. Vara-Ciruelos, D.; Russell, F.M.; Hardie, D.G. The strange case of AMPK and cancer: Dr Jekyll or Mr Hyde? (dagger). *Open Biol.* **2019**, *9*, 190099. [CrossRef] [PubMed]
64. Cairns, R.A.; Harris, I.S.; Mak, T.W. Regulation of cancer cell metabolism. *Nat. Rev. Cancer* **2011**, *11*, 85–95. [CrossRef]
65. Jeon, S.M.; Chandel, N.S.; Hay, N. AMPK regulates NADPH homeostasis to promote tumour cell survival during energy stress. *Nature* **2012**, *485*, 661–665. [CrossRef]
66. Zmijewski, J.W.; Banerjee, S.; Bae, H.; Friggeri, A.; Lazarowski, E.R.; Abraham, E. Exposure to hydrogen peroxide induces oxidation and activation of AMP-activated protein kinase. *J. Biol. Chem.* **2010**, *285*, 33154–33164. [CrossRef]
67. Li, L.; Chen, Y.; Gibson, S.B. Starvation-induced autophagy is regulated by mitochondrial reactive oxygen species leading to AMPK activation. *Cell. Signal.* **2013**, *25*, 50–65. [CrossRef]
68. Hinchy, E.C.; Gruszczyk, A.V.; Willows, R.; Navaratnam, N.; Hall, A.R.; Bates, G.; Bright, T.P.; Krieg, T.; Carling, D.; Murphy, M.P. Mitochondria-derived ROS activate AMP-activated protein kinase (AMPK) indirectly. *J. Biol. Chem.* **2018**, *293*, 17208–17217. [CrossRef]
69. Gordan, J.D.; Simon, M.C. Hypoxia-inducible factors: Central regulators of the tumor phenotype. *Curr. Opin. Genet. Dev.* **2007**, *17*, 71–77. [CrossRef]
70. Vaupel, P.; Harrison, L. Tumor hypoxia: causative factors, compensatory mechanisms, and cellular response. *Oncologist* **2004**, *9* (Suppl. S5), 4–9. [CrossRef]
71. Ghanbari Movahed, Z.; Rastegari-Pouyani, M.; Mohammadi, M.H.; Mansouri, K. Cancer cells change their glucose metabolism to overcome increased ROS: One step from cancer cell to cancer stem cell? *Biomed. Pharmacother.* **2019**, *112*, 108690. [CrossRef] [PubMed]
72. Kim, J.W.; Tchernyshyov, I.; Semenza, G.L.; Dang, C.V. HIF-1-mediated expression of pyruvate dehydrogenase kinase: A metabolic switch required for cellular adaptation to hypoxia. *Cell Metab.* **2006**, *3*, 177–185. [CrossRef] [PubMed]
73. Lu, C.W.; Lin, S.C.; Chen, K.F.; Lai, Y.Y.; Tsai, S.J. Induction of pyruvate dehydrogenase kinase-3 by hypoxia-inducible factor-1 promotes metabolic switch and drug resistance. *J. Biol. Chem.* **2008**, *283*, 28106–28114. [CrossRef] [PubMed]
74. Tello, D.; Balsa, E.; Acosta-Iborra, B.; Fuertes-Yebra, E.; Elorza, A.; Ordonez, A.; Corral-Escariz, M.; Soro, I.; Lopez-Bernardo, E.; Perales-Clemente, E.; et al. Induction of the mitochondrial NDUFA4L2 protein by HIF-1alpha decreases oxygen consumption by inhibiting Complex I activity. *Cell Metab.* **2011**, *14*, 768–779. [CrossRef] [PubMed]
75. Zhao, T.; Zhu, Y.; Morinibu, A.; Kobayashi, M.; Shinomiya, K.; Itasaka, S.; Yoshimura, M.; Guo, G.; Hiraoka, M.; Harada, H. HIF-1-mediated metabolic reprogramming reduces ROS levels and facilitates the metastatic colonization of cancers in lungs. *Sci. Rep.* **2014**, *4*, 3793. [CrossRef] [PubMed]

76. Xia, C.; Meng, Q.; Liu, L.Z.; Rojanasakul, Y.; Wang, X.R.; Jiang, B.H. Reactive oxygen species regulate angiogenesis and tumor growth through vascular endothelial growth factor. *Cancer Res.* **2007**, *67*, 10823–10830. [CrossRef]
77. Gao, N.; Shen, L.; Zhang, Z.; Leonard, S.S.; He, H.; Zhang, X.G.; Shi, X.; Jiang, B.H. Arsenite induces HIF-1alpha and VEGF through PI3K, Akt and reactive oxygen species in DU145 human prostate carcinoma cells. *Mol. Cell. Biochem.* **2004**, *255*, 33–45. [CrossRef]
78. Zhou, Q.; Liu, L.Z.; Fu, B.; Hu, X.; Shi, X.; Fang, J.; Jiang, B.H. Reactive oxygen species regulate insulin-induced VEGF and HIF-1alpha expression through the activation of p70S6K1 in human prostate cancer cells. *Carcinogenesis* **2007**, *28*, 28–37. [CrossRef]
79. Liu, B.; Chen, Y.; St Clair, D.K. ROS and p53: A versatile partnership. *Free Radic. Biol. Med.* **2008**, *44*, 1529–1535. [CrossRef]
80. Wu, G.S. The functional interactions between the p53 and MAPK signaling pathways. *Cancer Biol. Ther.* **2004**, *3*, 156–161. [CrossRef]
81. Iorio, M.V.; Croce, C.M. MicroRNA dysregulation in cancer: diagnostics, monitoring and therapeutics. A comprehensive review. *EMBO Mol. Med.* **2017**, *9*, 852. [CrossRef] [PubMed]
82. Jin, L.H.; Wei, C. Role of microRNAs in the Warburg effect and mitochondrial metabolism in cancer. *Asian Pac. J. Cancer Prev.* **2014**, *15*, 7015–7019. [CrossRef] [PubMed]
83. He, J.; Jiang, B.H. Interplay between Reactive oxygen Species and MicroRNAs in Cancer. *Curr. Pharmacol. Rep.* **2016**, *2*, 82–90. [CrossRef]
84. Kim, H.R.; Roe, J.S.; Lee, J.E.; Cho, E.J.; Youn, H.D. p53 regulates glucose metabolism by miR-34a. *Biochem. Biophys. Res. Commun.* **2013**, *437*, 225–231. [CrossRef] [PubMed]
85. Gorrini, C.; Harris, I.S.; Mak, T.W. Modulation of oxidative stress as an anticancer strategy. *Nat. Rev. Drug Discov.* **2013**, *12*, 931–947. [CrossRef] [PubMed]
86. Vafa, O.; Wade, M.; Kern, S.; Beeche, M.; Pandita, T.K.; Hampton, G.M.; Wahl, G.M. c-Myc can induce DNA damage, increase reactive oxygen species, and mitigate p53 function: A mechanism for oncogene-induced genetic instability. *Mol. Cell* **2002**, *9*, 1031–1044. [CrossRef]
87. Sullivan, L.B.; Chandel, N.S. Mitochondrial reactive oxygen species and cancer. *Cancer Metab* **2014**, *2*, 17. [CrossRef]
88. Goldman, R.D.; Kaplan, N.O.; Hall, T.C. Lactic Dehydrogenase in Human Neoplastic Tissues. *Cancer Res.* **1964**, *24*, 389–399.
89. Le, A.; Cooper, C.R.; Gouw, A.M.; Dinavahi, R.; Maitra, A.; Deck, L.M.; Royer, R.E.; Vander Jagt, D.L.; Semenza, G.L.; Dang, C.V. Inhibition of lactate dehydrogenase A induces oxidative stress and inhibits tumor progression. *Proc. Natl. Acad. Sci. USA* **2010**, *107*, 2037–2042. [CrossRef]
90. Arseneault, R.; Chien, A.; Newington, J.T.; Rappon, T.; Harris, R.; Cumming, R.C. Attenuation of LDHA expression in cancer cells leads to redox-dependent alterations in cytoskeletal structure and cell migration. *Cancer Lett.* **2013**, *338*, 255–266. [CrossRef]
91. Frederiks, W.M.; Vizan, P.; Bosch, K.S.; Vreeling-Sindelarova, H.; Boren, J.; Cascante, M. Elevated activity of the oxidative and non-oxidative pentose phosphate pathway in (pre)neoplastic lesions in rat liver. *Int. J. Exp. Pathol.* **2008**, *89*, 232–240. [CrossRef] [PubMed]
92. Bannasch, P.; Benner, U.; Hacker, H.J.; Klimek, F.; Mayer, D.; Moore, M.; Zerban, H. Cytochemical and biochemical microanalysis of carcinogenesis. *Histochem. J.* **1981**, *13*, 799–820. [CrossRef]
93. Van Driel, B.E.; De Goeij, A.F.; Song, J.Y.; De Bruine, A.P.; Van Noorden, C.J. Development of oxygen insensitivity of the quantitative histochemical assay of G6PDH activity during colorectal carcinogenesis. *J. Pathol.* **1997**, *182*, 398–403. [CrossRef]
94. Van Noorden, C.J.; Bahns, S.; Kohler, A. Adaptational changes in kinetic parameters of G6PDH but not of PGDH during contamination-induced carcinogenesis in livers of North Sea flatfish. *Biochim. Biophys. Acta* **1997**, *1342*, 141–148. [CrossRef]
95. Ju, H.Q.; Lu, Y.X.; Wu, Q.N.; Liu, J.; Zeng, Z.L.; Mo, H.Y.; Chen, Y.; Tian, T.; Wang, Y.; Kang, T.B.; et al. Disrupting G6PD-mediated Redox homeostasis enhances chemosensitivity in colorectal cancer. *Oncogene* **2017**, *36*, 6282–6292. [CrossRef] [PubMed]
96. Tian, W.N.; Braunstein, L.D.; Pang, J.; Stuhlmeier, K.M.; Xi, Q.C.; Tian, X.; Stanton, R.C. Importance of glucose-6-phosphate dehydrogenase activity for cell growth. *J. Biol. Chem.* **1998**, *273*, 10609–10617. [CrossRef] [PubMed]

97. Kuo, W.; Lin, J.; Tang, T.K. Human glucose-6-phosphate dehydrogenase (G6PD) gene transforms NIH 3T3 cells and induces tumors in nude mice. *Int. J. Cancer* **2000**, *85*, 857–864. [CrossRef]
98. Ruwende, C.; Hill, A. Glucose-6-phosphate dehydrogenase deficiency and malaria. *J. Mol. Med. (Berl.)* **1998**, *76*, 581–588. [CrossRef]
99. Zhang, Z.; Liew, C.W.; Handy, D.E.; Zhang, Y.; Leopold, J.A.; Hu, J.; Guo, L.; Kulkarni, R.N.; Loscalzo, J.; Stanton, R.C. High glucose inhibits glucose-6-phosphate dehydrogenase, leading to increased oxidative stress and beta-cell apoptosis. *FASEB J.* **2010**, *24*, 1497–1505. [CrossRef]
100. Leopold, J.A.; Walker, J.; Scribner, A.W.; Voetsch, B.; Zhang, Y.Y.; Loscalzo, A.J.; Stanton, R.C.; Loscalzo, J. Glucose-6-phosphate dehydrogenase modulates vascular endothelial growth factor-mediated angiogenesis. *J. Biol. Chem.* **2003**, *278*, 32100–32106. [CrossRef]
101. Leopold, J.A.; Zhang, Y.Y.; Scribner, A.W.; Stanton, R.C.; Loscalzo, J. Glucose-6-phosphate dehydrogenase overexpression decreases endothelial cell oxidant stress and increases bioavailable nitric oxide. *Arterioscler. Thromb. Vasc. Biol.* **2003**, *23*, 411–417. [CrossRef] [PubMed]
102. Boveris, A.; Cadenas, E.; Stoppani, A.O. Role of ubiquinone in the mitochondrial generation of hydrogen peroxide. *Biochem. J.* **1976**, *156*, 435–444. [CrossRef] [PubMed]
103. Turrens, J.F.; Alexandre, A.; Lehninger, A.L. Ubisemiquinone is the electron donor for superoxide formation by complex III of heart mitochondria. *Arch Biochem. Biophys.* **1985**, *237*, 408–414. [CrossRef]
104. Fleury, C.; Mignotte, B.; Vayssiere, J.L. Mitochondrial reactive oxygen species in cell death signaling. *Biochimie* **2002**, *84*, 131–141. [CrossRef]
105. Beckman, J.S.; Koppenol, W.H. Nitric oxide, superoxide, and peroxynitrite: The good, the bad, and ugly. *Am. J. Physiol.* **1996**, *271*, C1424–C1437. [CrossRef] [PubMed]
106. Radi, R.; Cassina, A.; Hodara, R.; Quijano, C.; Castro, L. Peroxynitrite reactions and formation in mitochondria. *Free Radic. Biol. Med.* **2002**, *33*, 1451–1464. [CrossRef]
107. Droge, W. Free radicals in the physiological control of cell function. *Physiol. Rev.* **2002**, *82*, 47–95. [CrossRef]
108. Stadtman, E.R.; Levine, R.L. Protein oxidation. *Ann. N. Y. Acad. Sci.* **2000**, *899*, 191–208. [CrossRef]
109. Rubbo, H.; Radi, R.; Trujillo, M.; Telleri, R.; Kalyanaraman, B.; Barnes, S.; Kirk, M.; Freeman, B.A. Nitric oxide regulation of superoxide and peroxynitrite-dependent lipid peroxidation. Formation of novel nitrogen-containing oxidized lipid derivatives. *J. Biol. Chem.* **1994**, *269*, 26066–26075.
110. Schulz, T.J.; Thierbach, R.; Voigt, A.; Drewes, G.; Mietzner, B.; Steinberg, P.; Pfeiffer, A.F.; Ristow, M. Induction of oxidative metabolism by mitochondrial frataxin inhibits cancer growth: Otto Warburg revisited. *J. Biol. Chem.* **2006**, *281*, 977–981. [CrossRef]
111. Gross, M.I.; Demo, S.D.; Dennison, J.B.; Chen, L.; Chernov-Rogan, T.; Goyal, B.; Janes, J.R.; Laidig, G.J.; Lewis, E.R.; Li, J.; et al. Antitumor activity of the glutaminase inhibitor CB-839 in triple-negative breast cancer. *Mol. Cancer Ther.* **2014**, *13*, 890–901. [CrossRef] [PubMed]
112. Shroff, E.H.; Eberlin, L.S.; Dang, V.M.; Gouw, A.M.; Gabay, M.; Adam, S.J.; Bellovin, D.I.; Tran, P.T.; Philbrick, W.M.; Garcia-Ocana, A.; et al. MYC oncogene overexpression drives renal cell carcinoma in a mouse model through glutamine metabolism. *Proc. Natl. Acad. Sci. USA* **2015**, *112*, 6539–6544. [CrossRef] [PubMed]
113. Matre, P.; Velez, J.; Jacamo, R.; Qi, Y.; Su, X.; Cai, T.; Chan, S.M.; Lodi, A.; Sweeney, S.R.; Ma, H.; et al. Inhibiting glutaminase in acute myeloid leukemia: metabolic dependency of selected AML subtypes. *Oncotarget* **2016**, *7*, 79722–79735. [CrossRef] [PubMed]
114. Wakil, S.J.; Porter, J.W.; Gibson, D.M. Studies on the mechanism of fatty acid synthesis. I. Preparation and purification of an enzymes system for reconstruction of fatty acid synthesis. *Biochim. Biophys. Acta* **1957**, *24*, 453–461. [CrossRef]
115. Geisbrecht, B.V.; Gould, S.J. The human PICD gene encodes a cytoplasmic and peroxisomal NADP(+)-dependent isocitrate dehydrogenase. *J. Biol. Chem.* **1999**, *274*, 30527–30533. [CrossRef]
116. Wise, D.R.; Ward, P.S.; Shay, J.E.; Cross, J.R.; Gruber, J.J.; Sachdeva, U.M.; Platt, J.M.; DeMatteo, R.G.; Simon, M.C.; Thompson, C.B. Hypoxia promotes isocitrate dehydrogenase-dependent carboxylation of alpha-ketoglutarate to citrate to support cell growth and viability. *Proc. Natl. Acad. Sci. USA* **2011**, *108*, 19611–19616. [CrossRef]
117. Stoddard, B.L.; Dean, A.; Koshland, D.E., Jr. Structure of isocitrate dehydrogenase with isocitrate, nicotinamide adenine dinucleotide phosphate, and calcium at 2.5-A resolution: a pseudo-Michaelis ternary complex. *Biochemistry* **1993**, *32*, 9310–9316. [CrossRef]

118. Ramachandran, N.; Colman, R.F. Chemical characterization of distinct subunits of pig heart DPN-specific isocitrate dehydrogenase. *J. Biol. Chem.* **1980**, *255*, 8859–8864.
119. Lee, S.H.; Jo, S.H.; Lee, S.M.; Koh, H.J.; Song, H.; Park, J.W.; Lee, W.H.; Huh, T.L. Role of NADP+-dependent isocitrate dehydrogenase (NADP+-ICDH) on cellular defence against oxidative injury by gamma-rays. *Int. J. Radiat. Biol.* **2004**, *80*, 635–642. [CrossRef]
120. Calvert, A.E.; Chalastanis, A.; Wu, Y.; Hurley, L.A.; Kouri, F.M.; Bi, Y.; Kachman, M.; May, J.L.; Bartom, E.; Hua, Y.; et al. Cancer-Associated IDH1 Promotes Growth and Resistance to Targeted Therapies in the Absence of Mutation. *Cell Rep.* **2017**, *19*, 1858–1873. [CrossRef]
121. Cancer Genome Atlas Research, N. Comprehensive genomic characterization defines human glioblastoma genes and core pathways. *Nature* **2008**, *455*, 1061–1068. [CrossRef] [PubMed]
122. Zarei, M.; Lal, S.; Parker, S.J.; Nevler, A.; Vaziri-Gohar, A.; Dukleska, K.; Mambelli-Lisboa, N.C.; Moffat, C.; Blanco, F.F.; Chand, S.N.; et al. Posttranscriptional Upregulation of IDH1 by HuR Establishes a Powerful Survival Phenotype in Pancreatic Cancer Cells. *Cancer Res.* **2017**, *77*, 4460–4471. [CrossRef] [PubMed]
123. Ma, Q.L.; Wang, J.H.; Wang, Y.G.; Hu, C.; Mu, Q.T.; Yu, M.X.; Wang, L.; Wang, D.M.; Yang, M.; Yin, X.F.; et al. High IDH1 expression is associated with a poor prognosis in cytogenetically normal acute myeloid leukemia. *Int. J. Cancer* **2015**, *137*, 1058–1065. [CrossRef]
124. Beljaards, R.C.; Meijer, C.J.; Scheffer, E.; van der Valk, P.; Willemze, R. Differential diagnosis of cutaneous large cell lymphomas using monoclonal antibodies reactive in paraffin-embedded skin biopsy specimens. *Am. J. Dermatopathol.* **1991**, *13*, 342–349. [CrossRef] [PubMed]
125. Figueroa, M.E.; Abdel-Wahab, O.; Lu, C.; Ward, P.S.; Patel, J.; Shih, A.; Li, Y.; Bhagwat, N.; Vasanthakumar, A.; Fernandez, H.F.; et al. Leukemic IDH1 and IDH2 mutations result in a hypermethylation phenotype, disrupt TET2 function, and impair hematopoietic differentiation. *Cancer Cell* **2010**, *18*, 553–567. [CrossRef] [PubMed]
126. Liu, X.; Kato, Y.; Kaneko, M.K.; Sugawara, M.; Ogasawara, S.; Tsujimoto, Y.; Naganuma, Y.; Yamakawa, M.; Tsuchiya, T.; Takagi, M. Isocitrate dehydrogenase 2 mutation is a frequent event in osteosarcoma detected by a multi-specific monoclonal antibody MsMab-1. *Cancer Med.* **2013**, *2*, 803–814. [CrossRef]
127. Murugan, A.K.; Bojdani, E.; Xing, M. Identification and functional characterization of isocitrate dehydrogenase 1 (IDH1) mutations in thyroid cancer. *Biochem. Biophys. Res. Commun.* **2010**, *393*, 555–559. [CrossRef]
128. Shi, J.; Sun, B.; Shi, W.; Zuo, H.; Cui, D.; Ni, L.; Chen, J. Decreasing GSH and increasing ROS in chemosensitivity gliomas with IDH1 mutation. *Tumour Biol.* **2015**, *36*, 655–662. [CrossRef]
129. Zhao, S.; Lin, Y.; Xu, W.; Jiang, W.; Zha, Z.; Wang, P.; Yu, W.; Li, Z.; Gong, L.; Peng, Y.; et al. Glioma-derived mutations in IDH1 dominantly inhibit IDH1 catalytic activity and induce HIF-1alpha. *Science* **2009**, *324*, 261–265. [CrossRef]
130. Losman, J.A.; Looper, R.E.; Koivunen, P.; Lee, S.; Schneider, R.K.; McMahon, C.; Cowley, G.S.; Root, D.E.; Ebert, B.L.; Kaelin, W.G., Jr. (R)-2-hydroxyglutarate is sufficient to promote leukemogenesis and its effects are reversible. *Science* **2013**, *339*, 1621–1625. [CrossRef]
131. Kaminska, B.; Czapski, B.; Guzik, R.; Krol, S.K.; Gielniewski, B. Consequences of IDH1/2 Mutations in Gliomas and an Assessment of Inhibitors Targeting Mutated IDH Proteins. *Molecules* **2019**, *24*, 968. [CrossRef] [PubMed]
132. Murai, S.; Ando, A.; Ebara, S.; Hirayama, M.; Satomi, Y.; Hara, T. Inhibition of malic enzyme 1 disrupts cellular metabolism and leads to vulnerability in cancer cells in glucose-restricted conditions. *Oncogenesis* **2017**, *6*, e329. [CrossRef] [PubMed]
133. Frenkel, R. Regulation and physiological functions of malic enzymes. *Curr. Top. Cell. Regul.* **1975**, *9*, 157–181. [CrossRef] [PubMed]
134. Pongratz, R.L.; Kibbey, R.G.; Shulman, G.I.; Cline, G.W. Cytosolic and mitochondrial malic enzyme isoforms differentially control insulin secretion. *J. Biol. Chem.* **2007**, *282*, 200–207. [CrossRef] [PubMed]
135. Nakashima, C.; Yamamoto, K.; Fujiwara-Tani, R.; Luo, Y.; Matsushima, S.; Fujii, K.; Ohmori, H.; Sasahira, T.; Sasaki, T.; Kitadai, Y.; et al. Expression of cytosolic malic enzyme (ME1) is associated with disease progression in human oral squamous cell carcinoma. *Cancer Sci.* **2018**, *109*, 2036–2045. [CrossRef]
136. Fernandes, L.M.; Al-Dwairi, A.; Simmen, R.C.M.; Marji, M.; Brown, D.M.; Jewell, S.W.; Simmen, F.A. Malic Enzyme 1 (ME1) is pro-oncogenic in Apc(Min/+) mice. *Sci. Rep.* **2018**, *8*, 14268. [CrossRef]
137. Lu, Y.X.; Ju, H.Q.; Liu, Z.X.; Chen, D.L.; Wang, Y.; Zhao, Q.; Wu, Q.N.; Zeng, Z.L.; Qiu, H.B.; Hu, P.S.; et al. ME1 Regulates NADPH Homeostasis to Promote Gastric Cancer Growth and Metastasis. *Cancer Res.* **2018**, *78*, 1972–1985. [CrossRef]

138. Robey, R.W.; Pluchino, K.M.; Hall, M.D.; Fojo, A.T.; Bates, S.E.; Gottesman, M.M. Revisiting the role of ABC transporters in multidrug-resistant cancer. *Nat. Rev. Cancer* **2018**, *18*, 452–464. [CrossRef]
139. Fletcher, J.I.; Williams, R.T.; Henderson, M.J.; Norris, M.D.; Haber, M. ABC transporters as mediators of drug resistance and contributors to cancer cell biology. *Drug Resist. Updat.* **2016**, *26*, 1–9. [CrossRef]
140. Thews, O.; Gassner, B.; Kelleher, D.K.; Schwerdt, G.; Gekle, M. Impact of extracellular acidity on the activity of P-glycoprotein and the cytotoxicity of chemotherapeutic drugs. *Neoplasia* **2006**, *8*, 143–152. [CrossRef]
141. Wojtkowiak, J.W.; Verduzco, D.; Schramm, K.J.; Gillies, R.J. Drug resistance and cellular adaptation to tumor acidic pH microenvironment. *Mol. Pharm.* **2011**, *8*, 2032–2038. [CrossRef] [PubMed]
142. Vukovic, V.; Tannock, I.F. Influence of low pH on cytotoxicity of paclitaxel, mitoxantrone and topotecan. *Br. J. Cancer* **1997**, *75*, 1167–1172. [CrossRef] [PubMed]
143. Seebacher, N.A.; Richardson, D.R.; Jansson, P.J. Glucose modulation induces reactive oxygen species and increases P-glycoprotein-mediated multidrug resistance to chemotherapeutics. *Br. J. Pharmacol.* **2015**, *172*, 2557–2572. [CrossRef] [PubMed]
144. Bhatt, A.N.; Chauhan, A.; Khanna, S.; Rai, Y.; Singh, S.; Soni, R.; Kalra, N.; Dwarakanath, B.S. Transient elevation of glycolysis confers radio-resistance by facilitating DNA repair in cells. *BMC Cancer* **2015**, *15*, 335. [CrossRef]
145. Wagner, W.; Ciszewski, W.M.; Kania, K.D. L- and D-lactate enhance DNA repair and modulate the resistance of cervical carcinoma cells to anticancer drugs via histone deacetylase inhibition and hydroxycarboxylic acid receptor 1 activation. *Cell Commun. Signal.* **2015**, *13*, 36. [CrossRef]
146. Zhao, J.G.; Ren, K.M.; Tang, J. Overcoming 5-Fu resistance in human non-small cell lung cancer cells by the combination of 5-Fu and cisplatin through the inhibition of glucose metabolism. *Tumour Biol.* **2014**, *35*, 12305–12315. [CrossRef]
147. Ma, S.; Jia, R.; Li, D.; Shen, B. Targeting Cellular Metabolism Chemosensitizes the Doxorubicin-Resistant Human Breast Adenocarcinoma Cells. *BioMed Res. Int.* **2015**, *2015*, 453986. [CrossRef]
148. Wintzell, M.; Lofstedt, L.; Johansson, J.; Pedersen, A.B.; Fuxe, J.; Shoshan, M. Repeated cisplatin treatment can lead to a multiresistant tumor cell population with stem cell features and sensitivity to 3-bromopyruvate. *Cancer Biol. Ther.* **2012**, *13*, 1454–1462. [CrossRef]
149. Sullivan, E.J.; Kurtoglu, M.; Brenneman, R.; Liu, H.; Lampidis, T.J. Targeting cisplatin-resistant human tumor cells with metabolic inhibitors. *Cancer Chemother. Pharmacol.* **2014**, *73*, 417–427. [CrossRef]
150. Zhang, X.Y.; Zhang, M.; Cong, Q.; Zhang, M.X.; Zhang, M.Y.; Lu, Y.Y.; Xu, C.J. Hexokinase 2 confers resistance to cisplatin in ovarian cancer cells by enhancing cisplatin-induced autophagy. *Int. J. Biochem. Cell Biol.* **2018**, *95*, 9–16. [CrossRef]
151. Vartanian, A.; Agnihotri, S.; Wilson, M.R.; Burrell, K.E.; Tonge, P.D.; Alamsahebpour, A.; Jalali, S.; Taccone, M.S.; Mansouri, S.; Golbourn, B.; et al. Targeting hexokinase 2 enhances response to radio-chemotherapy in glioblastoma. *Oncotarget* **2016**, *7*, 69518–69535. [CrossRef] [PubMed]
152. Fan, K.; Fan, Z.; Cheng, H.; Huang, Q.; Yang, C.; Jin, K.; Luo, G.; Yu, X.; Liu, C. Hexokinase 2 dimerization and interaction with voltage-dependent anion channel promoted resistance to cell apoptosis induced by gemcitabine in pancreatic cancer. *Cancer Med.* **2019**, *8*, 5903–5915. [CrossRef] [PubMed]
153. Jiang, J.X.; Gao, S.; Pan, Y.Z.; Yu, C.; Sun, C.Y. Overexpression of microRNA-125b sensitizes human hepatocellular carcinoma cells to 5-fluorouracil through inhibition of glycolysis by targeting hexokinase II. *Mol. Med. Rep.* **2014**, *10*, 995–1002. [CrossRef] [PubMed]
154. Li, W.C.; Huang, C.H.; Hsieh, Y.T.; Chen, T.Y.; Cheng, L.H.; Chen, C.Y.; Liu, C.J.; Chen, H.M.; Huang, C.L.; Lo, J.F.; et al. Regulatory Role of Hexokinase 2 in Modulating Head and Neck Tumorigenesis. *Front. Oncol.* **2020**, *10*, 176. [CrossRef]
155. Kwon, O.H.; Kang, T.W.; Kim, J.H.; Kim, M.; Noh, S.M.; Song, K.S.; Yoo, H.S.; Kim, W.H.; Xie, Z.; Pocalyko, D.; et al. Pyruvate kinase M2 promotes the growth of gastric cancer cells via regulation of Bcl-xL expression at transcriptional level. *Biochem. Biophys. Res. Commun.* **2012**, *423*, 38–44. [CrossRef]
156. Martinez-Balibrea, E.; Plasencia, C.; Gines, A.; Martinez-Cardus, A.; Musulen, E.; Aguilera, R.; Manzano, J.L.; Neamati, N.; Abad, A. A proteomic approach links decreased pyruvate kinase M2 expression to oxaliplatin resistance in patients with colorectal cancer and in human cell lines. *Mol. Cancer Ther.* **2009**, *8*, 771–778. [CrossRef]

157. Yoo, B.C.; Ku, J.L.; Hong, S.H.; Shin, Y.K.; Park, S.Y.; Kim, H.K.; Park, J.G. Decreased pyruvate kinase M2 activity linked to cisplatin resistance in human gastric carcinoma cell lines. *Int. J. Cancer* **2004**, *108*, 532–539. [CrossRef]
158. Wang, Y.; Hao, F.; Nan, Y.; Qu, L.; Na, W.; Jia, C.; Chen, X. PKM2 Inhibitor Shikonin Overcomes the Cisplatin Resistance in Bladder Cancer by Inducing Necroptosis. *Int. J. Biol. Sci.* **2018**, *14*, 1883–1891. [CrossRef]
159. Wang, X.; Zhang, F.; Wu, X.R. Inhibition of Pyruvate Kinase M2 Markedly Reduces Chemoresistance of Advanced Bladder Cancer to Cisplatin. *Sci. Rep.* **2017**, *7*, 45983. [CrossRef]
160. Liu, Y.; He, C.; Huang, X. Metformin partially reverses the carboplatin-resistance in NSCLC by inhibiting glucose metabolism. *Oncotarget* **2017**, *8*, 75206–75216. [CrossRef]
161. Guo, C.Y.; Yan, C.; Luo, L.; Goto, S.; Urata, Y.; Xu, J.J.; Wen, X.M.; Kuang, Y.K.; Tou, F.F.; Li, T.S. Enhanced expression of PKM2 associates with the biological properties of cancer stem cells from A549 human lung cancer cells. *Oncol. Rep.* **2017**, *37*, 2161–2166. [CrossRef] [PubMed]
162. Doi, K.; Onodera, T.; Tsuda, T.; Matsuzaki, H.; Mitsuoka, T. Histopathology of BALB/c mice infected with the D variant of encephalomyocarditis virus. *Br. J. Exp. Pathol.* **1988**, *69*, 395–401. [PubMed]
163. Shin, Y.K.; Yoo, B.C.; Hong, Y.S.; Chang, H.J.; Jung, K.H.; Jeong, S.Y.; Park, J.G. Upregulation of glycolytic enzymes in proteins secreted from human colon cancer cells with 5-fluorouracil resistance. *Electrophoresis* **2009**, *30*, 2182–2192. [CrossRef] [PubMed]
164. Sirover, M.A. Subcellular dynamics of multifunctional protein regulation: mechanisms of GAPDH intracellular translocation. *J. Cell. Biochem.* **2012**, *113*, 2193–2200. [CrossRef]
165. Butera, G.; Mullappilly, N.; Masetto, F.; Palmieri, M.; Scupoli, M.T.; Pacchiana, R.; Donadelli, M. Regulation of Autophagy by Nuclear GAPDH and Its Aggregates in Cancer and Neurodegenerative Disorders. *Int. J. Mol. Sci.* **2019**, *20*, 62. [CrossRef]
166. Phadke, M.; Krynetskaia, N.; Krynetskiy, E. Cytotoxicity of chemotherapeutic agents in glyceraldehyde-3-phosphate dehydrogenase-depleted human lung carcinoma A549 cells with the accelerated senescence phenotype. *Anticancer Drugs* **2013**, *24*, 366–374. [CrossRef]
167. Phadke, M.S.; Krynetskaia, N.F.; Mishra, A.K.; Krynetskiy, E. Glyceraldehyde 3-phosphate dehydrogenase depletion induces cell cycle arrest and resistance to antimetabolites in human carcinoma cell lines. *J. Pharmacol. Exp. Ther.* **2009**, *331*, 77–86. [CrossRef]
168. Dando, I.; Pacchiana, R.; Pozza, E.D.; Cataldo, I.; Bruno, S.; Conti, P.; Cordani, M.; Grimaldi, A.; Butera, G.; Caraglia, M.; et al. UCP2 inhibition induces ROS/Akt/mTOR axis: Role of GAPDH nuclear translocation in genipin/everolimus anticancer synergism. *Free Radic. Biol. Med.* **2017**, *113*, 176–189. [CrossRef]
169. Chen, R.W.; Saunders, P.A.; Wei, H.; Li, Z.; Seth, P.; Chuang, D.M. Involvement of glyceraldehyde-3-phosphate dehydrogenase (GAPDH) and p53 in neuronal apoptosis: evidence that GAPDH is upregulated by p53. *J. Neurosci.* **1999**, *19*, 9654–9662. [CrossRef]
170. Fiucci, G.; Beaucourt, S.; Duflaut, D.; Lespagnol, A.; Stumptner-Cuvelette, P.; Geant, A.; Buchwalter, G.; Tuynder, M.; Susini, L.; Lassalle, J.M.; et al. Siah-1b is a direct transcriptional target of p53: identification of the functional p53 responsive element in the siah-1b promoter. *Proc. Natl. Acad. Sci. USA* **2004**, *101*, 3510–3515. [CrossRef]
171. Hara, M.R.; Agrawal, N.; Kim, S.F.; Cascio, M.B.; Fujimuro, M.; Ozeki, Y.; Takahashi, M.; Cheah, J.H.; Tankou, S.K.; Hester, L.D.; et al. S-nitrosylated GAPDH initiates apoptotic cell death by nuclear translocation following Siah1 binding. *Nat. Cell Biol.* **2005**, *7*, 665–674. [CrossRef] [PubMed]
172. Hara, M.R.; Snyder, S.H. Nitric oxide-GAPDH-Siah: a novel cell death cascade. *Cell Mol. Neurobiol.* **2006**, *26*, 527–538. [CrossRef] [PubMed]
173. Thangima Zannat, M.; Bhattacharjee, R.B.; Bag, J. In the absence of cellular poly (A) binding protein, the glycolytic enzyme GAPDH translocated to the cell nucleus and activated the GAPDH mediated apoptotic pathway by enhancing acetylation and serine 46 phosphorylation of p53. *Biochem. Biophys. Res. Commun.* **2011**, *409*, 171–176. [CrossRef] [PubMed]
174. Butera, G.; Pacchiana, R.; Mullappilly, N.; Margiotta, M.; Bruno, S.; Conti, P.; Riganti, C.; Donadelli, M. Mutant p53 prevents GAPDH nuclear translocation in pancreatic cancer cells favoring glycolysis and 2-deoxyglucose sensitivity. *Biochim. Biophys. Acta Mol. Cell. Res.* **2018**, *1865*, 1914–1923. [CrossRef] [PubMed]
175. Li, K.; Huang, M.; Xu, P.; Wang, M.; Ye, S.; Wang, Q.; Zeng, S.; Chen, X.; Gao, W.; Chen, J.; et al. Microcystins-LR induced apoptosis via S-nitrosylation of GAPDH in colorectal cancer cells. *Ecotoxicol. Environ. Saf.* **2020**, *190*, 110096. [CrossRef] [PubMed]

176. Tamada, M.; Nagano, O.; Tateyama, S.; Ohmura, M.; Yae, T.; Ishimoto, T.; Sugihara, E.; Onishi, N.; Yamamoto, T.; Yanagawa, H.; et al. Modulation of glucose metabolism by CD44 contributes to antioxidant status and drug resistance in cancer cells. *Cancer Res.* **2012**, *72*, 1438–1448. [CrossRef]
177. Lucarelli, G.; Galleggiante, V.; Rutigliano, M.; Sanguedolce, F.; Cagiano, S.; Bufo, P.; Lastilla, G.; Maiorano, E.; Ribatti, D.; Giglio, A.; et al. Metabolomic profile of glycolysis and the pentose phosphate pathway identifies the central role of glucose-6-phosphate dehydrogenase in clear cell-renal cell carcinoma. *Oncotarget* **2015**, *6*, 13371–13386. [CrossRef]
178. Polimeni, M.; Voena, C.; Kopecka, J.; Riganti, C.; Pescarmona, G.; Bosia, A.; Ghigo, D. Modulation of doxorubicin resistance by the glucose-6-phosphate dehydrogenase activity. *Biochem. J.* **2011**, *439*, 141–149. [CrossRef]
179. Catanzaro, D.; Gaude, E.; Orso, G.; Giordano, C.; Guzzo, G.; Rasola, A.; Ragazzi, E.; Caparrotta, L.; Frezza, C.; Montopoli, M. Inhibition of glucose-6-phosphate dehydrogenase sensitizes cisplatin-resistant cells to death. *Oncotarget* **2015**, *6*, 30102–30114. [CrossRef]
180. Hong, W.; Cai, P.; Xu, C.; Cao, D.; Yu, W.; Zhao, Z.; Huang, M.; Jin, J. Inhibition of Glucose-6-Phosphate Dehydrogenase Reverses Cisplatin Resistance in Lung Cancer Cells via the Redox System. *Front. Pharmacol.* **2018**, *9*, 43. [CrossRef]
181. Zhang, R.; Tao, F.; Ruan, S.; Hu, M.; Hu, Y.; Fang, Z.; Mei, L.; Gong, C. The TGFbeta1-FOXM1-HMGA1-TGFbeta1 positive feedback loop increases the cisplatin resistance of non-small cell lung cancer by inducing G6PD expression. *Am. J. Transl. Res.* **2019**, *11*, 6860–6876. [PubMed]
182. Liu, R.; Li, W.; Tao, B.; Wang, X.; Yang, Z.; Zhang, Y.; Wang, C.; Liu, R.; Gao, H.; Liang, J.; et al. Tyrosine phosphorylation activates 6-phosphogluconate dehydrogenase and promotes tumor growth and radiation resistance. *Nat. Commun.* **2019**, *10*, 991. [CrossRef] [PubMed]
183. Ma, L.; Cheng, Q. Inhibiting 6-phosphogluconate dehydrogenase reverses doxorubicin resistance in anaplastic thyroid cancer via inhibiting NADPH-dependent metabolic reprogramming. *Biochem. Biophys. Res. Commun.* **2018**, *498*, 912–917. [CrossRef] [PubMed]
184. Sarfraz, I.; Rasul, A.; Hussain, G.; Shah, M.A.; Zahoor, A.F.; Asrar, M.; Selamoglu, Z.; Ji, X.Y.; Adem, S.; Sarker, S.D. 6-Phosphogluconate dehydrogenase fuels multiple aspects of cancer cells: From cancer initiation to metastasis and chemoresistance. *Biofactors* **2020**. [CrossRef]
185. Li, Q.; Qin, T.; Bi, Z.; Hong, H.; Ding, L.; Chen, J.; Wu, W.; Lin, X.; Fu, W.; Zheng, F.; et al. Rac1 activates non-oxidative pentose phosphate pathway to induce chemoresistance of breast cancer. *Nat. Commun.* **2020**, *11*, 1456. [CrossRef]
186. Shukla, S.K.; Purohit, V.; Mehla, K.; Gunda, V.; Chaika, N.V.; Vernucci, E.; King, R.J.; Abrego, J.; Goode, G.D.; Dasgupta, A.; et al. MUC1 and HIF-1alpha Signaling Crosstalk Induces Anabolic Glucose Metabolism to Impart Gemcitabine Resistance to Pancreatic Cancer. *Cancer Cell* **2017**, *32*, 71–87. [CrossRef]
187. Yang, H.; Wu, X.L.; Wu, K.H.; Zhang, R.; Ju, L.L.; Ji, Y.; Zhang, Y.W.; Xue, S.L.; Zhang, Y.X.; Yang, Y.F.; et al. MicroRNA-497 regulates cisplatin chemosensitivity of cervical cancer by targeting transketolase. *Am. J. Cancer Res.* **2016**, *6*, 2690–2699.
188. Ko, Y.H.; Verhoeven, H.A.; Lee, M.J.; Corbin, D.J.; Vogl, T.J.; Pedersen, P.L. A translational study "case report" on the small molecule "energy blocker" 3-bromopyruvate (3BP) as a potent anticancer agent: from bench side to bedside. *J. Bioenerg. Biomembr.* **2012**, *44*, 163–170. [CrossRef]
189. Zhang, Q.; Zhang, Y.; Zhang, P.; Chao, Z.; Xia, F.; Jiang, C.; Zhang, X.; Jiang, Z.; Liu, H. Hexokinase II inhibitor, 3-BrPA induced autophagy by stimulating ROS formation in human breast cancer cells. *Genes Cancer* **2014**, *5*, 100–112. [CrossRef]

International Journal of
Molecular Sciences

Review

Oxidative Damage in Sporadic Colorectal Cancer: Molecular Mapping of Base Excision Repair Glycosylases in Colorectal Cancer Patients

Pavel Vodicka [1,2,3,*], Marketa Urbanova [3], Pavol Makovicky [4], Kristyna Tomasova [1,2], Michal Kroupa [1,2], Rudolf Stetina [5], Alena Opattova [1,2,3], Klara Kostovcikova [6], Anna Siskova [1,3], Michaela Schneiderova [7], Veronika Vymetalkova [1,2,3] and Ludmila Vodickova [1,2,3]

1 Department of Molecular Biology of Cancer, Institute of Experimental Medicine of the Czech Academy of Sciences, Videnska 1083, 142 20 Prague, Czech Republic; Tomasova.Kristyna@seznam.cz (K.T.); misa.kroupa@gmail.com (M.K.); alenaopattova@gmail.com (A.O.); anna.siskova21@gmail.com (A.S.); veronika.vymetalkova@iem.cas.cz (V.V.); ludovod@seznam.cz (L.V.)
2 Biomedical Centre, Faculty of Medicine in Pilsen, Charles University, Alej Svobody 1655, 323 00 Pilsen, Czech Republic
3 Institute of Biology and Medical Genetics, First Faculty of Medicine, Charles University, Albertov 4, 128 00 Prague, Czech Republic; marketk@seznam.cz
4 Department of Biology, Faculty of Education, J Selye University, Bratislavska 3322, 945 01 Komarno, Slovakia; makovicky.pavol@gmail.com
5 Department of Toxicology and Military Pharmacy, Faculty of Military Health Sciences, University of Defence, Trebesska 1575, 500 01 Hradec Kralove, Czech Republic; r.stetina@tiscali.cz
6 Laboratory of Cellular and Molecular Immunology, Institute of Microbiology of the Czech Academy of Sciences, Videnska 1083, 142 20 Prague, Czech Republic; klimesov@biomed.cas.cz
7 Department of Surgery, University Hospital Kralovske Vinohrady in Prague, Third Medical Faculty, Charles University, Ruska 87, 100 00 Prague, Czech Republic; schneider77@seznam.cz
* Correspondence: pavel.vodicka@iem.cas.cz; Tel.: +420-241062694

Received: 13 February 2020; Accepted: 31 March 2020; Published: 2 April 2020

Abstract: Oxidative stress with subsequent premutagenic oxidative DNA damage has been implicated in colorectal carcinogenesis. The repair of oxidative DNA damage is initiated by lesion-specific DNA glycosylases (hOGG1, NTH1, MUTYH). The direct evidence of the role of oxidative DNA damage and its repair is proven by hereditary syndromes (MUTYH-associated polyposis, NTHL1-associated tumor syndrome), where germline mutations cause loss-of-function in glycosylases of base excision repair, thus enabling the accumulation of oxidative DNA damage and leading to the adenoma-colorectal cancer transition. Unrepaired oxidative DNA damage often results in G:C>T:A mutations in tumor suppressor genes and proto-oncogenes and widespread occurrence of chromosomal copy-neutral loss of heterozygosity. However, the situation is more complicated in complex and heterogeneous disease, such as sporadic colorectal cancer. Here we summarized our current knowledge of the role of oxidative DNA damage and its repair on the onset, prognosis and treatment of sporadic colorectal cancer. Molecular and histological tumor heterogeneity was considered. Our study has also suggested an additional important source of oxidative DNA damage due to intestinal dysbiosis. The roles of base excision repair glycosylases (hOGG1, MUTYH) in tumor and adjacent mucosa tissues of colorectal cancer patients, particularly in the interplay with other factors (especially microenvironment), deserve further attention. Base excision repair characteristics determined in colorectal cancer tissues reflect, rather, a disease prognosis. Finally, we discuss the role of DNA repair in the treatment of colon cancer, since acquired or inherited defects in DNA repair pathways can be effectively used in therapy.

Keywords: oxidative DNA damage; DNA repair; base excision repair (BER)glycosylases; colorectal cancer

1. Introduction

Colorectal cancer (CRC) represents significant social and public health problems, particularly in developed countries worldwide. At the most recent overview, colon cancer accounted for 1,096,601 new cases and 551,269 deaths in 2018, whereas rectal cancer was less frequent (704,376 newly diagnosed cases with 310,394 patients who died) [1]. Particular dietary and lifestyle habits and age constitute major risk factors in sporadic CRC, as recently reviewed [2,3].

Sporadic (non-hereditary) CRC (70–75% of CRCs cases) occurs in people without genetic predisposition or family history of CRC [4]. CRC often develops in genetically susceptible individuals as a consequence of the co-inheritance of multiple low-risk variants. Whereas up to 35% of interindividual variability in CRC risk has been attributed to genetic factors, high-risk germline mutations in *APC*, *MMR*, *MUTYH (MYH)*, *SMAD4*, *BMPR1A* and *STK11/LKB1* genes account for about 6% of all cases [5,6].

Colorectal carcinogenesis includes three major genetic and epigenetic pathways: chromosomal instability (CIN), CpG island methylator phenotype (CIMP) and microsatellite instability (MSI). MSI is driven by functional impairment of DNA mismatch repair (MMR) genes and it is characterized by alterations in the length of microsatellites [4,7,8]. CIN is hallmarked by changes in chromosomal copy numbers. CIMP, as a form of epigenetic modification, refers to hypermethylation at repetitive CpG dinucleotides (so-called CpG islands) in the promoter regions of tumor suppressor genes (such as *MLH1*, *MINT1*, *MINT2* and *MINT3*) that silences gene expression [4]. Ample and convincing evidence has been accumulated on the role of inflammation, lipid peroxidation, oxidative stress and metabolic dysfunction in CRC onset and development (reviewed in [3]). Several physiological and pathological processes, closely linked to CRC development in the human body (i.e., obesity, diabetes, inflammatory bowel diseases), stimulate the formation of reactive oxygen species (ROS) and subsequent DNA damage [9]. For instance, obesity increases inflammatory factors and adipokines (TNF, leptin, IL-1β and IL-6), subsequently promoting oxidative stress and suppressing the immune system. These alterations often end up in aberrant cell signaling, increased cell growth and angiogenesis [10,11]. Disturbances in DNA damage levels, antioxidant status and capacity for DNA repair result in the accumulation of mutations and genomic instability. The involvement of dietary factors in the etiology of CRC suggests that this disease may be preventable by prudent dietary adjustments, e.g., by antioxidant-rich food [12,13] and optimal selenium uptake [14].

In this study, we intended to summarize our current knowledge on the role of oxidative DNA damage and its repair on the onset and prognosis and treatment of sporadic CRC, taking into account tumor heterogeneity. We also addressed the roles of glycosylases (hOGG1, MUTYH) involved in the base excision repair (BER) of oxidative damage.

2. Colorectal Cancer and Oxidative DNA Damage

2.1. DNA Damage and Colorectal Cancer Pathogenesis

Chronic human inflammatory diseases, diabetes, aging and various malignant diseases, including CRC, are hallmarked by an increase in oxidative DNA damage [15]. Oxidative stress belongs to the ubiquitous events attacking biologic systems. It has been postulated that oxidative stress is responsible for steadily increasing oxidative damage burden from early adenoma to CRC progression [16,17]. In general, unrepaired DNA damage and subsequent disruption in DNA damage response (DDR) pathway have been recorded in many cancer types and are responsible for genomic instability, a pivotal feature of cancer [18]. Indeed, Pearl et al. [19] have documented complex functional impairment in DDR in several cancer types. Importance of DNA repair pathways (a constituent part of DDR) in maintaining genomic instability and cancer etiology is highlighted in familial cancers with known high-penetrance germline mutations in DNA repair genes: *BRCA1/BRCA2* in breast cancer, MMR and polymerase deficiency (*MLH1, MSH2, MSH6, PMS2 and POLE* genes) in CRC and ovarian cancers, deleterious mutations in *RAD51C* and *RAD51D* and *BRCA1* mutation in ovarian cancers [20–26].

2.2. Oxidative DNA Damage, Characteristics, Biologic Properties and Relevance

ROS are engaged in many redox-governing processes of the cells in order to maintain cellular homeostasis and they pose potential signaling molecules to control several physiological cellular functions (for review see [27]). Its overproduction results in oxidative stress, responsible for a bulk of oxidative damage in DNA. ROS comprise a group of highly reactive chemical ions and molecules that includes oxygen radicals, non-radicals and hydrogen peroxide [28]. They are produced either endogenously during normal aerobic cellular metabolism or exogenously by agents such as ionizing radiation, chemotherapeutic drugs and transition metals. Elevated levels of ROS or depressed antioxidant defense lead to the imbalance in cellular DNA damage formation. ROS attack biologic macromolecules, resulting in DNA base and sugar damage, apurinic or apyrimidinic sites, DNA–protein cross-links and strand breaks, all contributing to genomic instability [29–31]. Once ROS reach DNA, the oxidation of nucleophilic DNA bases and the ribose sugar ring leads to base loss and strand breaks. Guanine is the most prominent target, giving rise to 8-oxo-7,8-dihydro-2′deoxyguanosine (8-oxo-dG) and 2,6-diamino-4-hydroxy-5-formamidopyrimidine (FAPY). ROS also react with adenine (8-oxo-7,8-dihydro-2′deoxyadenosine, 2-hydroxyadenine)—and to a lesser extent with thymine and cytosine. Eight-oxo-dG is the most pro-mutagenic consequence of ROS, causing G > T transversion [32,33] and is commonly measured either as the base in DNA or as the nucleoside 8-oxo-dG in urine [29]. Oxidative DNA damage triggers multiple pathways that include DNA repair, cell cycle arrest and apoptosis. Under physiological conditions, the steady-state level between DNA damage, antioxidant status; capacity for DNA repair is established and critical mutations in cancer-related genes are rather rare events.

2.3. The Repair of Oxidative DNA Damage

Altered DNA repair, comprising BER, nucleotide excision repair (NER), MMR, direct DNA repair, homologous recombination repair (HR) and non-homologous end-joining repair, acts as an important player involved in both cancer initiation and progression [34,35]. Moreover, modulations in DNA repair processes contribute to genetic heterogeneity and cancer evolution (genomic/chromosomal instabilities). Relevance of DNA repair and DDR in cancer onset, its progression and patients′ therapeutic response has recently been reviewed in 33 cancer types. Genetic changes (mutations, loss of heterozygosity) were observed in 33% of DNA repair and DDR genes highlighting the participation of these pathways in tumorigenesis [36].

BER pathway is the main mechanism involved in the removal and repair of oxidized DNA bases (for reviews see [37,38]). The repair of oxidative DNA damage (including premutagenic 8-oxo-dG) is initiated by lesion-specific DNA glycosylases, such as hOGG1 and MUTYH, which are the first enzymes in this pathway responsible for locating and removing DNA single damaged base. Eleven DNA glycosylases have been identified in human BER so far (Table 1) [39]. The redundancy in the substrate specificities of the glycosylases that recognize and remove oxidized DNA bases supports robust and efficient cell defense against oxidative stress. It has developed in organisms to protect the genome from the perpetual attacks of oxygen radicals both under pathologies and physiological conditions.

Human 8-oxo-dG DNA N-glycosylase 1 (hOGG1) removes 8-oxo-dG from the DNA and mutY DNA glycosylase (MUTYH, also termed MYH) excises misincorporated adenines opposite to 8-oxo-dG via replicative DNA polymerases α, δ and ε (reviewed in [55]). Both glycosylases suppress tumorigenesis by preventing mutagenic G:C > T:A transversions, as well as by inducing MUTYH-dependent cell death (reviewed by [56]). It should be noted that hOGG1 glycosylase acts as a counter partner of MUTYH (Figure 1). Both glycosylases stimulate consequent steps in BER (action of AP endonuclease I and Polymerase β) to complete the repair process by incision, gap-filling and ligation. However, other BER enzymes also participate to protect DNA against oxidative damage, such as NTH1, MTH1, NEIL1-3, XRCC1 and PARP-1 [57–60]).

Table 1. List of human DNA glycosylases and their function.

Glycosylase Name	Gene	Enzyme Commission Number	Biologic Function	Reference
Adenine DNA glycosylase	*MUTYH*	3.2.2.31	MUTYH is a monofunctional DNA glycosylase which, after the replication, removes adenines mispaired with 8-oxo-dG.	Koger et al., 2019 [40]
N-glycosylase/DNA lyase	*OGG1*	4.2.99.18	OGG1 acts in cooperation with MUTYH. It is a major glycosylase for the removal of 8-oxo-dG. It possesses also an intrinsic AP lyase activity at abasic sites.	Wang et al., 2018 [41]
DNA-3-methyladenine glycosylase	*MPG*	3.2.2.21	MPG removes a variety of alkylated (3-methyladenine, 7-methylguanine) and deaminated (hypoxanthine) purines. It also recognizes and removes secondary oxidative lesions such as 1,N6-ethenoadenine.	Leitner-Dagan et al., 2012 [42]
Methyl-CpG-binding domain protein 4	*MBD4*	3.2.2.-	MBD4 preferentially binds to CpG sites and guards DNA against deamination of cytosine to uracil or 5-methylcytosine to thymine.	Sjolund et al., 2013 [43]
Single-strand selective monofunctional uracil DNA glycosylase	*SMUG1*	3.2.2.-	SMUG1 belongs to the uracil DNA glycosylase superfamily. It is a back-up uracil DNA glycosylase removing a wide variety of oxidized pyrimidines such as 5-hydroxyuracil, 5-hydroxymethyluracil, 5-formyluracil and 5-carboxyuracil In addition to that, SMUG1 has also an activity towards 5-fluorouracil, a commonly used chemotherapeutic agent to treat CRC.	Nagaria et al., 2013 [44], Alexeeva et al., 2019 [45]
Endonuclease III-like protein 1	*NTH1*	4.2.99.18	NTH1 cleaves a broad range of lesions such as thymine glycol, 5-hydroxyuracil, 5-formyluracil, 5-hydroxycytosine, 5-hydroxy-6-hydrothymine, 5,6-dihydroxycytosine, 5,6-dihydrouracil and formamidopyrimidine.	Shinmura et al., 2019 [46]
Endonuclease VIII-like 1	*NEIL1*	4.2.99.18	NEIL1 acts at the replication fork and it is implicated in direct removal of the 5-carboxylcytosine. Further, it stimulates TDG-mediated excision of 5-formylcytosine and 5-carboxylcytosine.	Slyvka et al., 2017 [47]

Table 1. *Cont.*

Glycosylase Name	Gene	Enzyme Commission Number	Biologic Function	Reference
Endonuclease VIII-like 2	*NEIL2*	4.2.99.18	NEIL2 takes part in the transcription-coupled BER. It excises 8-oxoguanine, thymine glycol, formamidopyrimidine lesions and oxidative products of cytosine, particularly 5-hydroxyuracil and 5-hydroxycytosine.	Sarker et al., 2014 [48], Han et al., 2019 [49], Minko et al., 2019 [50]
Endonuclease VIII-like 3	*NEIL3*	4.2.99.18	NEIL3 acts preferentially on ssDNA. It removes spiroiminodihydantoin and guanidinohydantoin, further oxidation products of 8-oxo-7,8-dihydroguanine. It is also implicated in the repair of formamidopyrimidine DNA adducts.	Massaad et al., 2016 [51], Minko et al., 2019 [50]
G/T mismatch-specific thymine DNA glycosylase	TDG	3.2.2.29	TDG recognizes U-G or T-G mismatches caused by the deamination of the cytosine or 5-methylcytosine. Therefore, it prevents the formation of a C→T mutation. Further, it excises oxidized products of the 5-methylcytosine and 5-hydroxymethylcytosine, such as the 5-formylcytosine and 5-carboxycytosine.	Da et al. 2018 [52], Fu et al., 2019 [53]
Uracil-DNA glycosylase	UNG	3.2.2.27	UNG hydrolyzes uracil from both ss and dsDNA, leaving an apyrimidinic site. Such lesions can arise due to deamination of cytosine or due to misincorporation of dUMPs during replication or repair.	Weiser et al., 2018 [54]

Another DNA glycosylase involved in the excision of a wide spectrum of oxidized pyrimidines is endonuclease VIII-like 1 encoded by the *NEIL1* gene, that figures as a back-up for DNA glycosylase NTH1 [61]. *NTH1* germ-line variant D239Y (G > T substitution) has been found to induce genomic instability and cellular transformation in non-transformed human and mouse mammary epithelial cells [61,62].

One of the most important nucleases removing oxidized deoxynucleotide triphosphates (dNTPs) from the cellular pool is mutY homolog named human mutT homolog 1 (MTH1). Nucleotides in the pool are particularly vulnerable to oxidation and incorporated and unrepaired 8-oxo-dGTPs further contribute to G:C > T:A transversion. MTH1 preserves genomic integrity by preventing the incorporation of mutagenic purines into nuclear and/or mitochondrial DNA [63]. Depletion or inhibition of MTH1 also results in DNA strand breaks [64]. Additionally, MTH1 helps to protect telomeres, the essential structures at the end of chromosomes, from their oxidation and shortening and prevents the induction of genomic instability [65].

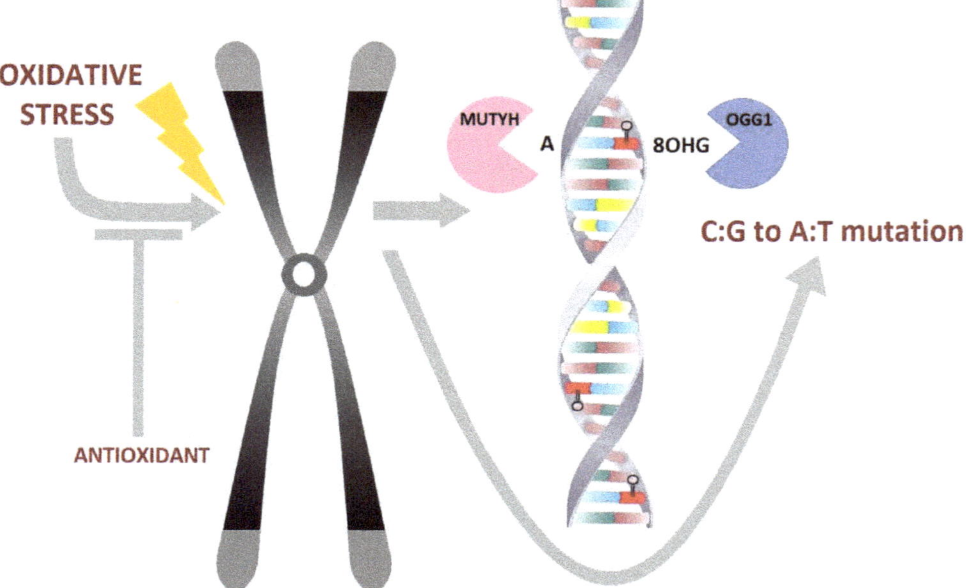

Figure 1. MUTYH and hOOG1 cooperate to prevent C:G to A:T transversion mutations under oxidative stress.

After the initial step of recognition and excision of oxidized bases by DNA glycosylases, DNA nicks occur that are sealed by DNA ligases. Several major types of DNA ligases (such as LIG1, LIG3, LIG4) have been discovered in human cells so far [66]. DNA nicks occurring during DNA replication or as the intermediate of BER are sealed by human DNA ligase I (LigI) or DNA ligase III (LigIII) along with XRCC1 DNA repair enzyme. High expression of *LigI* has been described in many human solid cancers [67]. The inhibition of LigI may therefore potentially block DNA replication and may also sensitize cancer cells to chemotherapeutic agents [68] leading to apoptosis [62,69].

Studies on laboratory animals have been intended to solve the question of whether oxidative damage occurs before or in the early stages of carcinogenesis or appears as a consequence of this process. The 8-oxo-dG has been detected in experimental animals treated with chemical carcinogens [70,71] or irradiated by X-rays [72] suggesting secondary oxidative stress that accompanies exposures to specific genotoxicants. In a comprehensive study of Olinski et al. the oxidized DNA bases were found in target organs of animals treated with various carcinogens (e.g., heavy metals) long before the tumor appeared [73]. Although the higher presence of 8-oxo-dG was not recorded in MUTYH knockout mice [74], simultaneous knocking out of both MUTYH and hOGG1 glycosylases resulted in a synergistic increase in G > T transversions [75]. All available evidence points to the enhanced oxidative DNA damage in tumors as a result of malignant transformation.

2.4. Oxidative DNA Damage, its Repair and Implications in Colorectal Carcinogenesis

2.4.1. Hereditary Syndromes with Defects in Glycosylases Predisposing Colorectal Cancer

Hereditary syndromes with germline mutations in selected repair genes predispose to complete loss-of-function of BER proteins and thus facilitate the inactivation of oxidative DNA damage removal process which results in accumulation of oxidative DNA damage in the transition from early adenoma to CRC. These comprise MUTYH-associated polyposis (MAP) and NTHL1-associated tumor syndrome (NATS) [26,56]. Although these hereditary syndromes account for less than 2% of all CRC, they represent a substantial (80%) risk of CRC development in MAP patients and directly connect the oxidative damage and its repair with CRC development. MAP-associated CRC exhibits G:C>T:A mutations in tumor suppressor genes and proto-oncogenes and widespread occurrence of chromosomal copy-neutral loss of heterozygosity [76]. Recessively inherited mutations in the *NTHL1* gene cause a polyposis and CRC syndrome [77], about five times less frequent than MAP. This NATS syndrome is based on a unique, clearly distinct mutational signature, G:C > A:T transition at the non-CpG site [56]. The above findings raise the question of whether other BER glycosylases could be candidate genes for new, yet undiscovered polyposis syndromes.

2.4.2. Sporadic Colorectal Cancer

In sporadic CRC, unrepaired 8-oxo-dG adducts induce mutations in proto-oncogenes, such as *KRAS* and tumor suppressor genes and, also, both *MUTYH* and *hOGG1* were found to be downregulated in neoplastic human colon tissues compared to adjacent tissues [78]. The concerted action of *MUTYH* and *hOGG1* is illustrated in Figure 1. This is in accordance with the suggestion that reduced BER capacity elevates susceptibility to oxidative DNA damage in various cancer types, including their invasiveness [79]. Unrepaired oxidative lesions, such as thymine glycol, can also stall the progression of a replicative fork and therefore contribute to genomic instability. Yet, another mechanism, based on DNA global demethylation mediated by BER in early colorectal tumorigenesis, has recently been postulated. The study of Furlan et al. found that MUTYH-associated polyposis adenomas exhibited strikingly pronounced hypomethylation than familial adenomatous and sporadic polyps. The authors concluded that DNA demethylation, together with specific KRAS/NRAS mutations, drives the early steps of oxidative damage related colorectal tumorigenesis [80]. In this context, we have not recorded any aberrant methylation in relevant DNA repair genes in tumor tissues from colorectal cancer patients [81]. Less information is currently available on MTH1 nuclease with regard to sporadic CRC. High expression of *MTH1* has been observed in many human malignancies [82,83], including CRC, where the enzyme bolsters survival of the malignant cells and acts therefore in procarcinogenic manner.

2.4.3. Base Excision Repair Capacity in Sporadic Colorectal Cancer

Recent years witnessed attempts to determine individual DNA excision repair capacities (DRC) both in healthy subjects and CRC patients [38]. DRC emerges as one of the most complex biomarkers since it integrates a plethora of factors such as gene variants, gene expressions, the interplay of relevant glycosylases, the stability of gene products, the effect of inhibitors/stimulators, lifestyle and environmental factors. Such a biomarker is of key importance in the identification of cancer risk in sporadic malignancies that are substantially affected by gene-environment interactions including oxidative DNA damage [34,84]. Functional DNA repair assays also provide fundamental information about the capacity of the organism to cope with chronic exposure to numerous environmental and dietary genotoxicants. Oxidative DNA damage, corresponding base excision repair capacity (BER-DRC) and relevant gene variants were addressed in 182 CRC patients and 245 controls. Whereas the 326Ser/Cys OGG1 and the 324Gln/His as well as the 324His/His MUTYH genotypes were associated with an increased CRC risk, the decreased efficiency of DNA repair was correlated with the 399Gln/Gln XRCC1 and the 324His/His MUTYH genotypes occurrence in CRC patients. Due to the missing information for some enrolled CRC patients, the validation study is warranted [85]. In our laboratory, we have measured both NER- and BER-DRC in tumor tissues and adjacent bowel mucosa of 70 incident CRC patients. In another study, BER-DRC in tumor tissue did not differ from that in adjacent mucosa. There was a good correlation between BER-DRC in tumor tissue, adjacent mucosa and peripheral blood lymphocytes. BER-DRC was not influenced by sex and age and, most importantly, did not differ between the colon and rectal tumors. No statistical significance was found in BER-DRC based on the pathologic stage of the tumors and the expression levels of BER genes did not correlate with BER-DRC [86]. Current reports suggest that oxidative DNA damage may be removed from DNA also via NER [87,88] and the authors postulate that loss of NER function shares common features arising from BER defects, including cancer predisposition [89]. Despite we have recorded significantly higher NER DRC in tumor tissue from CRC patients in comparison to adjacent mucosa, we fail to ascribe this change to any specific DNA damage [86]. Most recently, we have addressed BER-DRC in paired samples of tumor tissue and non-malignant adjacent mucosa of 123 incident colon cancer patients concerning 5-fluorouracil (5FU) therapy. Interestingly, BER-DRC in non-malignant adjacent mucosa was positively associated with overall and relapse-free survival. Moreover, the overall survival (OS) of these patients was further improved in patients with a decreased BER-DRC in tumor tissue. The ratio of BER-DRC in tumor tissue and adjacent mucosa positively correlated with advanced tumor stage [79].

2.4.4. Sporadic Colorectal Cancer and Gene Variants in Base Excision Repair

It was postulated that gene variants including single nucleotide polymorphisms (SNPs) in DNA repair genes may alter DNA repair function, including the function of BER glycosylases, modulate its capacity, induce genetic instability or deregulate cell growth and propagate cancer [90–92]. Earlier studies indicated that *hOGG1* Ser326Cys (rs1052133) SNP significantly affected BER-DRC [92]. However, the association of this SNP with CRC risk remains inconclusive: Recently the authors [93] reported on 727 CRC cases and 736 healthy controls from Taiwan significant association between *OGG1* Ser326Cys and *APE1* Asp148Glu (rs1130409) SNPs and an increased CRC risk. The authors concluded that *OGG1* and *APE1* SNPs are associated with stage- and sex-specific risk of CRC. An early study on 532 CRC cases and 532 matched controls [94] found an enhanced risk of CRC in smokers with *hOGG1* Ser326Cys polymorphism. Increased CRC risk was also reported in individuals simultaneously homozygous for the variant alleles of *APE1* Asn148Glu and *hOGG1* Ser326Cys [94]. A meta-analysis comprising 4174 cases and 6196 controls did not reveal any robust association between *hOGG1* Ser326Cys polymorphism and CRC, the authors however recommended further investigation [95]. Another rare nonsynonymous variant in *hOGG1* Gly308Glu (rs113561019) has been discussed in relation to the susceptibility to CRC. In a recent well-powered study, the authors found no evidence for the association of the above *hOGG1* polymorphism with CRC risk [96]. A meta-analysis by Zhang et al. evaluated 12 association studies and concluded that *hOGG1* Ser326Cys polymorphism does not associate with CRC risk [97]. However, its role in gene-gene interactions may not be ruled out.

In the study by Pardini et al. the authors investigated in 1098 sporadic CRC patients for prognostic effects of 3´-untranslated region polymorphisms (representing microRNA binding site) in BER genes. Interestingly, *NEIL2* rs6997097 polymorphism was associated with shorter survival and *NEIL3* rs1055678 polymorphism with CRC recurrence. The altered epigenetic regulation of the specialized glycosylases *NEIL2* and *NEIL3*, involved in the recognition of oxidized pyrimidines and transcription process, may further add to the understanding of the effect of oxidative DNA damage in colorectal carcinogenesis [98].

Single nucleotide polymorphism rs7689099 in the *NEIL3* gene was reported to modulate significantly survival of CRC patients. The NEIL3 encodes a DNA glycosylase involved in the first step of the BER pathway. Significantly elevated expression levels in tumors, compared to corresponding non-malignant tissues, were reported in 20 cancer sites, including CRC [99].

In a meta-analysis (comprising more than 8000 CRC cases and 6000 controls) Picelli et al. revisited the associations of rs3219484:G-A (MUTYH V22M) and rs3219489:G-C (MUTYH Q338H) polymorphisms with the risk of sporadic CRC. The associations with studied polymorphisms were, however, negative for all CRC as well as for colon and rectal cancer separately [100].

2.5. Colorectal Cancer, Oxidative Damage and Intestinal Microenvironment

Intestinal epithelial and immune cells are in permanent contact (interaction) with variable microbial inhabitants; these interactions result in modulations of numerous physiological and pathological processes [101]. Recent discoveries revealed that gut microbiome and CRC are tightly connected and during the disease, the composition and function of microbes can significantly differ [102,103].

Species unambiguously associated with colorectal carcinogenesis are reported in [104]. Intestinal bacteria induce proinflammatory and pro-carcinogenic pathways in colonic epithelium, produce genotoxins and ROS, promote host immune response disturbance and chronic inflammation and mediate the conversion of procarcinogens into carcinogens [105]. There are currently two hypotheses explaining the role of bacteria in colorectal carcinogenesis: (a) "driver-passenger" theory suggesting that certain intestinal bacteria (bacteria drivers) induce epithelial DNA damage and tumorigenesis; (b) dysbiotic microbial community with pro-carcinogenic characteristics via remodeling the whole microbiome initiates pro-inflammatory cascades and subsequent cellular transformation [106,107]. The former is illustrated by phylogenic group B2 of *Escherichia coli* identified as producers of colibactin (pks^+ *E. coli*), a peptide-polyketide, which induces several DNA adducts including those with bulky character, ultimately leading to inter-strand crosslinks or double-strand breaks [108]. The latter relates to a generation of ROS resulting in oxidative DNA damage [109,110]. Additionally, ROS are often produced in high amounts by tumor cells and influence local microenvironment and immune response [109]. The tumor microenvironment is composed of myeloid (innate immunity) and lymphoid (adaptive immunity) lineages. Infiltrating immune cells can function to control tumor growth or to help create an immunosuppressive environment in which the tumor can thrive. Recent understanding points to the fact that carcinogenesis shows many similarities to chronic inflammatory processes [111]. For instance, ulcerative colitis and CRC in relation to ROS were studied in mice deficient for epithelial-to-mesenchymal transition factor ZEB1 and DNA glycosylase MPG. Zeb1-deficient mice were partially protected from experimental colitis and, in a model of inflammatory CRC, they developed fewer tumors and exhibited lower levels of DNA damage (8-oxo-dG) and higher expression of *MPG* encoding DNA-3-methyladenine glycosylase [112]. The dynamic of microenvironment has been documented by different microbiota in relation to histology of adenoma/polyps. More strikingly, normalization of the microbiota has been recorded after colorectal cancer treatment. The intestinal microenvironment is further modulated by different bacterial strains due to the impacted generation of bacterial metabolites and toxins [113].

Oxidative Damage, Intestinal Microenvironment and CRC Prevention

Under physiological condition, a dynamic steady-state between ROS generation, antioxidant status, the formation of DNA damage and capacity for DNA repair, is constantly influenced by diet. However, the mechanisms by which nutritional components influence colorectal carcinogenesis are not yet clear. Since the first line of defense against ROS is the cellular antioxidant system, supplementation of volunteers' diets with antioxidants or antioxidant-rich foods has led in many trials to decreases in the level of endogenous oxidation of DNA bases [12,114] and increased resistance to oxidative damage ex vivo [115]. For example, fruit consumption has been suggested to increase DNA repair capacity and decrease DNA damage, likely due to antioxidants and bioactive compounds in fruits [13]. Soymilk in a form of yogurt also exerts substantial antioxidant potential [116]. Experimental and observational evidence indicate that sub-optimal dietary intakes of selenium may contribute to increased risk for several tumors including CRC, through oxidative and inflammatory response selenoproteins which require selenium for their biosynthesis [14,117]. Moreover, the diet may also significantly modulate BER and NER capacities, for instance, fruit- and vegetable-rich diet stimulates the repair of oxidative DNA damage [9]. It has been documented that diet substantially affects the composition of gut microbiota [118–120]. Various diets not only alter the abundances of several bacterial strains but also change the metabolic profile of whole microbiota, e.g., increase in the biotransformation of pro-carcinogenic polycyclic aromatic hydrocarbons, formed during meat processing [121]. Microbiota respond differently to dietary components: for instance, protein-rich diet correlates with *Bacteroides*, diet rich in fiber correlates with *Prevotella* [122] and consumption of dietary fiber is associated with increased fermentation of indigestible plant polysaccharides [122]. Therefore, certain dietary intervention impacts the gut microbiota and could promote changes that are either harmful or beneficial to health, and thus could influence cancer incidence by limiting the development/relapse of the disease.

3. Possible Utilization of Oxidative DNA Damage in Colorectal Cancer Therapy

The therapy of CRC has comprehensively been reviewed [38,123]. Despite all the efforts in CRC therapy over the years, 5-year survival remains unsatisfactory. The prognosis for CRC patients decreases with increasing TNM staging, the five-years survival rate is up to 90% for stage I, but only less than 15% for stage IV [99]. CRC treatment usually involves complete primary tumor resection and appropriate chemotherapy, which often causes severe adverse effects [124]. There are considerable interindividual differences among CRC patients in the response to the therapy, probably due to inherited genetic susceptibility, acquired resistance of tumor cells and the role of DNA damage/DNA repair in chemotherapy [125,126].

Cancer cells are due to their hypermetabolic activity highly sensitive to the oxidative balance and have a high antioxidant capacity [127]. Therefore, anticancer therapy targeting antioxidant defense of cancer cells or generating ROS represents an interesting approach in CRC treatment strategy. Oxaliplatin in combination with piperlongumine, a natural product constituent of the fruit of the Long pepper (*Piper longum*), has been shown to act synergically and induce CRC cell apoptosis via mitochondrial dysfunction and endoplasmic reticulum stress [128]. The mode of piperlongumine action in experimental colon cancer has recently been updated [129]. Our results showed that co-treatment of CRC cells with 5FU and *Ganoderma Lucidum* induces oxidative DNA damage in CRC cell lines. Moreover, the non-malignant cells were protected against oxidative DNA damage [130]. Co-treatment with paclitaxel and lentinan exerts synergistic apoptotic effects in A549 cells through inducing ROS production [131]. The anticancer effects of natural compounds and their tentative modes of action have recently been reviewed [118].

Efficient DNA repair often confers poor response to chemotherapy and worse prognosis [79]. DNA-alkylating agents used in CRC therapy—such as temozolomide (treatment of metastatic CRC [132])—induce DNA lesions repaired by BER [132]. Suppression of the BER pathway by inhibiting polymerase β activity may also represent a tool for improving therapy response [133].

More interestingly, *MTH1* overexpression has been previously associated with distinct cancer stages and survival of the cancer patients [134,135]. The important role of MTH1 during malignant transformation and an increasing number of articles on this topic resulted in the discovery of potent and selective MTH1 inhibitor [136], currently in phase I clinical testing. Unfortunately, selective inhibition of MTH1 in lung cancer cells showed increased oxidative DNA damage which indicates that MTH1 inhibition will likely not be utilized as an across-the-board therapeutic strategy [137].

4. Discussion

There are currently serious disputes and unresolved questions regarding the etiology of sporadic CRC: is genomic and/or chromosomal instability a cause or consequence of tumorigenesis, is the alteration of microbiota preceding CRC onset or it appears as a consequence of it. Another enigmatic aspect is whether high levels of the oxidative damage found in CRC tumors are associated with early stages of carcinogenesis or rather with its consequences. The outcomes from laboratory animals addressing oxidative bases in carcinogenesis are rather inconclusive, so is the issue of oxidative DNA damage in target/surrogate tissues. The bulk of studies so far reported DNA damage and DNA repair in CRC considered as a single entity. However, recent investigations highlighted the differences in embryogenesis, etiology, anatomy, genetics and treatment response between the colon and rectal cancers with additional impact on prognosis for patients and different treatment strategies [138–140].

Recent studies on hereditary polyposis syndromes (MAP polyposis, NTH1 polyposis; [56,141]) leading to CRC provided unambiguous evidence on the role of oxidative DNA damage and lack of function BER glycosylases in CRC etiopathogenesis. However, some mechanistic aspects have not fully been clarified yet, e.g., *NEIL1*, *NEIL2* and *NEIL3* triple-knockout mice were not prone to cancer and do not have increased mutational frequency produced by defective BER [142]. Further, recent studies disclosed the role of MTH1, which prevents incorporation of oxidized purines into DNA, in malignant transformation [143]. However, high expression of *MTH1* has been observed in many human malignancies [82,83], including CRC, where the enzyme bolster survival of the malignant cells. These seeming discrepancies may be explainable by the versatile role of oxidative DNA damage in carcinogenesis: its higher extent may trigger a malignant transformation in very early stages, whereas in developed tumors, its lower level (or efficient BER) may give additional survival/growth advantage to cancer cells [144]. Studies aimed at the comparison of BER capacity in tumor tissues and adjacent mucosa from sporadic CRC patients did not show major differences in BER between these tissues [86]. However, the recent study by Vodenkova et al. pointed to the importance of the ratio between BER capacity in tumor tissue and adjacent mucosa among CRC patients. Low BER in tumor and higher BER capacity in adjacent mucosa conferred to significantly longer survival and vice versa. Additionally, the ratio of BER capacity in tumor tissue over BER capacity in mucosa correlated positively with the advanced tumor stage [79]. Although relatively well-characterized at present, oxidative DNA damage and its repair warrants further investigation in complex diseases. Regarding sporadic CRC, last years witnessed an advent of additional questions closely related to oxidative DNA damage and its repair. One of those concerns the target and surrogate tissues, in which this damage occurs. Peripheral blood lymphocytes have been taken for long merely as a surrogate in biomonitoring studies. However, these cells represent an important player in the immune system, the last gatekeeper of cancer progression. Further, the generation of oxidative DNA damage is significantly affected by the intestinal microenvironment (microbiota) and this microbiota have been shown to determine immune response. On the other hand, tumor cells produce in high amounts ROS, with subsequent effects on the above systems [109].

Several association studies addressed the role of low penetrance BER gene variants in sporadic CRC with often controversial results. The recent whole genome association studies failed to identify these variants in large CRC patient cohorts [5,145]. Despite it was shown that h*OGG1* 326Cys SNP significantly affected BER DRC, the overall effect is rather minor, and these gene variants may rather find their relevance in interactions [146,147].

In our recent review, we have summarized the role of DNA repair in the treatment of colon cancer [148]. Acquired or inherited defects in DNA repair pathways can be effectively utilized in the therapy; for instance, CRC patients bearing deficiency in *RAD51C* or *CHEK2* genes, belonging to HR pathway, benefited from the treatment with olaparib [149]. This drug inhibits PARP1, a protein implicated in BER, which is often over-expressed in various types of cancers. Tumors with dysfunctional HR may be dependent on PARP1-facilitated DNA repair and are sensitive for its inhibition [150]. However, olaparib-based therapies were tested mainly for breast, ovarian, pancreatic and prostate cancer cases with a mutation in high-penetrance genes BRCA1 and BRCA2 [151]. As evidenced in pre-clinical studies, hLigI may represent an additional attractive target for inhibition in rapidly dividing malignant cells [68]. Despite some promising results, not a single inhibitor is currently applied in clinical practice [152].

Oxidative DNA damage and activities of glycosylases and ligases involved in its repair still await their full application in therapeutic strategies of CRC.

5. Conclusions

The pivotal role of ROS in both health and disease has been recognized. Arising oxidative DNA damage represents an important factor in the etiopathogenesis of CRC and further effort should be dedicated to its monitoring. Similarly, it may also represent a significant marker of prognosis and its level may contribute to treatment outcome.

With increasing knowledge on the role of microenvironment in colorectal carcinogenesis, proper attention should be given to the dynamic of DNA damage formation (and oxidative DNA damage in particular) and its repair.

However, binary roles of ROS and emerging oxidative DNA damage may be utilized in cancer therapy by exploiting combinations of conventional chemotherapeutics with substances leading to oxidative DNA damage in CRC cells. Our recent study suggests that higher capacity of BER in adjacent mucosa and lower in tumor cells accompanied longer survival and good prognosis (and vice versa) of CRC patients. Scarce data are available on the extent of oxidative DNA damage in colorectal tumor tissues and adjacent mucosa.

Author Contributions: Conceptualization P.V., L.V.; writing—review and editing P.V., L.V., V.V.; writing—original draft preparation M.K., K.T., P.V., R.S.; data curation and investigation K.K., M.U., M.S., A.O., P.M., A.S.; funding acquisition P.V.; project administration L.V.; supervision P.V., V.V. All authors have read and agreed to the published version of the manuscript.

Funding: The authors acknowledge the support from the Czech Science Foundation Grants 18-09709S, 19-10543S, Czech Health Research Council 15-27580A and NV18/03/00199. The work was also supported by the Charles University Research Center program [UNCE/MED/006]; by the Charles University Research Fund [Progress Q28/LF1] and by National Sustainability Program I (NPU I) L01503.

Conflicts of Interest: The authors declare no conflict of interest.

Abbreviations

5FU	5-fluorouracil
8-oxo-dG	8-oxo-7,8-dihydro-2′deoxyguanosine
BER	base excision repair
CIMP	CpG island methylator phenotype
CIN	chromosomal instability
CRC	colorectal cancer
DDR	DNA damage response
dNTPs	deoxynucleotide triphosphates
DRC	DNA excision repair capacity
EFS	event-free survival
FAPY	2,6-diamino-4-hydroxy-5-formamidopyrimidine
HR	homologous recombination repair
LigI	human DNA ligase I
LigIII	human DNA ligase III
LOH	loss of heterozygosity
hOOG1	human 8-oxo-dG DNA N-glycosylase 1
MAP	MUTYH-associated polyposis
MMR	mismatch repair
MSI	microsatellite instability
MSS	microsatellite stable
MTH1	human mutT homolog 1
MUTYH, MYH	mutY DNA glycosylase
NATS	NTHL1-associated tumor syndrome
NER	nucleotide excision repair
OS	overall survival
ROS	reactive oxygen species
SNP	single nucleotide polymorphism

References

1. Bray, F.; Ferlay, J.; Soerjomataram, I.; Siegel, R.L.; Torre, L.A.; Jemal, A. Global cancer statistics 2018: GLOBOCAN estimates of incidence and mortality worldwide for 36 cancers in 185 countries. *Cancer J. Clin.* **2018**, *68*, 394–424. [CrossRef] [PubMed]
2. Brenner, H.; Chen, C. The colorectal cancer epidemic: Challenges and opportunities for primary, secondary and tertiary prevention. *Br. J. Cancer* **2018**, *119*, 785–792. [CrossRef] [PubMed]
3. Murphy, N.; Moreno, V.; Hughes, D.J.; Vodicka, L.; Vodicka, P.; Aglago, E.K.; Gunter, M.J.; Jenab, M. Lifestyle and dietary environmental factors in colorectal cancer susceptibility. *Mol. Asp. Med.* **2019**, *69*, 2–9. [CrossRef] [PubMed]
4. Keum, N.; Giovannucci, E. Global burden of colorectal cancer: Emerging trends, risk factors and prevention strategies. *Nat. Rev. Gastroenterol. Hepatol.* **2019**, *16*, 713–732. [CrossRef]

5. Huyghe, J.R.; Bien, S.A.; Harrison, T.A.; Kang, H.M.; Chen, S.; Schmit, S.L.; Conti, D.V.; Qu, C.; Jeon, J.; Edlund, C.K.; et al. Discovery of common and rare genetic risk variants for colorectal cancer. *Nat. Genet.* **2019**, *51*, 76–87. [CrossRef]
6. Medina Pabon, M.A.; Babiker, H.M. A Review Of Hereditary Colorectal Cancers. In *StatPearls*; StatPearls Publishing LLC.: Treasure Island, FL, USA, 2019.
7. Pawlik, T.M.; Raut, C.P.; Rodriguez-Bigas, M.A. Colorectal carcinogenesis: MSI-H versus MSI-L. *Dis. Markers* **2004**, *20*, 199–206. [CrossRef]
8. Yamagishi, H.; Kuroda, H.; Imai, Y.; Hiraishi, H. Molecular pathogenesis of sporadic colorectal cancers. *Chin. J. Cancer* **2016**, *35*, 4. [CrossRef]
9. Collins, A.R.; Azqueta, A.; Langie, S.A.S. Effects of micronutrients on DNA repair. *Eur. J. Nutr.* **2012**, *51*, 261–279. [CrossRef]
10. Kompella, P.; Vasquez, K.M. Obesity and cancer: A mechanistic overview of metabolic changes in obesity that impact genetic instability. *Mol. Carcinog.* **2019**, *58*, 1531–1550. [CrossRef]
11. Murphy, N.; Jenab, M.; Gunter, M.J. Adiposity and gastrointestinal cancers: Epidemiology, mechanisms and future directions. *Nat. Rev. Gastroenterol. Hepatol.* **2018**, *15*, 659–670. [CrossRef]
12. Chakraborty, A.; Ferk, F.; Simić, T.; Brantner, A.; Dusinská, M.; Kundi, M.; Hoelzl, C.; Nersesyan, A.; Knasmüller, S. DNA-protective effects of sumach (*Rhus coriaria* L.), a common spice: Results of human and animal studies. *Mutat. Res.* **2009**, *661*, 10–17. [CrossRef] [PubMed]
13. Slyskova, J.; Lorenzo, Y.; Karlsen, A.; Carlsen, M.H.; Novosadova, V.; Blomhoff, R.; Vodicka, P.; Collins, A.R. Both genetic and dietary factors underlie individual differences in DNA damage levels and DNA repair capacity. *DNA Repair* **2014**, *16*, 66–73. [CrossRef] [PubMed]
14. Méplan, C.; Hesketh, J. Selenium and cancer: A story that should not be forgotten-insights from genomics. *Cancer Treat. Res.* **2014**, *159*, 145–166. [CrossRef]
15. Simonelli, V.; Mazzei, F.; D'Errico, M.; Dogliotti, E. Gene susceptibility to oxidative damage: From single nucleotide polymorphisms to function. *Mutat. Res.* **2012**, *731*, 1–13. [CrossRef]
16. Tudek, B.; Speina, E. Oxidatively damaged DNA and its repair in colon carcinogenesis. *Mutat. Res.* **2012**, *736*, 82–92. [CrossRef] [PubMed]
17. Guo, C.; Ding, P.; Xie, C.; Ye, C.; Ye, M.; Pan, C.; Cao, X.; Zhang, S.; Zheng, S. Potential application of the oxidative nucleic acid damage biomarkers in detection of diseases. *Oncotarget* **2017**, *8*, 75767–75777. [CrossRef] [PubMed]
18. Hanahan, D.; Weinberg, R.A. Hallmarks of cancer: The next generation. *Cell* **2011**, *144*, 646–674. [CrossRef]
19. Pearl, L.H.; Schierz, A.C.; Ward, S.E.; Al-Lazikani, B.; Pearl, F.M. Therapeutic opportunities within the DNA damage response. *Nat. Rev. Cancer* **2015**, *15*, 166–180. [CrossRef]
20. Carethers, J.M.; Jung, B.H. Genetics and Genetic Biomarkers in Sporadic Colorectal Cancer. *Gastroenterology* **2015**, *149*, 1177–1190.e1173. [CrossRef]
21. Grady, W.M.; Markowitz, S.D. The molecular pathogenesis of colorectal cancer and its potential application to colorectal cancer screening. *Dig. Dis. Sci.* **2015**, *60*, 762–772. [CrossRef]
22. Nielsen, F.C.; van Overeem Hansen, T.; Sørensen, C.S. Hereditary breast and ovarian cancer: New genes in confined pathways. *Nat. Rev. Cancer* **2016**, *16*, 599–612. [CrossRef]
23. Hernández, G.; Ramírez, M.J.; Minguillón, J.; Quiles, P.; Ruiz de Garibay, G.; Aza-Carmona, M.; Bogliolo, M.; Pujol, R.; Prados-Carvajal, R.; Fernández, J.; et al. Decapping protein EDC4 regulates DNA repair and phenocopies BRCA1. *Nat. Commun.* **2018**, *9*, 967. [CrossRef] [PubMed]
24. Song, H.; Dicks, E.; Ramus, S.J.; Tyrer, J.P.; Intermaggio, M.P.; Hayward, J.; Edlund, C.K.; Conti, D.; Harrington, P.; Fraser, L.; et al. Contribution of Germline Mutations in the RAD51B, RAD51C, and RAD51D Genes to Ovarian Cancer in the Population. *J. Clin. Oncol.* **2015**, *33*, 2901–2907. [CrossRef] [PubMed]
25. Niskakoski, A.; Pasanen, A.; Lassus, H.; Renkonen-Sinisalo, L.; Kaur, S.; Mecklin, J.-P.; Bützow, R.; Peltomäki, P. Molecular changes preceding endometrial and ovarian cancer: A study of consecutive endometrial specimens from Lynch syndrome surveillance. *Mod. Pathol.* **2018**, *31*, 1291–1301. [CrossRef] [PubMed]
26. Markowitz, S.D.; Bertagnolli, M.M. Molecular origins of cancer: Molecular basis of colorectal cancer. *N. Engl. J. Med.* **2009**, *361*, 2449–2460. [CrossRef] [PubMed]
27. Bardaweel, S.K.; Gul, M.; Alzweiri, M.; Ishaqat, A.; HA, A.L.; Bashatwah, R.M. Reactive Oxygen Species: The Dual Role in Physiological and Pathological Conditions of the Human Body. *Eurasian J. Med.* **2018**, *50*, 193–201. [CrossRef]

28. Schieber, M.; Chandel, N.S. ROS function in redox signaling and oxidative stress. *Curr. Biol.* **2014**, *24*, R453–R462. [CrossRef]
29. Manini, P.; De Palma, G.; Andreoli, R.; Marczynski, B.; Hanova, M.; Mozzoni, P.; Naccarati, A.; Vodickova, L.; Hlavac, P.; Mutti, A.; et al. Biomarkers of nucleic acid oxidation, polymorphism in, and expression of, hOGG1 gene in styrene-exposed workers. *Toxicol. Lett.* **2009**, *190*, 41–47. [CrossRef]
30. Naccarati, A.; Pardini, B.; Hemminki, K.; Vodicka, P. Sporadic colorectal cancer and individual susceptibility: A review of the association studies investigating the role of DNA repair genetic polymorphisms. *Mutat. Res.* **2007**, *635*, 118–145. [CrossRef]
31. Vodicka, P.; Kumar, R.; Stetina, R.; Sanyal, S.; Soucek, P.; Haufroid, V.; Dusinska, M.; Kuricova, M.; Zamecnikova, M.; Musak, L.; et al. Genetic polymorphisms in DNA repair genes and possible links with DNA repair rates, chromosomal aberrations and single-strand breaks in DNA. *Carcinogenesis* **2004**, *25*, 757–763. [CrossRef]
32. Ohno, M.; Sakumi, K.; Fukumura, R.; Furuichi, M.; Iwasaki, Y.; Hokama, M.; Ikemura, T.; Tsuzuki, T.; Gondo, Y.; Nakabeppu, Y. 8-oxoguanine causes spontaneous de novo germline mutations in mice. *Sci. Rep.* **2014**, *4*, 4689. [CrossRef] [PubMed]
33. Dizdaroglu, M. Oxidatively induced DNA damage and its repair in cancer. *Mutat. Res. Rev. Mutat. Res.* **2015**, *763*, 212–245. [CrossRef] [PubMed]
34. Roos, W.P.; Thomas, A.D.; Kaina, B. DNA damage and the balance between survival and death in cancer biology. *Nat. Rev. Cancer* **2016**, *16*, 20–33. [CrossRef] [PubMed]
35. Curtin, N.J. DNA repair dysregulation from cancer driver to therapeutic target. *Nat. Rev. Cancer* **2012**, *12*, 801–817. [CrossRef] [PubMed]
36. Knijnenburg, T.A.; Wang, L.; Zimmermann, M.T.; Chambwe, N.; Gao, G.F.; Cherniack, A.D.; Fan, H.; Shen, H.; Way, G.P.; Greene, C.S.; et al. Genomic and Molecular Landscape of DNA Damage Repair Deficiency across The Cancer Genome Atlas. *Cell Rep.* **2018**, *23*, 239–254.e6. [CrossRef] [PubMed]
37. Wallace, S.S. DNA glycosylases search for and remove oxidized DNA bases. *Environ. Mol. Mutagenesis* **2013**, *54*, 691–704. [CrossRef]
38. Vodicka, P.; Vodenkova, S.; Opattova, A.; Vodickova, L. DNA damage and repair measured by comet assay in cancer patients. *Mutat. Res.* **2019**, *843*, 95–110. [CrossRef]
39. Radom, M.; Machnicka, M.A.; Krwawicz, J.; Bujnicki, J.M.; Formanowicz, P. Petri net-based model of the human DNA base excision repair pathway. *PLoS ONE* **2019**, *14*, e0217913. [CrossRef]
40. Köger, N.; Brieger, A.; Hinrichsen, I.M.; Zeuzem, S.; Plotz, G. Analysis of MUTYH alternative transcript expression, promoter function, and the effect of human genetic variants. *Hum. Mutat.* **2019**, *40*, 472–482. [CrossRef]
41. Wang, R.; Hao, W.; Pan, L.; Boldogh, I.; Ba, X. The roles of base excision repair enzyme OGG1 in gene expression. *Cell. Mol. Life Sci.* **2018**, *75*, 3741–3750. [CrossRef]
42. Leitner-Dagan, Y.; Sevilya, Z.; Pinchev, M.; Kramer, R.; Elinger, D.; Roisman, L.C.; Rennert, H.S.; Schechtman, E.; Freedman, L.; Rennert, G.; et al. N-methylpurine DNA glycosylase and OGG1 DNA repair activities: Opposite associations with lung cancer risk. *J. Natl. Cancer Inst.* **2012**, *104*, 1765–1769. [CrossRef] [PubMed]
43. Sjolund, A.B.; Senejani, A.G.; Sweasy, J.B. MBD4 and TDG: Multifaceted DNA glycosylases with ever expanding biological roles. *Mutat. Res.* **2013**, *743–744*, 12–25. [CrossRef] [PubMed]
44. Nagaria, P.; Svilar, D.; Brown, A.R.; Wang, X.-h.; Sobol, R.W.; Wyatt, M.D. SMUG1 but not UNG DNA glycosylase contributes to the cellular response to recovery from 5-fluorouracil induced replication stress. *Mutat. Res.* **2013**, *743–744*, 26–32. [CrossRef] [PubMed]
45. Alexeeva, M.; Moen, M.N.; Grøsvik, K.; Tesfahun, A.N.; Xu, X.M.; Muruzábal-Lecumberri, I.; Olsen, K.M.; Rasmussen, A.; Ruoff, P.; Kirpekar, F.; et al. Excision of uracil from DNA by hSMUG1 includes strand incision and processing. *Nucleic Acids Res.* **2019**, *47*, 779–793. [CrossRef] [PubMed]
46. Shinmura, K.; Kato, H.; Kawanishi, Y.; Goto, M.; Tao, H.; Yoshimura, K.; Nakamura, S.; Misawa, K.; Sugimura, H. Defective repair capacity of variant proteins of the DNA glycosylase NTHL1 for 5-hydroxyuracil, an oxidation product of cytosine. *Free Radic. Biol. Med.* **2019**, *131*, 264–273. [CrossRef] [PubMed]
47. Slyvka, A.; Mierzejewska, K.; Bochtler, M. Nei-like 1 (NEIL1) excises 5-carboxylcytosine directly and stimulates TDG-mediated 5-formyl and 5-carboxylcytosine excision. *Sci. Rep.* **2017**, *7*, 9001. [CrossRef]

48. Sarker, A.H.; Chatterjee, A.; Williams, M.; Lin, S.; Havel, C.; Jacob, P., 3rd; Boldogh, I.; Hazra, T.K.; Talbot, P.; Hang, B. NEIL2 protects against oxidative DNA damage induced by sidestream smoke in human cells. *PLoS ONE* **2014**, *9*, e90261. [CrossRef]
49. Han, D.; Schomacher, L.; Schüle, K.M.; Mallick, M.; Musheev, M.U.; Karaulanov, E.; Krebs, L.; von Seggern, A.; Niehrs, C. NEIL1 and NEIL2 DNA glycosylases protect neural crest development against mitochondrial oxidative stress. *Elife* **2019**, *8*, e49044. [CrossRef]
50. Minko, I.G.; Christov, P.P.; Li, L.; Stone, M.P.; McCullough, A.K.; Lloyd, R.S. Processing of N(5)-substituted formamidopyrimidine DNA adducts by DNA glycosylases NEIL1 and NEIL3. *DNA Repair* **2019**, *73*, 49–54. [CrossRef]
51. Massaad, M.J.; Zhou, J.; Tsuchimoto, D.; Chou, J.; Jabara, H.; Janssen, E.; Glauzy, S.; Olson, B.G.; Morbach, H.; Ohsumi, T.K.; et al. Deficiency of base excision repair enzyme NEIL3 drives increased predisposition to autoimmunity. *J. Clin. Investig.* **2016**, *126*, 4219–4236. [CrossRef]
52. Da, L.-T.; Shi, Y.; Ning, G.; Yu, J. Dynamics of the excised base release in thymine DNA glycosylase during DNA repair process. *Nucleic Acids Res.* **2018**, *46*, 568–581. [CrossRef] [PubMed]
53. Fu, T.; Liu, L.; Yang, Q.-L.; Wang, Y.; Xu, P.; Zhang, L.; Liu, S.; Dai, Q.; Ji, Q.; Xu, G.-L.; et al. Thymine DNA glycosylase recognizes the geometry alteration of minor grooves induced by 5-formylcytosine and 5-carboxylcytosine. *Chem. Sci.* **2019**, *10*, 7407–7417. [CrossRef]
54. Weiser, B.P.; Rodriguez, G.; Cole, P.A.; Stivers, J.T. N-terminal domain of human uracil DNA glycosylase (hUNG2) promotes targeting to uracil sites adjacent to ssDNA-dsDNA junctions. *Nucleic Acids Res.* **2018**, *46*, 7169–7178. [CrossRef] [PubMed]
55. Bai, H.; Grist, S.; Gardner, J.; Suthers, G.; Wilson, T.M.; Lu, A.L. Functional characterization of human MutY homolog (hMYH) missense mutation (R231L) that is linked with hMYH-associated polyposis. *Cancer Lett.* **2007**, *250*, 74–81. [CrossRef] [PubMed]
56. Schubert, S.A.; Morreau, H.; de Miranda, N.F.C.C.; van Wezel, T. The missing heritability of familial colorectal cancer. *Mutagenesis* **2019**. [CrossRef] [PubMed]
57. Krokan, H.E.; Bjørås, M. Base excision repair. *Cold Spring Harb. Perspect. Biol.* **2013**, *5*, a012583. [CrossRef]
58. Hazra, T.K.; Izumi, T.; Kow, Y.W.; Mitra, S. The discovery of a new family of mammalian enzymes for repair of oxidatively damaged DNA, and its physiological implications. *Carcinogenesis* **2003**, *24*, 155–157. [CrossRef]
59. Sakumi, K.; Furuichi, M.; Tsuzuki, T.; Kakuma, T.; Kawabata, S.; Maki, H.; Sekiguchi, M. Cloning and expression of cDNA for a human enzyme that hydrolyzes 8-oxo-dGTP, a mutagenic substrate for DNA synthesis. *J. Biol. Chem.* **1993**, *268*, 23524–23530.
60. El-Khamisy, S.F.; Masutani, M.; Suzuki, H.; Caldecott, K.W. A requirement for PARP-1 for the assembly or stability of XRCC1 nuclear foci at sites of oxidative DNA damage. *Nucleic Acids Res.* **2003**, *31*, 5526–5533. [CrossRef]
61. Takao, M.; Kanno, S.-I.; Kobayashi, K.; Zhang, Q.-M.; Yonei, S.; van der Horst, G.T.J.; Yasui, A. A back-up glycosylase in Nth1 knock-out mice is a functional Nei (endonuclease VIII) homologue. *J. Biol. Chem.* **2002**, *277*, 42205–42213. [CrossRef]
62. Galick, H.A.; Kathe, S.; Liu, M.; Robey-Bond, S.; Kidane, D.; Wallace, S.S.; Sweasy, J.B. Germ-line variant of human NTH1 DNA glycosylase induces genomic instability and cellular transformation. *Proc. Natl. Acad. Sci. USA* **2013**, *110*, 14314–14319. [CrossRef]
63. Gad, H.; Koolmeister, T.; Jemth, A.-S.; Eshtad, S.; Jacques, S.A.; Ström, C.E.; Svensson, L.M.; Schultz, N.; Lundbäck, T.; Einarsdottir, B.O.; et al. MTH1 inhibition eradicates cancer by preventing sanitation of the dNTP pool. *Nature* **2014**, *508*, 215–221. [CrossRef]
64. Rai, P.; Sobol, R.W. Mechanisms of MTH1 inhibition-induced DNA strand breaks: The slippery slope from the oxidized nucleotide pool to genotoxic damage. *DNA Repair* **2019**, *77*, 18–26. [CrossRef] [PubMed]
65. Ahmed, W.; Lingner, J. PRDX1 and MTH1 cooperate to prevent ROS-mediated inhibition of telomerase. *Genes Dev.* **2018**, *32*, 658–669. [CrossRef] [PubMed]
66. Ellenberger, T.; Tomkinson, A.E. Eukaryotic DNA ligases: Structural and functional insights. *Annu. Rev. Biochem.* **2008**, *77*, 313–338. [CrossRef] [PubMed]
67. Sun, D.; Urrabaz, R.; Nguyen, M.; Marty, J.; Stringer, S.; Cruz, E.; Medina-Gundrum, L.; Weitman, S. Elevated expression of DNA ligase I in human cancers. *Clin. Cancer Res.* **2001**, *7*, 4143–4148. [PubMed]

68. Saquib, M.; Ansari, M.I.; Johnson, C.R.; Khatoon, S.; Kamil Hussain, M.; Coop, A. Recent advances in the targeting of human DNA ligase I as a potential new strategy for cancer treatment. *Eur. J. Med. Chem.* **2019**, *182*, 111657. [CrossRef]
69. McNally, J.R.; O'Brien, P.J. Kinetic analyses of single-stranded break repair by human DNA ligase III isoforms reveal biochemical differences from DNA ligase I. *J. Biol. Chem.* **2017**, *292*, 15870–15879. [CrossRef]
70. Adachi, S.; Takemoto, K.; Hirosue, T.; Hosogai, Y. Spontaneous and 2-nitropropane induced levels of 8-hydroxy-2′-deoxyguanosine in liver DNA of rats fed iron-deficient or manganese- and copper-deficient diets. *Carcinogenesis* **1993**, *14*, 265–268. [CrossRef]
71. Kasai, H.; Crain, P.F.; Kuchino, Y.; Nishimura, S.; Ootsuyama, A.; Tanooka, H. Formation of 8-hydroxyguanine moiety in cellular DNA by agents producing oxygen radicals and evidence for its repair. *Carcinogenesis* **1986**, *7*, 1849–1851. [CrossRef]
72. Wood, M.L.; Dizdaroglu, M.; Gajewski, E.; Essigmann, J.M. Mechanistic studies of ionizing radiation and oxidative mutagenesis: Genetic effects of a single 8-hydroxyguanine (7-hydro-8-oxoguanine) residue inserted at a unique site in a viral genome. *Biochemistry* **1990**, *29*, 7024–7032. [CrossRef] [PubMed]
73. Olinski, R.; Gackowski, D.; Foksinski, M.; Rozalski, R.; Roszkowski, K.; Jaruga, P. Oxidative DNA damage: Assessment of the role in carcinogenesis, atherosclerosis, and acquired immunodeficiency syndrome. *Free Radic. Biol. Med.* **2002**, *33*, 192–200. [CrossRef]
74. Robertson, A.B.; Klungland, A.; Rognes, T.; Leiros, I. DNA repair in mammalian cells: Base excision repair: The long and short of it. *Cell Mol. Life Sci.* **2009**, *66*, 981–993. [CrossRef] [PubMed]
75. Russo, M.T.; De Luca, G.; Degan, P.; Parlanti, E.; Dogliotti, E.; Barnes, D.E.; Lindahl, T.; Yang, H.; Miller, J.H.; Bignami, M. Accumulation of the oxidative base lesion 8-hydroxyguanine in DNA of tumor-prone mice defective in both the Myh and Ogg1 DNA glycosylases. *Cancer Res.* **2004**, *64*, 4411–4414. [CrossRef] [PubMed]
76. Pilati, C.; Shinde, J.; Alexandrov, L.B.; Assié, G.; André, T.; Hélias-Rodzewicz, Z.; Ducoudray, R.; Le Corre, D.; Zucman-Rossi, J.; Emile, J.-F.; et al. Mutational signature analysis identifies MUTYH deficiency in colorectal cancers and adrenocortical carcinomas. *J. Pathol.* **2017**, *242*, 10–15. [CrossRef] [PubMed]
77. Weren, R.D.A.; Ligtenberg, M.J.; Geurts van Kessel, A.; De Voer, R.M.; Hoogerbrugge, N.; Kuiper, R.P. NTHL1 and MUTYH polyposis syndromes: Two sides of the same coin? *J. Pathol.* **2018**, *244*, 135–142. [CrossRef]
78. Nascimento, E.F.R.; Ribeiro, M.L.; Magro, D.O.; Carvalho, J.; Kanno, D.T.; Martinez, C.A.R.; Coy, C.S.R. tissue expresion of the genes mutyh and ogg1 in patients with sporadic colorectal cancer. *Arq. Bras. Cir. Dig.* **2017**, *30*, 98–102. [CrossRef]
79. Vodenkova, S.; Jiraskova, K.; Urbanova, M.; Kroupa, M.; Slyskova, J.; Schneiderova, M.; Levy, M.; Buchler, T.; Liska, V.; Vodickova, L.; et al. Base excision repair capacity as a determinant of prognosis and therapy response in colon cancer patients. *DNA Repair* **2018**, *72*, 77–85. [CrossRef]
80. Koketsu, S.; Watanabe, T.; Nagawa, H. Expression of DNA repair protein: MYH, NTH1, and MTH1 in colorectal cancer. *Hepatogastroenterology* **2004**, *51*, 638–642.
81. Zhang, X.; Song, W.; Zhou, Y.; Mao, F.; Lin, Y.; Guan, J.; Sun, Q. Expression and function of MutT homolog 1 in distinct subtypes of breast cancer. *Oncol. Lett.* **2017**, *13*, 2161–2168. [CrossRef]
82. Furlan, D.; Trapani, D.; Berrino, E.; Debernardi, C.; Panero, M.; Libera, L.; Sahnane, N.; Riva, C.; Tibiletti, M.G.; Sessa, F.; et al. Oxidative DNA damage induces hypomethylation in a compromised base excision repair colorectal tumourigenesis. *Br. J. Cancer* **2017**, *116*, 793–801. [CrossRef]
83. Farkas, S.A.; Vymetalkova, V.; Vodickova, L.; Vodicka, P.; Nilsson, T.K. DNA methylation changes in genes frequently mutated in sporadic colorectal cancer and in the DNA repair and Wnt/beta-catenin signaling pathway genes. *Epigenomics* **2014**, *6*, 179–191. [CrossRef] [PubMed]
84. Cooke, M.S.; Evans, M.D.; Dizdaroglu, M.; Lunec, J. Oxidative DNA damage: Mechanisms, mutation, and disease. *FASEB J.* **2003**, *17*, 1195–1214. [CrossRef] [PubMed]
85. Przybylowska, K.; Kabzinski, J.; Sygut, A.; Dziki, L.; Dziki, A.; Majsterek, I. An association selected polymorphisms of XRCC1, OGG1 and MUTYH gene and the level of efficiency oxidative DNA damage repair with a risk of colorectal cancer. *Mutat. Res.* **2013**, *745–746*, 6–15. [CrossRef] [PubMed]
86. Slyskova, J.; Korenkova, V.; Collins, A.R.; Prochazka, P.; Vodickova, L.; Svec, J.; Lipska, L.; Levy, M.; Schneiderova, M.; Liska, V.; et al. Functional, genetic, and epigenetic aspects of base and nucleotide excision repair in colorectal carcinomas. *Clin. Cancer Res.* **2012**, *18*, 5878–5887. [CrossRef] [PubMed]

87. Lerner, L.K.; Moreno, N.C.; Rocha, C.R.R.; Munford, V.; Santos, V.; Soltys, D.T.; Garcia, C.C.M.; Sarasin, A.; Menck, C.F.M. XPD/ERCC2 mutations interfere in cellular responses to oxidative stress. *Mutagenesis* **2019**, *34*, 341–354. [CrossRef]
88. Melis, J.P.M.; van Steeg, H.; Luijten, M. Oxidative DNA damage and nucleotide excision repair. *Antioxid. Redox Signal.* **2013**, *18*, 2409–2419. [CrossRef]
89. Lee, T.-H.; Kang, T.-H. DNA Oxidation and Excision Repair Pathways. *Int. J. Mol. Sci.* **2019**, *20*, 6092. [CrossRef]
90. He, J.; Shi, T.-Y.; Zhu, M.-L.; Wang, M.-Y.; Li, Q.-X.; Wei, Q.-Y. Associations of Lys939Gln and Ala499Val polymorphisms of the XPC gene with cancer susceptibility: A meta-analysis. *Int. J. Cancer* **2013**, *133*, 1765–1775. [CrossRef]
91. Slyskova, J.; Naccarati, A.; Pardini, B.; Polakova, V.; Vodickova, L.; Smerhovsky, Z.; Levy, M.; Lipska, L.; Liska, V.; Vodicka, P. Differences in nucleotide excision repair capacity between newly diagnosed colorectal cancer patients and healthy controls. *Mutagenesis* **2012**, *27*, 225–232. [CrossRef]
92. Vodicka, P.; Stetina, R.; Polakova, V.; Tulupova, E.; Naccarati, A.; Vodickova, L.; Kumar, R.; Hanova, M.; Pardini, B.; Slyskova, J.; et al. Association of DNA repair polymorphisms with DNA repair functional outcomes in healthy human subjects. *Carcinogenesis* **2007**, *28*, 657–664. [CrossRef] [PubMed]
93. Lai, C.-Y.; Hsieh, L.-L.; Tang, R.; Santella, R.M.; Chang-Chieh, C.R.; Yeh, C.-C. Association between polymorphisms of APE1 and OGG1 and risk of colorectal cancer in Taiwan. *World J. Gastroenterol.* **2016**, *22*, 3372–3380. [CrossRef]
94. Pardini, B.; Naccarati, A.; Novotny, J.; Smerhovsky, Z.; Vodickova, L.; Polakova, V.; Hanova, M.; Slyskova, J.; Tulupova, E.; Kumar, R.; et al. DNA repair genetic polymorphisms and risk of colorectal cancer in the Czech Republic. *Mutat. Res.* **2008**, *638*, 146–153. [CrossRef] [PubMed]
95. Guo, C.-L.; Han, F.-F.; Wang, H.-Y.; Wang, L. Meta-analysis of the association between hOGG1 Ser326Cys polymorphism and risk of colorectal cancer based on case–control studies. *J. Cancer Res. Clin. Oncol.* **2012**, *138*, 1443–1448. [CrossRef] [PubMed]
96. Kinnersley, B.; Buch, S.; Castellví-Bel, S.; Farrington, S.M.; Forsti, A.; Hampe, J.; Hemminki, K.; Hofstra, R.M.W.; Northwood, E.; Palles, C.; et al. Re: Role of the oxidative DNA damage repair gene OGG1 in colorectal tumorigenesis. *J. Natl. Cancer Inst.* **2014**, *106*, dju086. [CrossRef] [PubMed]
97. Zhang, Y.; He, B.-S.; Pan, Y.-Q.; Xu, Y.-Q.; Wang, S.-K. Association of OGG1 Ser326Cys polymorphism with colorectal cancer risk: A meta-analysis. *Int. J. Colorectal Dis.* **2011**, *26*, 1525–1530. [CrossRef] [PubMed]
98. Pardini, B.; Rosa, F.; Barone, E.; Di Gaetano, C.; Slyskova, J.; Novotny, J.; Levy, M.; Garritano, S.; Vodickova, L.; Buchler, T.; et al. Variation within 3′-UTRs of base excision repair genes and response to therapy in colorectal cancer patients: A potential modulation of microRNAs binding. *Clin. Cancer Res.* **2013**, *19*, 6044–6056. [CrossRef] [PubMed]
99. Jiraskova, K.; Hughes, D.J.; Brezina, S.; Gumpenberger, T.; Veskrnova, V.; Buchler, T.; Schneiderova, M.; Levy, M.; Liska, V.; Vodenkova, S.; et al. Functional Polymorphisms in DNA Repair Genes Are Associated with Sporadic Colorectal Cancer Susceptibility and Clinical Outcome. *Int. J. Mol. Sci.* **2018**, *20*, 97. [CrossRef] [PubMed]
100. Picelli, S.; Lorenzo Bermejo, J.; Chang-Claude, J.; Hoffmeister, M.; Fernández-Rozadilla, C.; Carracedo, A.; Castells, A.; Castellví-Bel, S.; The EPICOLON Consortium Members of the EPICOLON Consortium; Naccarati, A.; et al. Meta-analysis of mismatch repair polymorphisms within the cogent consortium for colorectal cancer susceptibility. *PLoS ONE* **2013**, *8*, e72091. [CrossRef] [PubMed]
101. Tlaskalová-Hogenová, H.; Stěpánková, R.; Kozáková, H.; Hudcovic, T.; Vannucci, L.; Tučková, L.; Rossmann, P.; Hrnčíř, T.; Kverka, M.; Zákostelská, Z.; et al. The role of gut microbiota (commensal bacteria) and the mucosal barrier in the pathogenesis of inflammatory and autoimmune diseases and cancer: Contribution of germ-free and gnotobiotic animal models of human diseases. *Cell. Mol. Immunol.* **2011**, *8*, 110–120. [CrossRef]
102. Bajer, L.; Kverka, M.; Kostovcik, M.; Macinga, P.; Dvorak, J.; Stehlikova, Z.; Brezina, J.; Wohl, P.; Spicak, J.; Drastich, P. Distinct gut microbiota profiles in patients with primary sclerosing cholangitis and ulcerative colitis. *World J. Gastroenterol.* **2017**, *23*, 4548–4558. [CrossRef] [PubMed]
103. Flemer, B.; Lynch, D.B.; Brown, J.M.R.; Jeffery, I.B.; Ryan, F.J.; Claesson, M.J.; O'Riordain, M.; Shanahan, F.; O'Toole, P.W. Tumour-associated and non-tumour-associated microbiota in colorectal cancer. *Gut* **2017**, *66*, 633–643. [CrossRef] [PubMed]

104. Park, H.E.; Kim, J.H.; Cho, N.-Y.; Lee, H.S.; Kang, G.H. Intratumoral Fusobacterium nucleatum abundance correlates with macrophage infiltration and CDKN2A methylation in microsatellite-unstable colorectal carcinoma. *Virchows Arch.* **2017**, *471*, 329–336. [CrossRef] [PubMed]
105. Arthur, J.C.; Perez-Chanona, E.; Mühlbauer, M.; Tomkovich, S.; Uronis, J.M.; Fan, T.-J.; Campbell, B.J.; Abujamel, T.; Dogan, B.; Rogers, A.B.; et al. Intestinal inflammation targets cancer-inducing activity of the microbiota. *Science* **2012**, *338*, 120–123. [CrossRef] [PubMed]
106. Klimesova, K.; Jiraskova Zakostelska, Z.; Tlaskalova-Hogenova, H. Oral Bacterial and Fungal Microbiome Impacts Colorectal Carcinogenesis. *Front. Microbiol.* **2018**, *9*, 774. [CrossRef] [PubMed]
107. Gagnière, J.; Raisch, J.; Veziant, J.; Barnich, N.; Bonnet, R.; Buc, E.; Bringer, M.-A.; Pezet, D.; Bonnet, M. Gut microbiota imbalance and colorectal cancer. *World J. Gastroenterol.* **2016**, *22*, 501–518. [CrossRef]
108. Wilson, M.R.; Jiang, Y.; Villalta, P.W.; Stornetta, A.; Boudreau, P.D.; Carrá, A.; Brennan, C.A.; Chun, E.; Ngo, L.; Samson, L.D.; et al. The human gut bacterial genotoxin colibactin alkylates DNA. *Science* **2019**, *363*, eaar7785. [CrossRef]
109. Panieri, E.; Santoro, M.M. ROS homeostasis and metabolism: A dangerous liason in cancer cells. *Cell Death Dis.* **2016**, *7*, e2253. [CrossRef]
110. Hsu, C.W.; Sowers, M.L.; Hsu, W.; Eyzaguirre, E.; Qiu, S.; Chao, C.; Mouton, C.P.; Fofanov, Y.; Singh, P.; Sowers, L.C. How does inflammation drive mutagenesis in colorectal cancer? *Trends Cancer Res.* **2017**, *12*, 111–132.
111. Shalapour, S.; Karin, M. Pas de Deux: Control of Anti-tumor Immunity by Cancer-Associated Inflammation. *Immunity* **2019**, *51*, 15–26. [CrossRef]
112. De Barrios, O.; Sanchez-Moral, L.; Cortés, M.; Ninfali, C.; Profitós-Pelejà, N.; Martínez-Campanario, M.C.; Siles, L.; Del Campo, R.; Fernández-Aceñero, M.J.; Darling, D.S.; et al. ZEB1 promotes inflammation and progression towards inflammation-driven carcinoma through repression of the DNA repair glycosylase MPG in epithelial cells. *Gut* **2019**, *68*, 2129–2141. [CrossRef] [PubMed]
113. Weng, M.T.; Chiu, Y.T.; Wei, P.Y.; Chiang, C.W.; Fang, H.L.; Wei, S.C. Microbiota and gastrointestinal cancer. *J. Formos. Med. Assoc.* **2019**, *118* (Suppl. 1), S32–S41. [CrossRef]
114. Møller, P.; Loft, S. Dietary antioxidants and beneficial effect on oxidatively damaged DNA. *Free Radic. Biol. Med.* **2006**, *41*, 388–415. [CrossRef] [PubMed]
115. Hoelzl, C.; Knasmüller, S.; Misík, M.; Collins, A.; Dusinská, M.; Nersesyan, A. Use of single cell gel electrophoresis assays for the detection of DNA-protective effects of dietary factors in humans: Recent results and trends. *Mutat. Res.* **2009**, *681*, 68–79. [CrossRef] [PubMed]
116. Yamamoto, N.; Shoji, M.; Hoshigami, H.; Watanabe, K.; Watanabe, K.; Takatsuzu, T.; Yasuda, S.; Igoshi, K.; Kinoshita, H. Antioxidant capacity of soymilk yogurt and exopolysaccharides produced by lactic acid bacteria. *Biosci. Microbiota Food Health* **2019**, *38*, 97–104. [CrossRef]
117. Rejhová, A.; Opattová, A.; Čumová, A.; Slíva, D.; Vodička, P. Natural compounds and combination therapy in colorectal cancer treatment. *Eur. J. Med. Chem.* **2018**, *144*, 582–594. [CrossRef]
118. Claesson, M.J.; Jeffery, I.B.; Conde, S.; Power, S.E.; O'Connor, E.M.; Cusack, S.; Harris, H.M.B.; Coakley, M.; Lakshminarayanan, B.; O'Sullivan, O.; et al. Gut microbiota composition correlates with diet and health in the elderly. *Nature* **2012**, *488*, 178–184. [CrossRef]
119. Zoetendal, E.G.; de Vos, W.M. Effect of diet on the intestinal microbiota and its activity. *Curr. Opin. Gastroenterol.* **2014**, *30*, 189–195. [CrossRef]
120. Kau, A.L.; Ahern, P.P.; Griffin, N.W.; Goodman, A.L.; Gordon, J.I. Human nutrition, the gut microbiome and the immune system. *Nature* **2011**, *474*, 327–336. [CrossRef]
121. David, L.A.; Maurice, C.F.; Carmody, R.N.; Gootenberg, D.B.; Button, J.E.; Wolfe, B.E.; Ling, A.V.; Devlin, A.S.; Varma, Y.; Fischbach, M.A.; et al. Diet rapidly and reproducibly alters the human gut microbiome. *Nature* **2014**, *505*, 559–563. [CrossRef]
122. De Filippo, C.; Cavalieri, D.; Di Paola, M.; Ramazzotti, M.; Poullet, J.B.; Massart, S.; Collini, S.; Pieraccini, G.; Lionetti, P. Impact of diet in shaping gut microbiota revealed by a comparative study in children from Europe and rural Africa. *Proc. Natl. Acad. Sci. USA* **2010**, *107*, 14691–14696. [CrossRef] [PubMed]
123. Vodenkova, S.; Buchler, T.; Cervena, K.; Veskrnova, V.; Vodicka, P.; Vymetalkova, V. 5-fluorouracil and other fluoropyrimidines in colorectal cancer: Past, present and future. *Pharmacol. Ther.* **2019**. [CrossRef] [PubMed]

124. Gustavsson, B.; Carlsson, G.; Machover, D.; Petrelli, N.; Roth, A.; Schmoll, H.-J.; Tveit, K.-M.; Gibson, F. A review of the evolution of systemic chemotherapy in the management of colorectal cancer. *Clin. Colorectal. Cancer* **2015**, *14*, 1–10. [CrossRef] [PubMed]
125. Slyskova, J.; Cordero, F.; Pardini, B.; Korenkova, V.; Vymetalkova, V.; Bielik, L.; Vodickova, L.; Pitule, P.; Liska, V.; Matejka, V.M.; et al. Post-treatment recovery of suboptimal DNA repair capacity and gene expression levels in colorectal cancer patients. *Mol. Carcinog.* **2015**, *54*, 769–778. [CrossRef]
126. Wyatt, M.D.; Wilson, D.M., 3rd. Participation of DNA repair in the response to 5-fluorouracil. *Cell. Mol. Life Sci.* **2009**, *66*, 788–799. [CrossRef] [PubMed]
127. Gorrini, C.; Harris, I.S.; Mak, T.W. Modulation of oxidative stress as an anticancer strategy. *Nat. Rev. Drug Discov.* **2013**, *12*, 931–947. [CrossRef]
128. Chen, W.; Lian, W.; Yuan, Y.; Li, M. The synergistic effects of oxaliplatin and piperlongumine on colorectal cancer are mediated by oxidative stress. *Cell Death Dis.* **2019**, *10*, 600. [CrossRef]
129. Kumar, S.; Agnihotri, N. Piperlongumine, a piper alkaloid targets Ras/PI3K/Akt/mTOR signaling axis to inhibit tumor cell growth and proliferation in DMH/DSS induced experimental colon cancer. *Biomed. Pharmacother.* **2019**, *109*, 1462–1477. [CrossRef]
130. Opattova, A.; Horak, J.; Vodenkova, S.; Kostovcikova, K.; Cumova, A.; Macinga, P.; Galanova, N.; Rejhova, A.; Vodickova, L.; Kozics, K.; et al. Ganoderma Lucidum induces oxidative DNA damage and enhances the effect of 5-Fluorouracil in colorectal cancer in vitro and in vivo. *Mutat. Res.* **2019**, *845*, 403065. [CrossRef]
131. Liu, W.; Gu, J.; Qi, J.; Zeng, X.-N.; Ji, J.; Chen, Z.-Z.; Sun, X.-L. Lentinan exerts synergistic apoptotic effects with paclitaxel in A549 cells via activating ROS-TXNIP-NLRP3 inflammasome. *J. Cell Mol. Med.* **2015**, *19*, 1949–1955. [CrossRef]
132. Morano, F.; Corallo, S.; Niger, M.; Barault, L.; Milione, M.; Berenato, R.; Moretto, R.; Randon, G.; Antista, M.; Belfiore, A.; et al. Temozolomide and irinotecan (TEMIRI regimen) as salvage treatment of irinotecan-sensitive advanced colorectal cancer patients bearing MGMT methylation. *Ann. Oncol.* **2018**, *29*, 1800–1806. [CrossRef] [PubMed]
133. Jaiswal, A.S.; Banerjee, S.; Aneja, R.; Sarkar, F.H.; Ostrov, D.A.; Narayan, S. DNA polymerase β as a novel target for chemotherapeutic intervention of colorectal cancer. *PLoS ONE* **2011**, *6*, e16691. [CrossRef] [PubMed]
134. Fujishita, T.; Okamoto, T.; Akamine, T.; Takamori, S.; Takada, K.; Katsura, M.; Toyokawa, G.; Shoji, F.; Shimokawa, M.; Oda, Y.; et al. Association of MTH1 expression with the tumor malignant potential and poor prognosis in patients with resected lung cancer. *Lung Cancer* **2017**, *109*, 52–57. [CrossRef] [PubMed]
135. Akiyama, S.; Saeki, H.; Nakashima, Y.; Iimori, M.; Kitao, H.; Oki, E.; Oda, Y.; Nakabeppu, Y.; Kakeji, Y.; Maehara, Y. Prognostic impact of MutT homolog-1 expression on esophageal squamous cell carcinoma. *Cancer Med.* **2017**, *6*, 258–266. [CrossRef]
136. Zhou, W.; Ma, L.; Yang, J.; Qiao, H.; Li, L.; Guo, Q.; Ma, J.; Zhao, L.; Wang, J.; Jiang, G.; et al. Potent and specific MTH1 inhibitors targeting gastric cancer. *Cell Death Dis.* **2019**, *10*, 434. [CrossRef]
137. Abbas, H.H.K.; Alhamoudi, K.M.H.; Evans, M.D.; Jones, G.D.D.; Foster, S.S. MTH1 deficiency selectively increases non-cytotoxic oxidative DNA damage in lung cancer cells: More bad news than good? *BMC Cancer* **2018**, *18*, 423. [CrossRef]
138. Frattini, M.; Balestra, D.; Suardi, S.; Oggionni, M.; Alberici, P.; Radice, P.; Costa, A.; Daidone, M.G.; Leo, E.; Pilotti, S.; et al. Different genetic features associated with colon and rectal carcinogenesis. *Clin. Cancer Res.* **2004**, *10*, 4015–4021. [CrossRef]
139. Lee, G.H.; Malietzis, G.; Askari, A.; Bernardo, D.; Al-Hassi, H.O.; Clark, S.K. Is right-sided colon cancer different to left-sided colorectal cancer?-a systematic review. *Eur. J. Surg Oncol.* **2015**, *41*, 300–308. [CrossRef]
140. Punt, C.J.A.; Koopman, M.; Vermeulen, L. From tumour heterogeneity to advances in precision treatment of colorectal cancer. *Nat. Rev. Clin. Oncol.* **2017**, *14*, 235–246. [CrossRef]
141. Kuiper, R.P.; Hoogerbrugge, N. NTHL1 defines novel cancer syndrome. *Oncotarget* **2015**, *6*, 34069–34070. [CrossRef]
142. Rolseth, V.; Luna, L.; Olsen, A.K.; Suganthan, R.; Scheffler, K.; Neurauter, C.G.; Esbensen, Y.; Kuśnierczyk, A.; Hildrestrand, G.A.; Graupner, A.; et al. No cancer predisposition or increased spontaneous mutation frequencies in NEIL DNA glycosylases-deficient mice. *Sci. Rep.* **2017**, *7*, 4384. [CrossRef] [PubMed]
143. Samaranayake, G.J.; Huynh, M.; Rai, P. MTH1 as a Chemotherapeutic Target: The Elephant in the Room. *Cancers* **2017**, *9*, 47. [CrossRef] [PubMed]

144. Kay, J.; Thadhani, E.; Samson, L.; Engelward, B. Inflammation-induced DNA damage, mutations and cancer. *DNA Repair* **2019**, *83*, 102673. [CrossRef] [PubMed]
145. Pardini, B.; Corrado, A.; Paolicchi, E.; Cugliari, G.; Berndt, S.I.; Bezieau, S.; Bien, S.A.; Brenner, H.; Caan, B.J.; Campbell, P.T.; et al. DNA repair and cancer in colon and rectum: Novel players in genetic susceptibility. *Int. J. Cancer* **2020**, *146*, 363–372. [CrossRef]
146. Forsti, A.; Frank, C.; Smolkova, B.; Kazimirova, A.; Barancokova, M.; Vymetalkova, V.; Kroupa, M.; Naccarati, A.; Vodickova, L.; Buchancova, J.; et al. Genetic variation in the major mitotic checkpoint genes associated with chromosomal aberrations in healthy humans. *Cancer Lett.* **2016**, *380*, 442–446. [CrossRef]
147. Vodicka, P.; Musak, L.; Frank, C.; Kazimirova, A.; Vymetalkova, V.; Barancokova, M.; Smolkova, B.; Dzupinkova, Z.; Jiraskova, K.; Vodenkova, S.; et al. Interactions of DNA repair gene variants modulate chromosomal aberrations in healthy subjects. *Carcinogenesis* **2015**, *36*, 1299–1306. [CrossRef]
148. Vodicka, P.; Vodenkova, S.; Buchler, T.; Vodickova, L. DNA repair capacity and response to treatment of colon cancer. *Pharmacogenomics* **2019**, *20*, 1225–1233. [CrossRef]
149. Ghiringhelli, F.; Richard, C.; Chevrier, S.; Végran, F.; Boidot, R. Efficiency of olaparib in colorectal cancer patients with an alteration of the homologous repair protein. *World J. Gastroenterol.* **2016**, *22*, 10680–10686. [CrossRef]
150. Morales, J.; Li, L.; Fattah, F.J.; Dong, Y.; Bey, E.A.; Patel, M.; Gao, J.; Boothman, D.A. Review of poly (ADP-ribose) polymerase (PARP) mechanisms of action and rationale for targeting in cancer and other diseases. *Crit. Rev. Eukaryot Gene Expr.* **2014**, *24*, 15–28. [CrossRef]
151. Kamel, D.; Gray, C.; Walia, J.S.; Kumar, V. PARP Inhibitor Drugs in the Treatment of Breast, Ovarian, Prostate and Pancreatic Cancers: An Update of Clinical Trials. *Curr. Drug Targets* **2018**, *19*, 21–37. [CrossRef]
152. Tomkinson, A.E.; Howes, T.R.L.; Wiest, N.E. DNA ligases as therapeutic targets. *Transl. Cancer Res.* **2013**, *2*, 1219. [PubMed]

International Journal of
Molecular Sciences

Review

DNA Damage Response and Oxidative Stress in Systemic Autoimmunity

Vassilis L. Souliotis [1,2,†,*], Nikolaos I. Vlachogiannis [1,†], Maria Pappa [1], Alexandra Argyriou [1], Panagiotis A. Ntouros [1] and Petros P. Sfikakis [1]

1 First Department of Propaedeutic Internal Medicine and Joint Rheumatology Program, National and Kapodistrian University of Athens Medical School, 115 27 Athens, Greece; nivlachogiannis@gmail.com (N.I.V.); mariak.pappa@yahoo.com (M.P.); argyrioualex@gmail.com (A.A.); panagiotisd94@gmail.com (P.A.N.); psfikakis@med.uoa.gr (P.P.S.)

2 Institute of Chemical Biology, National Hellenic Research Foundation, 116 35 Athens, Greece

* Correspondence: vls@eie.gr

† These authors contributed equally to this work.

Received: 13 November 2019; Accepted: 18 December 2019; Published: 20 December 2019

Abstract: The DNA damage response and repair (DDR/R) network, a sum of hierarchically structured signaling pathways that recognize and repair DNA damage, and the immune response to endogenous and/or exogenous threats, act synergistically to enhance cellular defense. On the other hand, a deregulated interplay between these systems underlines inflammatory diseases including malignancies and chronic systemic autoimmune diseases, such as systemic lupus erythematosus, systemic sclerosis, and rheumatoid arthritis. Patients with these diseases are characterized by aberrant immune response to self-antigens with widespread production of autoantibodies and multiple-tissue injury, as well as by the presence of increased oxidative stress. Recent data demonstrate accumulation of endogenous DNA damage in peripheral blood mononuclear cells from these patients, which is related to (a) augmented DNA damage formation, at least partly due to the induction of oxidative stress, and (b) epigenetically regulated functional abnormalities of fundamental DNA repair mechanisms. Because endogenous DNA damage accumulation has serious consequences for cellular health, including genomic instability and enhancement of an aberrant immune response, these results can be exploited for understanding pathogenesis and progression of systemic autoimmune diseases, as well as for the development of new treatments.

Keywords: DNA damage response and repair network; immune response; autoimmunity; systemic lupus erythematosus; systemic sclerosis; rheumatoid arthritis; oxidative stress; abasic sites; chromatin organization; apoptosis

---

1. Introduction

The human genome confronts thousands of DNA lesions every day due to normal "mistakes" during DNA replication, or exposure to exogenous or endogenous "toxic" factors, which can block the replication process, lead to genomic instability, and threaten cell function and homeostasis [1]. To ensure proper cell function and viability, a well-organized mechanism, namely, DNA damage response and repair (DDR/R) network, has been evolved over the years. DDR/R is a hierarchically structured mechanism, the main aspects of which are conserved from prokaryotes and phages to humans [2], consisting of sensors, mediators, transducers, and effectors, which recognize any defects during the cell cycle and assign the proper repair process [1]. In case of unrepaired lesions and depending on the extent and type of damage, the cell either passes the mutated genome to its offspring or is neutralized by programmed cell death (apoptosis) or senescence [2].

The interplay between DDR/R and innate immune response has been increasingly recognized in the past years. Several studies have demonstrated that a shift in the balance of DDR/R network driven by either exposure to DNA-damaging agents or deregulation of DNA repair mechanisms results in the accumulation of cytosolic single-stranded DNAs (ssDNAs) and double-stranded DNAs (dsDNAs) that can act as potent immunostimulators through the induction of the cGAS-STING (stimulator of interferon genes)-IRF3 pathway and the production of type I interferon (IFN) [3–11]. Moreover, recent studies have shown that cell cycle progression through mitosis following DNA double-strand breaks (DSBs) formation and DDR/R induction leads to the generation of micronuclei, which precede activation of the immune system [12–14]. On the other hand, loss of immune homeostasis and prolonged inflammatory response generated by different sources (infection, radiation, toxins, autoimmunity, ageing, etc.) can lead to DNA damage and activate the DDR/R network [15–21], proposing a bi-directional relationship between DDR/R and immune response (ImmR) [2].

Systemic autoimmune disorders comprise a heterogeneous group of diseases characterized by aberrant immune response to self-antigens with widespread production of autoantibodies and multiple tissue injury, as well as by oxidative stress along with the excess production of reactive oxygen species (ROS) and reactive nitrogen species (RNS). Inappropriate activation of adaptive immunity and production of autoantibodies has been classically linked to autoimmunity, whereas innate immune activation and, specifically, the recognition of nucleic acids by Toll-like receptors and other cytoplasmic innate immune receptors, are considered as part of the pathophysiology of autoimmune diseases [22].

Aberrant DDR/R has been reported in patients with systemic autoimmune diseases, such as systemic lupus erythematosus (SLE) [3,4], systemic sclerosis (SSc) [23], and rheumatoid arthritis (RA) [24,25]. Polymorphisms of nucleases or molecules central in the DNA repair process have been detected with increased frequency among patients with autoimmune diseases, whereas gene/protein expression assays have shown downregulation of molecular components that are implicated in the DNA repair machinery and upregulation of apoptosis genes among patients with autoimmune disease [4]. However, only a few studies to date have examined the burden of DNA damage in patients with systemic autoimmune diseases and mechanistic aspects underlying this phenomenon. Whether aberrant DDR/R response precedes immune activation in autoimmune diseases or the chronic immune activation/inflammation leads to increased DNA damage formation and deregulation of DNA repair mechanisms remains largely unknown. Recently, our group suggested a role of deficient DNA repair and increased formation of endogenous DNA damage, at least partly due to the induction of oxidative stress, that may lead both to augmented apoptosis rates and subsequent autoantibody production in patients with SLE [3,4].

In the current review, we first briefly overview the normal DDR/R pathways and highlight DDR/R aberrations and critical endogenous factors/processes that lead to the intracellular formation of DNA damage, which are observed in systemic autoimmune diseases. The main goal of this review is to serve as a "tool" for the comprehensive presentation of up-to-date literature on the subject and thus help in the design of new mechanistic studies to better understand the involvement of the DDR/R network in the pathogenesis of systemic autoimmunity, as well as to suggest new therapeutic perspectives and potential targets.

2. Normal DNA Repair Pathways

To compensate for the many types of DNA damage that occur, cells have developed multiple repair mechanisms wherein each corrects a different subset of lesions. In general, there are six major DNA repair pathways, which will be presented below.

2.1. Nucleotide Excision Repair (NER)

NER is a fundamental DNA repair mechanism involved in the removal of bulky, helix-distorting lesions from DNA [26]. DNA adducts that are repaired by NER include cyclobutane pyrimidine dimers (CPDs) and 6-4 photoproducts (6-4 PPs) produced by UV radiation, DNA lesions generated by ROS or endogenous lipid peroxidation products, intrastrand cross-links and adducts produced by genotoxic drugs (melphalan, cisplatin), or environmental carcinogens (benzo[a]pyrene) [27,28]. There are two subpathways of NER, termed GGR (global genome repair) and TCR (transcription-coupled repair), where approximately 30 proteins are involved in both subpathways. The first step, the recognition of DNA damage, differs between the two subpathways. In GGR, the formation of a bulky DNA adduct induces an increase in helix distortion, which facilitates the recruitment of the damage recognition factor XPC/RAD23/CETN2 and UV-DDB. On the other hand, damage recognition in TCR is initiated when an elongating RNA polymerase II (RNAPII) is arrested upon encountering a site of DNA damage. Subsequently, two TCR-specific proteins, Cockayne syndrome A (CSA) and B (CSB), are thought to displace the stalled RNAPII to allow the access of the NER proteins to the lesion. Following damage recognition, both GGR and TCR proceed through common NER reactions. The biological importance of NER for human health is obvious by the fact that defects in this repair pathway cause several human genetic disorders, including Cockayne syndrome (CS), xeroderma pigmentosum (XP), and trichothiodystrophy (TTD), which are all associated with photosensitivity [29].

2.2. Base Excision Repair (BER)

BER is a conserved and ubiquitous DNA repair pathway, which recognizes and removes damaged DNA bases that do not significantly distort the structure of the DNA helix [30]. BER is used by the cell to correct DNA lesions that occur through the spontaneous deamination or hydroxylation of bases and by oxidation of nucleotides by ROS produced either by normal metabolism or environmental stresses such as smoking, oxidizing chemicals, or ionizing radiation [31]. In addition, BER is implicated in the repair of alkylated DNA bases generated by endogenous or exogenous factors (carcinogens, antineoplastic drugs, etc.), which if left unrepaired produce mutations in the cells [32]. BER consists of two subpathways, known as single-nucleotide or short-patch and long-patch; the activation of one or the other is predicated by the cause and type of damage, the type of abasic (AP; apurinic/apyrimidinic) site generated in the first repair step and the cell cycle phase in progress when the damage occurs. The short-patch pathway quickly repairs single-base damage during the G1 phase; the long-patch pathway handles lengthier repair during S or G2, when resynthesis of two to eight nucleotides surrounding the AP-site is required. Among the enzymes that take part in BER, DNA glycosylases, mono- or bi-functional, are the most important. They recognize and hydrolyze the *N*-glycosylic bond between the damaged base and the sugar phosphate backbone, creating an AP intermediate site.

2.3. Mismatch Repair (MMR)

MMR mechanism is a major contributor to replication fidelity, which removes base substitution and insertion/deletion mismatches that arise as a result of replication errors escaping the proofreading function of DNA polymerases [33]. The recognition of DNA lesions is accomplished by the complex Mutator Sα (MUTSα), a heterodimer of the DNA mismatch repair proteins Mutator S homolog 2 (MSH2) and Mutator S homolog 6 (MSH6). Another heterodimer complex, called MUTSβ, which consists of MSH2 and MSH3, is able to bind only to insertion/deletion mismatches. Lesion recognition is followed by the recruitment of Mutator Lα (MutLα) [MLH1/postmeiotic segregation increased 2 (PMS2)] or MutLβ (MLH1/MLH3), which have endonuclease activity that can incise DNA near the mismatch. The nick is used by the 5′ exonuclease 1 (Exo1) as an entry point to degrade DNA past the mismatch, and the resulting single-stranded DNA gap is filled in by polymerase δ and sealed with DNA ligase I [34,35]. Deficiencies in MMR lead to microsatellite instability (MSI), which is a pattern of hypermutation that occurs at genomic microsatellites, and is associated with unique clinical features, prognosis and response to therapy, and immune checkpoint blockade [36,37].

2.4. Double-Strand Breaks (DSBs) Repair

DSBs may occur as a result of exposure to both exogenous factors, including ionizing radiation, UV light and genotoxic drugs [38], and endogenous events, including oxidative stress, replication fork collapse, and telomere erosion [39]. Of interest, these lesions also occur as programmed events during meiosis, as well as during V(D)J recombination [the process by which T cells and B cells randomly assemble different gene segments—known as variable (V), diversity (D) and joining (J) genes—in order to generate unique receptors (known as antigen receptors) that can collectively recognize many different types of molecule] and class-switch recombination (CSR) required for immunoglobulin diversity and function [40]. DSBs, if left unrepaired, have severe adverse consequences for the cell including the generation of mutations, chromosomal aberrations, and cell death [41]. To maintain genomic integrity, cells have evolved several pathways to remove DSBs.

2.4.1. Homologous Recombination Repair (HRR)

HRR is an error-free DNA repair mechanism, which operates during the S and G2 phases of the cell cycle so that it can find a large area of homology on a sister chromatid to use as a template for resynthesizing damaged or lost bases [42]. HRR can be divided into several steps. During initiation, both the 5′-ends of the DSB are resected by the action of a specific nuclease to yield 3′-single-strand DNA (3′-ssDNA) tails. Then, one of these tails invades an intact homologous duplex and generates a D-loop structure, while the other could simply anneal with the displaced strand at the joint. Both 3′-ends then prime new DNA synthesis using the intact duplex as a template. This process, followed by ligation, leads to the formation of two Holliday junctions (four-stranded branched structures), which are finally cleaved by the action of a resolvase [43].

2.4.2. Canonical Non-Homologous End Joining (c-NHEJ)

c-NHEJ is an error-prone process, which is active throughout the entire cell cycle. C-NHEJ is initiated by the binding of the heterodimeric protein complex X-ray repair cross complementing 5/6 to both DNA ends. Then, DNA-dependent protein kinase (DNA-PK) is recruited, a DNA dependant protein kinase, which activates X-ray repair cross-complementing protein 4 (XRCC4)-ligase IV complex to link the broken DNA ends together. However, before re-ligation, MRN complex (MRE11-Rad50-NBS1), together with the Flap Endonuclease 1 (FEN1) and Artemis, are involved in processing DNA ends [44,45]. Aberrant c-NHEJ is a major source of genomic rearrangements and chromosomal translocations, leading to genomic instability [46]. Interestingly, deficient c-NHEJ is associated with defective V(D)J recombination and immune defects [47]. The choice between c-NHEJ and HRR pathways is regulated by complex regulatory mechanisms and involves competition between the p53-binding protein 1 (53BP1), which favors c-NHEJ, and BReast CAncer gene 1 (BRCA1), which promotes HRR. Methylation of histone H4 by Multiple Myeloma SET (MMSET) results in 53BP1 recruitment at the DSB site, which blocks DNA end resection by the MRN complex, C-terminal binding protein 1 interacting protein (CtIP), and BRCA1. On the other hand, histone H4 acetylation by the Tat-interactive protein (Tip60) blocks 53BP1 recruitment and promotes BRCA1 occupancy and HRR. Cell cycle-regulated proteins such as cyclin-dependent kinases also play a key role in the choice of pathway to resolve DSBs [38].

2.4.3. Alternative Non-Homologous End Joining (alt-NHEJ)

alt-NHEJ is a mechanistically distinct pathway of DSB repair that is frequently termed microhomology-mediated end-joining [48]. Indeed, the foremost distinguishing property of alt-NHEJ is the use of 5–25 base pair microhomologous sequences during the alignment of broken ends before joining, thereby resulting in deletions flanking the original break [49]. Thus, alt-NHEJ is frequently associated with chromosome abnormalities including translocations, deletions, and inversions [50]. The viewpoint that alt-NHEJ is the major DNA repair pathway to pathogenic chromosomal errors is further strengthened by the finding that c-NHEJ-deficient mice develop tumors with chromosomal translocations generated by alt-NHEJ [51].

2.4.4. Single-Strand Annealing (SSA)

SSA is a highly mutagenic but very efficient DSB repair mechanism [38,52]. This process involves a DSB between homologous repeats, followed by DSB end resection that generates 3′-ssDNA, which reveals flanking homologous sequences that are annealed together to form a synapsed intermediate. This intermediate is then processed for ligation, which requires endonucleolytic cleavage of nonhomologous 3′-ssDNA tails, and polymerase filling of the gaps. Genetically, SSA is distinct from other homologous recombination pathways, as it occurs independently of Rad51 recombinase. Instead, it depends on Rad59, the Rad52 paralog that is structurally homologous to the N-terminus of Rad52. Biochemically, both Rad52 and Rad59 can anneal ssDNA, but only Rad52 can anneal ssDNA coated with RPA proteins. Although relatively mutagenic in terms of causing a rearrangement between repeat elements, SSA is critical to restore a broken chromosome with DSB ends that have undergone extensive end resection, but are unable to be resolved by HRR or alt-NHEJ [53]. The importance of SSA in DNA repair depends on a number of factors, including the state of the cell cycle, the presence or absence of the sister chromatid, and the length of uninterrupted homology.

2.5. Interstrand Cross-Link (ICL) Repair

The formation of cross-links between the two strands of DNA is considered a critical event, causing cell cycle and replication arrest and eventually cell death if not repaired [54]. Cross-linking agents are exogenous chemicals, including the drugs cyclophosphamide, melphalan, cisplatin, mitomycin C, and psoralen [55], as well as endogenously formed aldehydes [56]. There are three routes for cross-link detection in mammalian cells. Adducts can be recognized in otherwise unperturbed duplex DNA by factors that recognize DNA damage. Cross-link detection might also occur via encounter with the transcription machinery. Finally, ICLs could block a replication fork, triggering a repair response that would remove the cross-link and restore replication. Interestingly, in non-replicating cells, the repair of ICL is mediated by the NER mechanism and by the DNA translocase FANCM, which facilitates the access of nucleases to the lesion. In S-phase cells, cross-link repair is coupled to DNA replication, features DSBs as repair intermediates, and depends on the homologous recombination machinery [57]. ICL repair in human cells is accomplished in four distinct steps: (a) unhooking of the ICL on one strand and induction of a DNA replication-dependent DSB, (b) translesion DNA synthesis using the DNA strand with the unhooked ICL as a template, (c) processing of the DSB and restoration of the stalled DNA replication fork, and (d) removal of the residual unhooked ICL [58]. Proteins implicated in the repair of ICLs have a critical role in the pathophysiology of several hereditary disorders, such as Fanconi anemia, xeroderma pigmentosum, Cockayne syndrome, cerebro-oculo-facio-skeletal syndrome, and trichothyodistrophy.

2.6. Direct Repair Pathway

The direct repair mechanism is a single step pathway, which is unique in that only one protein is implicated in the repair process [59]. Indeed, the sole protein involved, O6-methylguanine-DNA methyltransferase (MGMT), removes alkyl groups from the O6 position of guanine or to a lesser extent from the O4 position of thymine, such as those generated by treatment with alkylating drugs (procarbazine, dacarbazine, temozolomide), and transfers it to an internal cysteine residue of MGMT [60]. Because the alkyl group is covalently bound to the MGMT protein, MGMT is functionally inactivated after each reaction, and degraded through the ubiquitin proteolytic pathway. Without MGMT repair, alkyl adducts would cause thymine mispairing during replication, leading to G:C to A:T transitions or strand breaks [61]. Overactivity of MGMT is also considered responsible for chemoresistance; for example, >90% of recurrent gliomas show no response to a second cycle of chemotherapy. Conversely, inhibition of MGMT renders cancer cells sensitive to temozolomide, whereas MGMT promoter alkylation is a significant determinant in the sensitivity of drugs such as temozolomide. There is abundant evidence linking methylation of the MGMT promoter to loss of protein expression, resulting in increased sensitivity to chemotherapeutic agents and to the prognostic outcome of patients treated. Similarly, low MGMT expression appears to be a biomarker for slower tumor progression [62].

3. The Interplay between the DDR/R Network and the Immune Response: The Role of Oxidative Stress

Although not completely delineated to date, interplay between the DDR/R network and the ImmR has been suggested by a series of studies, nicely reviewed in [2]. A first hint that defective nucleic acid metabolism may trigger aberrant innate immune activation with serious consequences for life is derived from Aicardi-Goutières (AGS) syndrome. AGS is a childhood-onset, possibly fatal encephalopathy characterized by mutations of central molecules implicated in DNA and RNA metabolism, such as (a) RNaseH2, which is involved in excision of a single ribonucleotide embedded in genomic DNA and removal of an R-loop formed in cells [63–65], or (b) the 3′–5′ exonuclease TREX1 [66–68]. On the other hand, increasing data suggest that aberrant, chronic (auto)immune activation and chronic inflammation may cause DNA damage and trigger the DDR/R network [4]. Indeed, under inflammatory conditions, ROS and RNS are generated from inflammatory and epithelial cells and result in oxidative and nitrative DNA damage, such as 8-oxo-dG and 8-nitro-dG, as well as in the inhibition of key proteins of the DNA repair machinery, indicating the bi-directional interplay between DDR/R and ImmR via oxidative stress [2].

The central role of type I IFN pathway activation in the pathophysiology of systemic autoimmune diseases has been extensively studied since its first description almost 40 years ago [69,70]. The recognition of nucleic acids by innate immune receptors (Toll-like receptors (TLR) and non-TLRs) has a central role in autoimmunity, suggesting that abnormal DNA or RNA metabolism may initiate and/or perpetuate innate immune activation [71]. Innate immune receptors can either recognize pathogen-derived "non-self" DNA (pathogen-associated molecular patterns, PAMPs), for example, those derived from a DNA virus, but also damaged "self" DNA (damage-associated molecular patterns, DAMPs) at sites of inflammation, and can initiate an immune response [2]. In physiological conditions, DAMPs are found intracellularly; are invisible to the immune system; and serve metabolic, structural, or enzymatic functions [72]. On the other hand, DAMPS are exposed or released upon stress, injury, and cell death, thereby becoming able to bind appropriate receptors on immune cells. Of note, following treatment with some anticancer drugs, such as anthracyclines (doxorubicin, epirubicin, idarubicin), mitoxantrone, oxaliplatin, cyclophosphamide, and bortezomib, cancer cells undergo a form of cell death named immunogenic cell death, which is characterized by an increased immunogenic potential, owing to the emission of the DAMPs, which act as danger signals to produce immunostimulatory effects, such as the recruitment and activation of neutrophils, macrophages, and other immune cells [73]. DAMPs released during immunogenic cell death include plasma membrane exposure of endoplasmic reticulum chaperones such as calreticulin (CALR), secretion of ATP, release of double-stranded DNA resulting in activation of STING and release of type I IFN and proinflammatory cytokines, secretion CXCL10, as well as the release of high-mobility group box 1 (HMGB1) and annexin A1 (ANXA1) [74].

3.1. DNA Double-Strand Breaks Per Se Induce Innate Immune Activation

Apart from oxidative DNA damage, the presence of DSBs *per se* has been shown to induce type I IFN production [5]. Indeed, treatment of healthy donor-derived primary monocytes with etoposide, a chemotherapeutic agent that blocks topoisomerase II activity and leads to the accumulation of DSBs, up-regulated type I IFN-induced gene expression and type III IFNs (IFN-λ). Similarly, other DSB-inducing drugs (mitomycin C, adriamycin, etc.) were also able to induce type I and III IFNs in primary monocytes and various cell lines, suggesting that DDR-induced IFN expression is a universal mechanism that may underline different pathological processes [5]. In line with these results, treatment of breast cancer cell lines with DSB-inducing drugs, including ionizing radiation or therapeutic drugs (mitomycin C, cisplatin), led to the accumulation of cytoplasmic ssDNA and finally the activation of the STING-IRF3 pathway [10]. Furthermore, basic components of DSB repair were shown to be responsible for the production of cytoplasmic ssDNA, which seems to be the main immunostimulant [10]. Interestingly, TREX1 was found to be the main responsible nuclease for the restriction of this cytoplasmic ssDNA and prevention of aberrant innate immune activation [6,10]. An association between immune activation and TREX1 was also observed when TREX1 null mice developed inflammatory myocarditis due to an interferon-dependent autoimmune response leading to dilated cardiomyopathy and a significantly reduced survival [75]. Of note, type I IFN is indeed implicated in the inflammatory myocarditis and early mortality observed in this mouse model, as disease manifestations were strikingly attenuated in Trex1-deficient mice also lacking the type I IFN receptor (IFNαR1) [76], and were also improved in mice treated with an inhibitor of the downstream kinase TANK-binding kinase 1 [77].

3.2. Defective DNA Repair and Chronic Low-Level DNA Damage "Prime" the Innate Immune Response

Another clue that defective DNA repair primes innate immune response comes from ataxia–telengiectasia (AT), a neurodegenerative disorder associated with mutations of the central DNA repair kinase ATM [6]. In this study, Härtlova and colleagues showed that AT-derived fibroblasts had higher constitutive expression of type I and III IFNs and mounted a profoundly high response upon transfection with DNA virus or the intracellular microbe *Listeria* monocytogenes. These results suggested that loss of ATM, potentially leading to chronic accumulation of low-grade DNA damage, may prime the innate immune response [6]. Similar results were obtained from ATM-deficient mice and cell lines where ATM was silenced. Of interest, γ-irradiation or etoposide treatment of normal bone marrow-derived myeloid cells mimicked the elevated basal expression of IFN and hyper-sensitivity to PRR (TLR and non-TLR)-induced response observed in ATM-deficient cells, suggesting that the accumulation of damaged DNA underlined this phenomenon. A series of mechanistic studies with knock-out of various innate immune adaptors revealed that STING was mainly responsible for the observed phenotype [6]. In summary, ATM deficiency led to the accumulation of DNA damage, exportation of damaged ssDNA and dsDNA into the cytoplasm, activation of the cGAS-STING pathway, and finally type I IFN production that primed cells for response to exogenous or endogenous stimuli such as viral or bacterial infections (Figure 1).

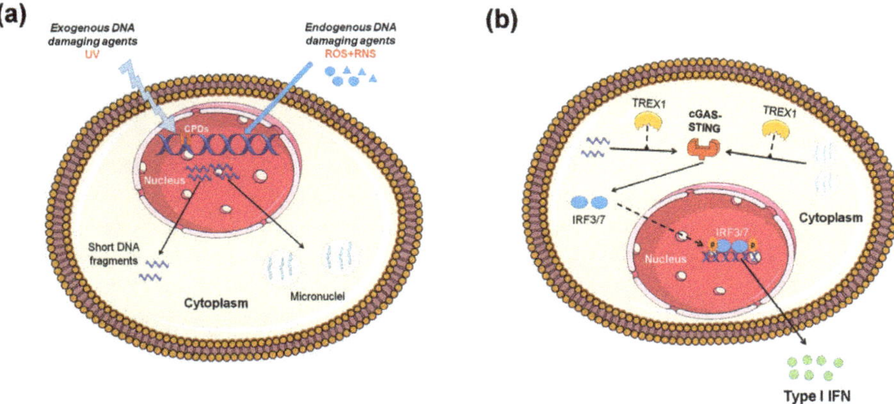

Figure 1. Induction of type I interferon (IFN) expression by endogenous DNA damage. (**A**) Exogenous and/or endogenous genotoxic agents may lead to the accumulation of DNA damage in the nucleus, followed by exportation of damaged DNA into the cytoplasm and the induction of micronuclei. (**B**) Damaged cytoplasmic DNA, if it is not cleared by the exonuclease Trex1, activates the cGAS-STING (stimulator of interferon genes)-IRF3 pathway and the production of type I IFN. ROS: reactive oxygen species, RNS: reactive nitrogen species, CPDs: cyclobutane pyrimidine dimers.

Moreover, Günther and colleagues suggested that defective ribonucleotide removal and accumulation of base lesions and low-grade DNA damage "primed" ImmR [69]. Indeed, they showed that fibroblasts from AGS and SLE patients with mutations in the DNA repair enzyme RNaseH2 produced increased levels of IFNβ upon stimulation with poly(I:C), a phenomenon that was enhanced when poly(I:C) treatment was combined with UVC irradiation. In the same study, patients' fibroblasts showed a decreased proliferation rate in vitro, increased p53 phosphorylation at Ser15, and senescence. Of interest, RNaseH2 deficiency in heterozygous carriers (parents of AGS patients) significantly increased the prevalence of antinuclear antibodies (ANAs), suggesting that defective ribonucleotide removal may promote formation of autoantibodies [69]. Type I IFN activation in RNaseH2-null cells was also shown to be mediated by STING [11]. That is, dermal fibroblasts isolated from AGS/SLE patients with RNaseH2 mutations and mouse embryonic fibroblasts (MEFs) isolated from RNaseH2-null mice showed significantly increased single-strand breaks (SSBs) and DSBs and were also more sensitive to UV-irradiation, as shown by increased CPD formation.

3.3. Micronuclei: Connecting Nuclear DNA Damage and Cytosolic Innate Immune Receptors

The strict compartmentalization of DNA in the cell's nucleus and mitochondria raises the question as to how damaged self DNA becomes accessible to STING, which resides in the cytoplasm. Recent sophisticated studies connected the dots featuring a new role for micronuclei [70]. Micronuclei are components of the nuclear membrane encompassing DNA, which are released in the cytoplasm during mitotic cell division. Two independent studies showed that RNaseH2-null cells, which have been previously shown to express higher levels of IFN-induced genes, probably through a STING-mediated pathway [11,69], have increased numbers of micronuclei in their cytoplasm [12,14]. Of note, the majority of micronuclei were enriched for cGAS, which is essential for the production of cGAMP, the activator of STING [71].

3.4. Oxidative Stress Causes DNA Damage That Activates the Immune System

Cellular oxidative damage is a general mechanism of cell and tissue injury, which is primarily caused by free radicals and ROS. ROS are chemically reactive molecules containing an oxygen atom. Although normally ROS are essential elements of the ImmR involved in cytokine production, microbial clearance, cell proliferation, and cell death, overproduction and/or inadequate removal of these species results in oxidative stress [78].

ROS are produced by both endogenous and exogenous sources. Endogenous sources include the generation of ROS from mitochondria; peroxisomes (intracellular organelles that are also called microbodies); activated inflammatory cells, such as macrophages, neutrophils, and eosinophils; as well as during the metabolism of xenobiotics mediated by cytochromes P450 oxidoreductases [79]. These endogenously induced DNA lesions can often reach a level much higher than the ones induced by environmental factors. ROS are constantly generated in mitochondria as respiration byproducts (1–5% of consumed oxygen), and in general are accepted as the major source of oxidative injury in aerobic organisms. Another source of constant generation of free radicals is the chronic exposure to viral infections. The high intracellular oxidation status in viral infections consists of decreased antioxidant enzymes such as catalase, glutathione peroxidase, glutathione reductase, as well as high levels of hydroxyl radicals. Of note, tumor growth and development is always accompanied by oxidative stress, which develops due to various inflammatory and immune reactions [80].

As for the extracellular sources of ROS, these include ionizing radiations such as X-, γ-, or cosmic rays and α-particles from radon decay, oxidizing chemicals, ultraviolet A (UVA) light, chemotherapeutics, environmental toxins, and other pollutants [78]. Exposure to extracellular sources of ROS is especially prevalent in skin cells, as they are constantly exposed to the environment. Radiation can react with oxygen and form superoxide anion radical, hydroxide anion, and hydroxyl radical that are able to destroy the structural integrity of DNA. Moreover, chronic exposure to cigarette smoke promotes lipid peroxidation and has detrimental effects for the cardiac and respiratory systems. Some xenobiotics appear to interfere with mitochondrial bioenergetics and promote superoxide production.

DNA lesions associated with ROS are oxidized purines and pyrimidines, SSBs, DSBs, and abasic sites. Two of the most common endogenous DNA base modifications are 8-oxo-7,8-dihydroguanine (8-oxoGua) and 2,6-diamino-4-hydroxy-5-formamido-pyrimidine. These lesions can be originated from the addition of the hydroxyl radical to the C8 position of the guanine ring producing a 8-hydroxy-7,8-dihydroguanyl radical, which can be either oxidized to 8-oxoGua or reduced to give the ring-opened 2,6-diamino-4-hydroxy-5-formamidopyrimidine (FapyGua) [81]. Moreover, interaction of hydroxyl radical with pyrimidines (thymine and cytosine) at positions 5 or 6 of the ring can produce several base lesions, such as 5,6-dihydroxy-5,6-dihydrothymine and 5,6-dihydroxy-5,6-dihydrocytosine. Two other pyrimidine lesions are the 5-(hydroxymethyl) uracil and the 5-formyluracil, which are often detected in humans as the result of the interaction of the hydroxyl radical with the methyl group of thymine. With the interaction of the hydroxyl radicals with DNA, SSBs may also occur, which in turn trigger the induction of DSBs. The mechanism consists of hydrogen abstraction from the 2-deoxyribose, leading to the formation of carbon-based radicals, which under the presence of oxygen can be converted to peroxyl radicals. The peroxyl radicals, through different reactions, can also abstract hydrogen atoms from sugar moieties, thus leading to DNA strand breaks. The most prevalent and characteristic abasic sites formed under oxidative stress are 2-deoxyribonolactone and the C4′ oxidized abasic site that arise from hydroxyl radical-mediated hydrogen abstraction at C1 and C4 of the 2-deoxyribose moiety of DNA, respectively [82]. Interestingly, peroxyl radical-mediated DNA adducts are potential precursors of apurinic sites, as the opening of the imidazole ring of the purine bases may lead to increased hydrolytic lability of their *N*-glycosidic bonds. This occurrence is very common and can occur spontaneously or enzymatically as "repair intermediates" of the BER pathway.

On exposure to oxidative stress, cells initiate a variety of defense mechanisms, including both enzymatic and non-enzymatic antioxidants. In mammalian cells, enzymatic antioxidants include superoxide dismutase, glutathione peroxidase, glutathione reductase, glutathione-S-transferase, and catalase, whereas non-enzymatic antioxidants contain ascorbic acid (vitamin C), α-tocopherol (vitamin E), total thiol, glutathione, carotenoids, and flavonoids [83]. Oxidative stress causes damage on the primary cellular components, including DNA, proteins, and lipids. In particular, ROS-induced DNA lesions include oxidized bases, abasic sites, single-strand breaks (SSBs), and DSBs, which during the replication process can lead to replication fork stalling, thus giving rise to mutations and genetic instability [84].

Accumulating evidence suggests that oxidative stress can participate in the pathogenesis, progression, and complications of many diseases, including cancer and autoimmunity [78]. Especially with regard to systemic autoimmune diseases, several studies have shown that SLE patients are characterized by increased oxidative stress, resulting in immune system dysregulation, abnormal activation and processing of cell-death signals, and autoantibody production [85]. Indeed, previous studies have shown that oxidative stress causes a significant delay in the apoptotic clearance, resulting in a prolonged interaction between ROS and nuclear residues, which in turn triggers neo-epitope production and autoantibody formation [86]. In addition, oxidative stress is involved in the pathogenesis of SSc [87]. That is, SSc patients are characterized by increased production of ROS in the skin, visceral fibroblasts, and endothelial cells, as well as by reduced concentrations of various antioxidants, including antioxidant vitamins (ascorbic acid, α-tocopherol, β-carotene) and minerals (zinc, selenium) [88]. Also, oxidative stress has been observed in patients with RA. In fact, these patients show augmented intracellular ROS, lipid peroxidation, protein oxidation, DNA damage, and deregulated antioxidant defense system of the body. Moreover, deficient MMR system was observed in RA patients, resulting in increased formation of DNA adducts in the joints and acceleration of the disease progression [89]. In line with these data, low levels of non-enzymatic antioxidants [reduced glutathione (GSH) and vitamin C] were found in RA patients, as compared with healthy individuals.

In an attempt to explain the immunogenicity of "self" DNA under inflammatory conditions, Gehrke and colleagues used a series of in vitro experiments and revealed a role for oxidized DNA as DAMP [90]. Indeed, they found that oxidized DNA, that is, from UV-induced damage or from ROS released during cell death, activated the innate immune receptor STING (stimulator of interferon genes), whereas "normal" DNA did not. Further exploring the mechanistic aspects of this "paradox", oxidized DNA was shown to be resistant to TREX1 degradation, thus accumulating in the cytoplasm and activating the cGAS-STING and the type I IFN pathway. Of note, this axis, that being oxidized DNA-cGAS-STING-type I IFN, was verified in a SLE mouse model [Murphy Roths Large lymphoproliferation (MRL/lpr) mice], as well as in skin biopsies of patients with SLE, where oxidized DNA co-localized with the type I IFN-induced gene myxovirus (influenza) resistance 1 (MX1) [90].

The immunogenicity of UV radiation and subsequent oxidative DNA damage, as observed in SLE flares following sun exposure, has also been studied in preclinical models [91]. The researchers showed that UV radiation potentiates STING-dependent activation of IFN regulatory factor 3 (IRF3; immune signaling transcription factor) in response to cytosolic DNA and cyclic dinucleotides in keratinocytes and other human cells. Furthermore, they found that stimulation of STING-dependent IRF3 by UV is due to apoptotic signaling-dependent disruption of ULK1 (Unc51-like kinase 1), a pro-autophagic protein that negatively regulates STING.

3.5. Oxidative Stress and Immune Senescence

It is generally accepted that oxidative stress induces the senescent phenotype. Cellular senescence is a cell state implicated in various physiological processes and a wide spectrum of age-related diseases [92]. There are four different molecular mechanisms of oxidative stress-induced cell senescence: (a) the DDR/R mechanism, in which oxidative damage stimulates the DDR/R network through activating p53 and up-regulating p21 expression to cause senescence [93]; (b) the nuclear factor kappa B (NF-κB) mechanism, in which oxidative stress activates the inhibitor of kappa B (IκBs) kinase, which phosphorylates IκB to activate NF-κB and makes it transfer into the nucleus to stimulate IL-8 expression and increase p53 protein stability and then induce cellular senescence [94]; (c) the p38 mitogen-activated protein kinase (MAPK) mechanism, which is activated by ROS, up-regulates p19 protein expression, and limits self-renewal (the process by which stem cells divide to make more stem cells) to induce cellular senescence [95]; and (d) the microRNA mechanism, in which oxidative stress affects the amount of microRNA and promotes senescence [96].

Interestingly, the senescence-induced decline of the immune system is known as immunosenescence and is implicated in impaired autoantigen recognition and vaccination in the elderly. Indeed, several functions of the cells involved in the innate and adaptive immune responses are seriously compromised with age progression, including chronic inflammatory state, changes in lymphocyte subsets, and decreased proliferative responses, among others. Recent data have shown that during senescence, the LINE-1 retrotransposon is transcriptionally expressed and stimulates the IFN-I response, thus contributing to the maintenance of the senescence-associated secretory phenotype, which determines the ability of senescent cells to express and secrete cytokines, chemokines, proteases, growth factors, and bioactive lipids [97].

Moreover, age-related transformations redesign the immune architecture and the balance between pro-inflammatory and anti-inflammatory protective factors, as well as between pro-apoptotic and anti-apoptotic signals. In fact, elderly people experience increased reactivity to autoantigens, loss of tolerance, and systemic inflammation, while at the same time they suffer from degenerative diseases, which, in turn, increase the risk of developing an autoimmune disease [98]. Moreover, epigenetic changes and the increase in inflammatory cytokines and chemokines such as TNF-α, C-reactive protein, IL-8, MCP1, and RANTES (Regulated on Activation, Normal T Cell Expressed and Secreted) that occur in the elderly play a crucial role in the onset of autoimmune diseases [99].

These alterations make older persons more prone, not only to autoimmune disease, but also to cancer, as well as metabolic, neurodegenerative, and infectious diseases [100]. In fact, infectious diseases account for roughly 20% of hospitalizations in the elderly, whereas one-third of deaths in persons aged >65 years has been reported to be due to infectious diseases [101]. In addition, immunosenescence also results in reduced responses to vaccination, a common phenomenon in the elderly [102] Growing interest in therapeutically targeting senescence to improve healthy aging and age-related disease with compounds known as senolytic drugs has recently led to the first clinical trials [92].

3.6. Defects in Degradation of Endogenous DNA and Immune Activation

Apart from increased DNA damage formation and defective repair, a third mechanism may also be implicated in the accumulation of immunogenic damaged DNA—defective DNA degradation. Indeed, degradation of cytosolic DNA by TREX1 is integral for the prevention of aberrant innate immune responses [76]. Although previous mechanistic studies have revealed TREX1 as the main nuclease for removal of oxidized DNA, DNAse II may also have an important role in prevention of misplaced innate immune response, as DNAse II-deficient mice have been shown to spontaneously develop polyarthritis mimicking RA [103].

4. The DDR/R Network in Systemic Autoimmune Diseases

4.1. Systemic Lupus Erythematosus

SLE is a prototypic autoimmune disease characterized by abnormal T and B cell responses, in which excessive antibody production and immune complex formation are considered central pathogenetic mechanisms [104]. In the past years, innate immunity and specifically the recognition of nucleic acids by TLRs and cytoplasmic receptors have also been gaining attention as critical components in SLE pathogenesis [105]. The first hint that abnormalities in DDR/R pathway may be involved in SLE pathophysiology comes from the increased frequency of polymorphisms of central molecules involved in the DDR/R pathway such as TREX1 [106]. Moreover, autoantibodies against components of the DDR pathway have been detected in approximately 10%–20% of patients with SLE [107]. Deficiencies in DNA repair have been shown to induce lupus-like disease in animals. Mice carrying the Y265C hypomorphic allele of POLB (DNA polymerase, beta; a key enzyme in BER mechanism) demonstrated several pathologies resembling lupus, such as nephritis and skin manifestations, along with high titers of anti-nuclear antibodies in serum [108]. Moreover, mice with compound deficiency in Gadd45β (Growth arrest and DNA-damage-inducible, beta) and Gadd45γ proteins, which are involved in DDR/R as well as in initiation of the type 1 helper T cell (Th1) response, showed features resembling lupus, such as antibodies against dsDNA and histones in sera, and immune complex deposits in renal glomeruli [109].

4.1.1. Gene Polymorphisms Associated with Impaired DNA Repair Machinery and SLE

Data from two cohort studies evaluated the association of the most common polymorphisms of XRCC1, a ligase protein involved in BER, with SLE susceptibility. The rs25487 single-nucleotide polymorphism (SNP), which encodes an arginine to glutamine substitution at position 399 (R399Q) was found to be associated with high titer of anti-dsDNA antibodies in Brazilian SLE patients, whereas the presence of two common SNPs was associated with neuropsychiatric manifestations and antiphospholipid syndrome [110]. At the same time, a Chinese Han population cohort study revealed that individuals with the aforementioned SNP are nearly two times more prone to develop SLE compared with healthy controls [111]. Polymorphisms in another key enzyme of BER, Polβ, have also been associated with SLE in two large, independent cohort studies of a Chinese Han population [112,113]. Poly (Adenosine diphosphate-ribose) polymerase 1 (PARP1), a core protein of BER and DSB repair mechanisms, is also implicated in the susceptibility for SLE development. Indeed, a genetic analysis of chromosome 1q41-q42 revealed that a specific allele of PARP1, with a length of approximately 85bp, confers defective DNA repair and abnormal apoptosis, thus predisposing to SLE [114]. The contribution of polymorphisms in DDR/R components to the development and progression of SLE has been recently reviewed here [115].

4.1.2. Increased Endogenous DNA Damage in SLE: Defective Repair or Increased Formation?

We have previously shown that peripheral blood mononuclear cells (PBMCs) from SLE patients display defects in two main DNA repair pathways, namely, NER and DSB repair [3,4]. Specifically, study of the formation of *N*-alkylpurine-monoadducts (almost exclusively repaired by NER) at the N-ras (neuroblastoma RAS) locus, the repair rate of which is representative of the total cellular NER capacity, and phosphorylated H2AX (γ-H2AX), a sensitive marker for DNA DSBs that was measured at the level of the whole cell, revealed that SLE patients with nephritis have approximately 3–5 times higher intrinsic DNA damage compared with healthy controls. Of interest, patients with quiescent disease also exhibited increased levels of DNA damage, although lower than patients with nephritis, suggesting that DNA damage levels may also be associated with disease activity. Following ex vivo treatment of PBMCs with genotoxic drugs such as melphalan or cisplatin, we also observed that SLE patients were defective in NER and DSB repair mechanisms [3,4]. Accordingly, genes involved in NER [DNA damage-binding protein 1 (DDB1), excision repair cross-complementation group 2 (ERCC2), Xeroderma pigmentosum complementation group A (XPA), Xeroderma pigmentosum complementation group C (XPC)] and DSBs repair [Bloom syndrome RecQ like helicase (BLM), checkpoint kinase 1 (CHEK1), HUS1 checkpoint clamp component (HUS1), Meiotic Recombination 11 Homolog A (MRE11A), Nijmegen Breakage Syndrome 1 (Nibrin; NBN), RAD50, RAD51, Replication Protein A1 (RPA1), tumor protein p53 binding protein 1 (TP53BP1), X-ray repair cross complementing 2 (XRCC2), X-ray repair cross complementing 6 (XRCC6)] were significantly downregulated in SLE patients compared with healthy controls [4]. In line with previous data showing that epigenetic dysregulation (particularly global hypomethylation in T cells) is well documented in SLE [116], we also found that SLE patients are characterized by more condensed chromatin structure at the N-ras locus than their matched controls [4]. Moreover, in accordance with previous data showing that the histone deacetylase inhibitor (HDACi) vorinostat reverses the abnormal chromatin compaction that impedes the access of DNA repair proteins to sites of DNA damage, we found that treatment of PBMCs from quiescent SLE patients with this drug resulted in increased efficiency of the DNA repair machinery and decreased DNA damage burden of these cells [4]. Also, B lymphoblastoid cell lines isolated from children with lupus provided adequate information of defects in the repair of DNA DSBs. Indeed, results from neutral comet assay and colony survival assay showed delayed DSBs repair that might contribute further to the progression of autoimmunity, according to the writers [117]. In addition, previous studies have reported that neutrophils from SLE patients are characterized by increased DNA damage, defective repair of oxidative DNA damage, and augmented apoptosis rates [118,119]. Of note, recent data suggest that the enhanced generation of neutrophil extracellular traps (NETosis) driven by mitochondrial ROS promotes externalization of pro-inflammatory oxidized mtDNA and subsequent activation of STING-dependent type I IFN signaling pathway in SLE [120].

In line with our results [4], another research group, which obtained gene expression profiles from SLE patients and healthy individuals, reported downregulation of genes classified in cell cycle sensors (ATPase/ATPase domain-containing genes) and in NER pathway (ERCC2/XPD and ERCC5/XPG) [121]. Indeed, the team concluded that ATP depletion in combination with downregulation of ATP-dependent genes ERCC2 and ERCC5 suggest insufficient DNA repair in SLE patients, resulting in increased apoptosis and perpetuation of autoimmunity. Moreover, in order to identify rare alleles associated with SLE, Delgado-Vega and colleagues performed whole exome sequencing in SLE patients from well-studied Icelandic SLE multi-case families [122]. They found rare, possibly pathogenic variants in 19 genes, including the X-ray repair cross-complementation group 6 binding protein 1 (XRCC6BP1), also termed Ku70-binding protein 3 (KUB3). Of note, the XRCC6 protein, which is involved in NHEJ required for DSB repair pathway and V(D)J recombination, is a well-established lupus autoantigen. Defective PARP1 activity has also been found in PBMCs from SLE patients [123]. In that particular study, Cerboni and colleagues showed that the activity of PARP1 after UV radiation was significantly lower in SLE patients than in healthy controls, suggesting that PARP1 is implicated in the susceptibility for SLE development.

4.1.3. Autoantibodies against DNA Repair Enzymes

The two subunits of Ku protein (Ku70 and Ku80) involved in NHEJ and the DNA-dependent protein kinase (DNA-PK), a pivotal component of the DNA repair machinery that governs the response to DNA damage and is also involved in V(D)J recombination, are known targets of autoantibodies in SLE. Moreover, ELISA and immunoblotting assays in sera from a total of 155 patients with systemic autoimmune diseases identified two more proteins of NHEJ pathway, namely, DNA ligase IV and XRCC4, as autoantibody targets in approximately 20% of SLE patients [124]. Another research group studied by immunoprecipitation the correlation between anti-Ku antibodies in SLE sera and antibodies against four different DNA repair proteins (DNA-PK, PARP, Mre11, and Werner protein) and found that more than 50% of anti-Ku positive sera contained at least one out of four autoantibodies, providing further evidence that abnormal DSB repair influences the development of certain autoimmune diseases [125]. Recently, Luo and colleagues, using a commercial human protein microarray platform bearing over 9400 antigens, compared the autoantibody profile of SLE patients with those of healthy controls and found novel autoantibodies that were related to DNA repair pathways and apoptosis [126]. Interestingly, they observed that the levels of autoantibodies against Apurinic/Apyrimidinic Endodeoxyribonuclease 1 (APEX1), High mobility group box 1 (HMGB1), vaccinia-related kinase 1 (VRK1), Aurora-A kinase (AURKA), peptidyl arginine deiminase 4 (PADI4), and signal recognition particle 19 (SRP19) (all involved in the DDR/R pathways) were positively correlated with the level of anti-dsDNA in SLE patients, suggesting that these autoantibodies may play a critical role in the pathogenesis of SLE.

4.1.4. Defects in Apoptosis and SLE Pathogenesis

Recent studies suggest that either dysregulated apoptosis or defects in dead cell clearance contribute to the perpetuation of autoimmunity and SLE pathogenesis [127]. GWAS studies in the previous years have identified at least eight different genes that function in the clearance of apoptotic cells, finding that their underexpression is related with the development of SLE or a similar phenotype of autoimmunity [128]. Analysis of bone marrow immune cells by immunochemistry from 14 SLE patients (5 of them presented active lupus nephritis at the moment of the bone marrow biopsy) revealed a significantly higher percentage of apoptotic cells than in controls, which was also positively correlated with the number of plasmacytoid dendritic cells, the major type I IFN-a producer [129]. Interestingly, our previous studies have shown that genotoxic drug-induced apoptosis rates were higher in PBMCs from quiescent SLE patients than healthy controls and correlated inversely with DNA repair efficiency, supporting the hypothesis that accumulation of DNA damage contributes to increased apoptosis [3,4]. Also, the same cells after vorinostat treatment showed a suppressed apoptotic rate through modifications in the degree of the chromatin condensation. Accordingly, several apoptosis-associated genes [Protein phosphatase 1 regulatory subunit 15A (PPP1R15A), cyclin-dependent kinase inhibitor 1A (CDKN1A), BRCA1-associated RING domain protein 1 (BARD1), RAD21, RAD9 checkpoint clamp component A (RAD9A), Protein Kinase, DNA-Activated, Catalytic Subunit (PRKDC), Calcium and integrin-binding protein 1 (CIB1), BRCA1, Abelson murine leukemia viral oncogene homolog 1 (ABL1), checkpoint kinase 2 (CHEK2), and Bcl-2-binding component 3 (BBC3)] were found to be significantly overexpressed in SLE compared with healthy controls [4]. Taken together, in these studies we proposed that SLE patients are characterized by lower DNA repair capacity, resulting in the accumulation of DNA damage and the induction of the apoptotic pathway.

Of note, studies in lupus animal models have proven an association between defective apoptosis and SLE development. Indeed, macrophages from mice lacking the intracellular receptor of the membrane tyrosine kinase c-mer (Merkd mice) present in vivo impaired clearance of apoptotic bodies [130]. Moreover, these mice develop a lupus-like autoimmunity phenotype with autoantibodies to ssDNA and dsDNA and renal pathology. In addition, Shao and Cohen reported that mice lacking the T cell immunoglobulin mucin 4 (TIM-4), a phosphatidylserine receptor that assists phagocytosis of apoptotic debris by macrophages, develop autoantibodies to dsDNA (hallmark of SLE), suggesting that this protein is an important component for dead cell clearance [25].

4.2. Systemic Sclerosis

Systemic sclerosis is a connective tissue disorder characterized by vascular alterations, autoantibody production, and fibrosis of skin and internal organs [131]. Although the pathophysiology of SSc remains largely unknown, oxidative stress has been implicated in the development and perpetuation of SSc [132]. In fibroblasts isolated from the skin of patients with diffuse SSc, levels of ROS and type I collagen are significantly higher and the amounts of free thiol are significantly lower when compared to normal fibroblasts [133]. Moreover, sera from patients with diffuse SSc and lung fibrosis contain elevated levels of advanced oxidation protein products (AOPPs) compared to sera from healthy individuals or from patients with limited SSc and no lung fibrosis [134]. AOPPS are able to induce hydrogen peroxide production by endothelial cells. Similarly, in vitro treatment of endothelial cells with sera from patients with either limited or diffuse SSc induced higher hydrogen peroxide production compared to sera from healthy individuals [134]. AOPPs are also able to induce proliferation of fibroblasts. Of interest, in vitro synthesized AOPPs, from DNA topoisomerase 1 oxidized by hypochlorous acid or hydroxyl radicals, increased the proliferation of fibroblasts and the production of hydrogen peroxide by endothelial cells compared to AOPP generated from other proteins [134]. In line with these results, oxidative stress induced by either immunoglobulins isolated from SSc patients or by oxidative DNA-damaging agents led to decreased Wingless inhibitory factor 1 (WIF-1) expression in SSc fibroblasts and was associated with higher collagen production. This effect, mediated by the central DDR kinase ATM, linked oxidative stress and DNA damage with fibrosis, suggesting an important role of the DDR/R pathway in pathogenesis of fibrosing conditions such as SSc [135]. On the other hand, inhibition of ATM in SSc fibroblasts with the competitive inhibitor KU55933 (KuDOS 55933) significantly increased the expression of the WIF-1 gene, suggesting a therapeutic benefit from targeting components of the DDR/R [135].

Moreover, increased DNA damage levels have also been detected in the peripheral blood of patients with SSc, regardless of disease subtype (diffuse or limited SSc) or treatment [23]. To examine whether DNA damage is a result of dysfunction of DNA repair enzymes, DNA damage and polymorphic sites in two genes encoding DNA repair enzymes XRCC1 [arginine to glutamine polymorphism at position 399 (Arg399Gln)] and XRCC4 [Isoleucine to threonine polymorphism at position 401 (Ile401Thr)] were evaluated. Regarding the XRCC1 gene, healthy individuals with the Arg399Gln allele presented higher levels of DNA damage compared with healthy individuals with the XRCC1 wild type, something that was not observed in SSc patients. However, SSc patients with either XRCC1 allele presented increased DNA damage compared to healthy individuals. Regarding the XRCC4 gene, both healthy individuals and SSc patients with the Ile401Thr allele presented higher levels of DNA damage compared to healthy individuals or SSc patients with the XRCC4 wild type allele [23]. Together, these results indicate that SSc patients with polymorphisms at genes of DNA repair enzymes are characterized by increased DNA damage. Of interest, XRCC4 was also found to be enriched in patients with diffuse SSc in a study using whole-exome sequencing (WES) in 32 diffuse cutaneous systemic sclerosis (dcSSc) patients and 17 healthy controls [136].

4.3. Rheumatoid Arthritis

The role of oxidative DNA damage and aberrations of the DDR/R network have been long studied in RA [137]. Initial studies reported overexpression and tissue-specific mutations of p53, a central molecule in DNA repair and regulator of apoptosis, in the synovium of patients with RA [138,139]. P53 mutations were characteristically detected at the lining region of the synovium [140], which mainly consists of fibroblast-like synoviocytes (FLS), the "maestro" of synovial inflammatory milieu [141], and macrophage-like synoviocytes. Immunohistochemical analysis of RA synovial tissues revealed compensatory up-regulation of MMR enzymes, especially in the synovial lining, which, however, did not completely invert the observed oxidative damage [142]. Of interest, neutrophils isolated from synovial fluid of RA patients also displayed increased DNA damage levels when compared to osteoarthritic controls [143]. Extracellular mitochondrial DNA and 8-oxo-2′-deoxyguanosine (8-oxodG) DNA was also detected in synovial fluid from RA patients but not in controls [144].

Further, we and others have detected increased endogenous DNA damage levels in peripheral blood (PBMCs or granulocytes) of patients with RA compared with healthy controls [24,89,143]. Of interest, a positive correlation of endogenous DNA damage levels in PBMCs/peripheral blood neutrophils with the disease activity index DAS-28 has also been observed [89,143]. Previous studies have reported lower levels of DNA damage in neutrophils and T cells of patients under treatment compared with treatment-naïve patients. [143,145]. In line with these results, in our recent study we examined paired samples from patients before and after 12 week antirheumatic treatment and observed a significant decrease in the endogenous DNA damage levels [24].

Accumulation of the endogenous DNA damage in cells can be mediated either by augmented endogenous DNA damage formation and/or delayed/decreased efficiency of the DNA repair mechanisms, two possibilities that are not mutually exclusive. Numerous studies have shown increased levels of oxidative stress in RA PBMCs/neutrophils in correlation with endogenous DNA damage levels [24,89,143]. Increased levels of 8-oxodG have also been found in DNA of peripheral blood lymphocytes, CD4+ T cells, and granulocytes of RA patients [145,146]. Further, we have recently shown that abasic site formation, the most common spontaneously occurring DNA lesion, is also increased in patients with RA [24].

On the other hand, previous studies have shown significant defects in the DNA repair capacity of RA patients in association with increased senescence and apoptosis [145,146]. When cultured, RA lymphocytes showed increased apoptotic rates in association with spontaneous accumulation of DNA damage [145], whereas they were also more sensitive to hydrogen peroxide-induced DNA damage and growth-arrest [146]. Further, RA T cells were more sensitive to ionizing radiation and showed delayed repair of DNA damage. Several sensors of DSBs were down-regulated in RA T cells (ATM, Rad50, MRE11, and NBS1) in the basal state and also failed to increase in response to radiation-induced DNA damage [145]. Of note, recent studies have shown that MRE11A, in addition to its DNA repair activity, plays a critical role in mitochondria protection, as the deficiency of MRE11A in RA T cells disrupted mitochondrial oxygen consumption; suppressed ATP generation; caused leakage of mitochondrial DNA into the cytosol; and induced inflammasome assembly, caspase-1 activation, and pyroptotic cell death [147]. In line with these results, we have recently shown that RA PBMCs are characterized by decreased repairing capacity, mainly due to defects in the global genome repair (GGR) pathway of NER, directly controlled by the degree of chromatin condensation, and we also showed that 3 month treatment with antirheumatic drugs reversed the observed phenotype [24].

Taken together, accumulation of endogenous DNA damage, derived from augmented formation of DNA damage and deregulated DDR/R signals, which are both reversible after therapy, is implicated in the pathogenesis of RA. Last but not least, although Shao and colleagues [145] failed to spot similar deficiencies in components of the DNA repair machinery in SLE-derived CD4+ T cells, we have previously shown strong down-regulation of both ATM and MRE11 complex in PBMCs of patients with quiescent SLE [4]. The possibility of a shared defective mechanism among autoimmune diseases is rather tempting, taking into consideration the central role of nucleic acid metabolism and recognition in initialization and perpetuation of autoimmunity (Figure 1) [22].

5. Conclusion and Future Directions

The DDR/R network and the ImmR act synergistically for the survival of all living organisms. Aberrant activation of each one of these systems often leads to chronic and potentially fatal systemic autoimmune diseases. As reviewed herein, a balance shift in DDR/R may negatively affect ImmR; evidently, the opposite also may occur. As depicted in Figure 2, we propose that epigenetically regulated functional abnormalities of DNA repair mechanisms (i.e., downregulation of DDR/R-related genes and condensed chromatin structure that result in defective repair) and increased endogenous DNA damage formation, partly due to the induction of oxidative stress, may result in the augmented accumulation of DNA damage (both SSBs and DSBs). This accumulation may trigger the induction of apoptosis, which facilitates autoantibody production, as well as the generation of damaged cytosolic DNA and micronuclei that both can act as potent immunostimulators through the induction of the cGAS-STING-IRF3 pathway and the production of type I IFN, leading to systemic autoimmune disease expression. Notably, some of the components are partially reversible following histone hyperacetylation.

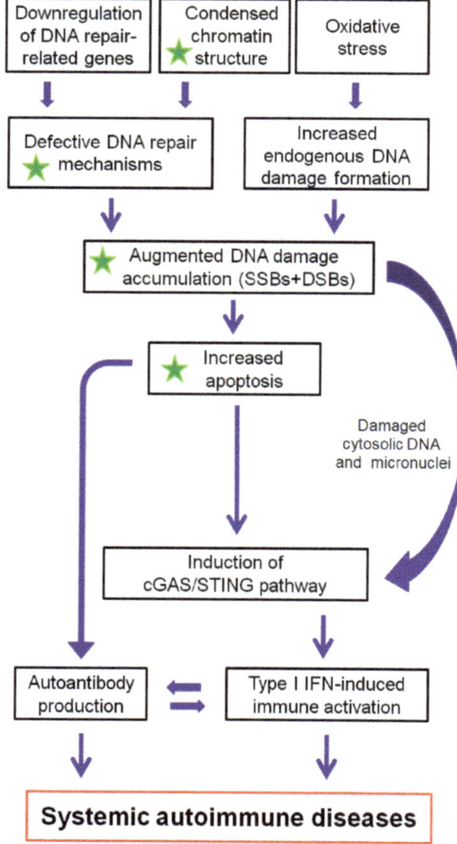

Figure 2. A proposed model of systemic autoimmune disease promotion by epigenetically regulated functional abnormalities of the DNA damage response and repair (DDR/R) network and oxidative stress. The green asterisk denotes partial reversibility following histone hyperacetylation. SSBs: single-strand breaks, DSBs: double-strand breaks.

Because targeting the DDR/R network can have an impact not only on cancer cells but also on host immunity, manipulation of molecular components of this network, alone or in combination with immune checkpoint inhibitors, has gained significant attention in cancer immunotherapy [148,149]. Although these approaches have been extensively studied in cancer, promising results have also been revealed in preclinical mouse models of autoimmunity. Therapeutic targeting of T lymphocytes from autoimmune disease patients with DDR/R inhibitors is based on their high proliferation rate and accumulation of DNA damage [150]. Indeed, in mice with experimental autoimmune encephalitis, the combination of p53 activators and CHK1/2 inhibitors led to the elimination of pathogenic, activated T lymphocytes with no side-toxicity of normal T cells. In addition, our recent data have shown that treatment of human SLE-derived PBMCs with the HDACi vorinostat results in hyperacetylation of histone H4, chromatin decondensation, restoration of the DNA repair capacity, and decreased apoptosis rates [4]. These results are in line with previous data, showing that HDACi ameliorate disease in lupus mouse models [151–153]. Also, treatment of lupus-prone Mrl/lpr mice with the HDACi panobinostat significantly reduced circulating naïve B and plasma cell numbers and the levels of autoantibodies [154]. More importantly, in children with systemic-onset juvenile idiopathic arthritis, the HDACi givinostat was found to be safe and beneficial, particularly in reducing the arthritic features, suggesting that HDACi may have important clinical applications in the treatment of systemic autoimmunity [155]. On the other hand, restoration of defective DNA repair factors, such as MRE11A, has also shown promising results in reducing the pro-inflammatory, pro-arthritogenic capacity of RA T-cells in vivo [156], whereas ATM overexpression in RA T cells was able to invert the observed apoptotic phenotype [145].

Taken together, the results reviewed herein suggest that the deregulated interplay between DDR/R and ImmR plays a crucial role in the pathogenesis and progression of systemic autoimmune diseases. Thus, unraveling the molecular mechanisms of this interplay can be exploited for understanding pathogenesis and progression of these diseases, as well as to discover new treatment opportunities in the field.

Author Contributions: Conceptualization, V.L.S., N.I.V., and P.P.S.; writing—review and editing, V.L.S., N.I.V., M.P., A.A., P.A.N., and P.P.S.; all authors have read and approved the submitted version. All authors have read and agreed to the published version of the manuscript.

Funding: This research was funded by an educational grant (National and Kapodistrian University of Athens ELKE, grant number 0974) and a PhD scholarship to M.P. by IKY (2018-050-0502-13136).

Conflicts of Interest: The authors declare no conflict of interest.

References

1. Jackson, S.P.; Bartek, J. The DNA-damage response in human biology and disease. *Nature* **2009**, *461*, 1071–1078. [CrossRef] [PubMed]
2. Pateras, I.S.; Havaki, S.; Nikitopoulou, X.; Vougas, K.; Townsend, P.A.; Panayiotidis, M.I.; Georgakilas, A.G.; Gorgoulis, V.G. The DNA damage response and immune signaling alliance: Is it good or bad? Nature decides when and where. *Pharmacol. Ther.* **2015**, *154*, 36–56. [CrossRef] [PubMed]
3. Souliotis, V.L.; Sfikakis, P.P. Increased DNA double-strand breaks and enhanced apoptosis in patients with lupus nephritis. *Lupus* **2015**, *24*, 804–815. [CrossRef] [PubMed]
4. Souliotis, V.L.; Vougas, K.; Gorgoulis, V.G.; Sfikakis, P.P. Defective DNA repair and chromatin organization in patients with quiescent systemic lupus erythematosus. *Arthritis Res. Ther.* **2016**, *18*, 182. [CrossRef] [PubMed]
5. Brzostek-Racine, S.; Gordon, C.; Van Scoy, S.; Reich, N.C. The DNA damage response induces IFN. *J. Immunol.* **2011**, *187*, 5336–5345. [CrossRef]
6. Härtlova, A.; Erttmann, S.F.; Raffi, F.A.; Schmalz, A.M.; Resch, U.; Anugula, S.; Lienenklaus, S.; Nilsson, L.M.; Kröger, A.; Nilsson, J.A.; et al. DNA damage primes the type I interferon system via the cytosolic DNA sensor STING to promote anti-microbial innate immunity. *Immunity* **2015**, *42*, 332–343.
7. Galluzzi, L.; Buqué, A.; Kepp, O.; Zitvogel, L.; Kroemer, G. Immunological Effects of Conventional Chemotherapy and Targeted Anticancer Agents. *Cancer Cell* **2015**, *28*, 690–714. [CrossRef]
8. Shen, Y.J.; Le Bert, N.; Chitre, A.A.; Koo, C.X.; Nga, X.H.; Ho, S.S.W.; Khatoo, M.; Tan, N.Y.; Ishii, K.J.; Gasser, S. Genome-derived cytosolic DNA mediates type I interferon-dependent rejection of B cell lymphoma cells. *Cell Rep.* **2015**, *11*, 460–473. [CrossRef]
9. Nakad, R.; Schumacher, B. DNA Damage Response and Immune Defense: Links and Mechanisms. *Front. Genet.* **2016**, *7*, 147. [CrossRef]

10. Erdal, E.; Haider, S.; Rehwinkel, J.; Harris, A.L.; McHugh, P.J. A prosurvival DNA damage-induced cytoplasmic interferon response is mediated by end resection factors and is limited by Trex1. *Genes Dev.* **2017**, *31*, 353–369. [CrossRef]
11. Mackenzie, K.J.; Carroll, P.; Lettice, L.; Tarnauskaitė, Ž.; Reddy, K.; Dix, F.; Revuelta, A.; Abbondati, E.; Rigby, R.E.; Rabe, B.; et al. Ribonuclease H2 mutations induce a cGAS/STING-dependent innate immune response. *EMBO J.* **2016**, *35*, 831–844. [CrossRef] [PubMed]
12. Mackenzie, K.J.; Carroll, P.; Martin, C.-A.; Murina, O.; Fluteau, A.; Simpson, D.J.; Olova, N.; Sutcliffe, H.; Rainger, J.K.; Leitch, A.; et al. cGAS surveillance of micronuclei links genome instability to innate immunity. *Nature* **2017**, *548*, 461–465. [CrossRef] [PubMed]
13. Harding, S.M.; Benci, J.L.; Irianto, J.; Discher, D.E.; Minn, A.J.; Greenberg, R.A. Mitotic progression following DNA damage enables pattern recognition within micronuclei. *Nature* **2017**, *548*, 466–470. [CrossRef] [PubMed]
14. Bartsch, K.; Knittler, K.; Borowski, C.; Rudnik, S.; Damme, M.; Aden, K.; Spehlmann, M.E.; Frey, N.; Saftig, P.; Chalaris, A.; et al. Absence of RNase H2 triggers generation of immunogenic micronuclei removed by autophagy. *Hum. Mol. Genet.* **2017**, *26*, 3960–3972. [CrossRef]
15. Meira, L.B.; Bugni, J.M.; Green, S.L.; Lee, C.-W.; Pang, B.; Borenshtein, D.; Rickman, B.H.; Rogers, A.B.; Moroski-Erkul, C.A.; McFaline, J.L. DNA damage induced by chronic inflammation contributes to colon carcinogenesis in mice. *J. Clin. Invest.* **2008**, *118*, 2516–2525. [CrossRef]
16. Ohnishi, S.; Ma, N.; Thanan, R.; Pinlaor, S.; Hammam, O.; Murata, M.; Kawanishi, S. DNA Damage in Inflammation-Related Carcinogenesis and Cancer Stem Cells. *Oxid. Med. Cell. Longev.* **2013**, *2013*, 1–9. [CrossRef]
17. Pálmai-Pallag, T.; Bachrati, C.Z. Inflammation-induced DNA damage and damage-induced inflammation: A vicious cycle. *Microbes Infect.* **2014**, *16*, 822–832. [CrossRef]
18. Kidane, D.; Chae, W.J.; Czochor, J.; Eckert, K.A.; Glazer, P.M.; Bothwell, A.L.; Sweasy, J.B. Interplay between DNA repair and inflammation, and the link to cancer. *Crit. Rev. Biochem. Mol. Biol.* **2014**, *49*, 116–139. [CrossRef]
19. Fontes, F.L.; Pinheiro, D.M.; Oliveira, A.H.; Oliveira, R.K.; Lajus, T.B.; Agnez-Lima, L.F. Role of DNA repair in host immune response and inflammation. *Mutat. Res. Rev. Mutat. Res.* **2015**, *763*, 246–257. [CrossRef]
20. Kiraly, O.; Gong, G.; Olipitz, W.; Muthupalani, S.; Engelward, B.P. Inflammation-Induced Cell Proliferation Potentiates DNA Damage-Induced Mutations In Vivo. *PLoS Genet.* **2015**, *11*, e1004901. [CrossRef]
21. Neves-Costa, A.; Moita, L.F. Modulation of inflammation and disease tolerance by DNA damage response pathways. *FEBS J.* **2017**, *284*, 680–698. [CrossRef] [PubMed]
22. Theofilopoulos, A.N.; Kono, D.H.; Baccala, R. The multiple pathways to autoimmunity. *Nat. Immunol.* **2017**, *18*, 716–724. [CrossRef] [PubMed]
23. Palomino, G.M.; Bassi, C.L.; Wastowski, I.J.; Xavier, D.J.; Lucisano-Valim, Y.M.; Crispim, J.C.O.; Rassi, D.M.; Marques-Netom, J.F.; Sakamoto-Hojo, E.T.; Moreau, P.; et al. Patients with systemic sclerosis present increased DNA damage differentially associated with DNA repair gene polymorphisms. *J. Rheumatol.* **2014**, *41*, 458–465. [CrossRef] [PubMed]
24. Souliotis, V.L.; Vlachogiannis, N.I.; Pappa, M.; Argyriou, A.; Sfikakis, P.P. DNA damage accumulation, defective chromatin organization and deficient DNA repair capacity in patients with rheumatoid arthritis. *Clin. Immunol.* **2019**, *203*, 28–36. [CrossRef] [PubMed]
25. Shao, W.-H.; Cohen, P.L. Disturbances of apoptotic cell clearance in systemic lupus erythematosus. *Arthritis Res. Ther.* **2011**, *13*, 202. [CrossRef] [PubMed]
26. Shuck, S.C.; Short, E.A.; Turchi, J.J. Eukaryotic nucleotide excision repair: From understanding mechanisms to influencing biology. *Cell Res.* **2008**, *18*, 64–72. [CrossRef] [PubMed]
27. Shah, P.; He, Y.-Y. Molecular regulation of UV-induced DNA repair. *Photochem. Photobiol.* **2015**, *91*, 254–264. [CrossRef]
28. Schärer, O.D. Nucleotide excision repair in eukaryotes. *Cold Spring Harb. Perspect. Biol.* **2013**, *5*, a012609. [CrossRef]
29. Cleaver, J.E.; Lam, E.T.; Revet, I. Disorders of nucleotide excision repair: The genetic and molecular basis of heterogeneity. *Nat. Rev. Genet.* **2009**, *10*, 756–768. [CrossRef]
30. Zharkov, D.O. Base excision DNA repair. *Cell. Mol. Life Sci.* **2008**, *65*, 1544–1565. [CrossRef]
31. Krokan, H.E.; Bjørås, M. Base excision repair. *Cold Spring Harb. Perspect. Biol.* **2013**, *5*, a012583. [CrossRef] [PubMed]

32. Luo, M.; He, H.; Kelley, M.R.; Georgiadis, M.M. Redox regulation of DNA repair: Implications for human health and cancer therapeutic development. *Antioxid. Redox Signal.* **2010**, *12*, 1247–1269. [CrossRef] [PubMed]
33. Martin, S.A.; Lord, C.J.; Ashworth, A. Therapeutic targeting of the DNA mismatch repair pathway. *Clin. Cancer Res.* **2010**, *16*, 5107–5113. [CrossRef] [PubMed]
34. Jiricny, J. The multifaceted mismatch-repair system. *Nat. Rev. Mol. Cell Biol.* **2006**, *7*, 335–346. [CrossRef] [PubMed]
35. Shaheen, M.; Allen, C.; Nickoloff, J.A.; Hromas, R. Synthetic lethality: Exploiting the addiction of cancer to DNA repair. *Blood* **2011**, *117*, 6074–6082. [CrossRef] [PubMed]
36. Diaz-Padilla, I.; Romero, N.; Amir, E.; Matias-Guiu, X.; Vilar, E.; Muggia, F.; Garcia-Donas, J. Mismatch repair status and clinical outcome in endometrial cancer: A systematic review and meta-analysis. *Crit. Rev. Oncol. Hematol.* **2013**, *88*, 154–167. [CrossRef]
37. Li, Y.-H.; Wang, X.; Pan, Y.; Lee, D.-H.; Chowdhury, D.; Kimmelman, A.C. Inhibition of non-homologous end joining repair impairs pancreatic cancer growth and enhances radiation response. *PLoS ONE* **2012**, *7*, e39588. [CrossRef]
38. Ciccia, A.; Elledge, S.J. The DNA Damage Response: Making It Safe to Play with Knives. *Mol. Cell* **2010**, *40*, 179–204. [CrossRef]
39. Dueva, R.; Iliakis, G. Alternative pathways of non-homologous end joining (NHEJ) in genomic instability and cancer. *Transl. Cancer Res.* **2013**, *2*, 163–177.
40. Scott, S.; Pandita, T.K. The cellular control of DNA double-strand breaks. *J. Cell. Biochem.* **2006**, *99*, 1463–1475. [CrossRef]
41. Mills, K.D.; Ferguson, D.O.; Alt, F.W. The role of DNA breaks in genomic instability and tumorigenesis. *Immunol. Rev.* **2003**, *194*, 77–95. [CrossRef] [PubMed]
42. Helleday, T. Homologous recombination in cancer development, treatment and development of drug resistance. *Carcinogenesis* **2010**, *31*, 955–960. [CrossRef] [PubMed]
43. Shah Punatar, R.; Martin, M.J.; Wyatt, H.D.; Chan, Y.W.; West, S.C. Resolution of single and double Holliday junction recombination intermediates by GEN1. *Proc. Natl. Acad. Sci. USA* **2017**, *114*, 443–450. [CrossRef] [PubMed]
44. Chakraborty, A.; Tapryal, N.; Venkova, T.; Horikoshi, N.; Pandita, R.K.; Sarker, A.H.; Pandita, T.K.; Hazra, T.K. Classical non-homologous end-joining pathway utilizes nascent RNA for error-free double-strand break repair of transcribed genes. *Nat. Commun.* **2016**, *7*, 13049. [CrossRef] [PubMed]
45. Davis, A.J.; Chen, D.J. DNA double strand break repair via non-homologous end-joining. *Transl. Cancer Res.* **2013**, *2*, 130–143.
46. Bunting, S.F.; Nussenzweig, A. End-joining, translocations and cancer. *Nat. Rev. Cancer* **2013**, *13*, 443–454. [CrossRef]
47. Zha, S.; Guo, C.; Boboila, C.; Oksenych, V.; Cheng, H.-L.; Zhang, Y.; Wesemann, D.R.; Yuen, G.; Patel, H.; Goff, P.H.; et al. ATM damage response and XLF repair factor are functionally redundant in joining DNA breaks. *Nature* **2011**, *469*, 250–254. [CrossRef]
48. Iliakis, G.; Murmann, T.; Soni, A. Alternative end-joining repair pathways are the ultimate backup for abrogated classical non-homologous end-joining and homologous recombination repair: Implications for the formation of chromosome translocations. *Mutat. Res. Genet. Toxicol. Environ. Mutagen.* **2015**, *793*, 166–175. [CrossRef]
49. Yu, A.M.; McVey, M. Synthesis-dependent microhomology-mediated end joining accounts for multiple types of repair junctions. *Nucleic Acids Res.* **2010**, *38*, 5706–5717. [CrossRef]
50. Simsek, D.; Jasin, M. Alternative end-joining is suppressed by the canonical NHEJ component Xrcc4-ligase IV during chromosomal translocation formation. *Nat. Struct. Mol. Biol.* **2010**, *17*, 410–416. [CrossRef]
51. Deriano, L.; Roth, D.B. Modernizing the nonhomologous end-joining repertoire: Alternative and classical NHEJ share the stage. *Annu. Rev. Genet.* **2013**, *47*, 433–455. [CrossRef] [PubMed]
52. Hartlerode, J.; Scully, R. Mechanisms of double-strand break repair in somatic mammalian cells. *Biochem. J.* **2009**, *423*, 157–168. [CrossRef] [PubMed]
53. Bhargava, R.; Onyango, D.O.; Stark, J.M. Regulation of Single-Strand Annealing and its Role in Genome Maintenance. *Trends Genet.* **2016**, *32*, 566–575. [CrossRef] [PubMed]
54. Hashimoto, S.; Anai, H.; Hanada, K. Mechanisms of interstrand DNA crosslink repair and human disorders. *Genes Environ.* **2016**, *38*, 9. [CrossRef]

55. Deans, A.J.; West, S.C. DNA interstrand crosslink repair and cancer. *Nat. Rev. Cancer* **2011**, *11*, 467–480. [CrossRef]
56. Garaycoechea, J.I.; Crossan, G.P.; Langevin, F.; Daly, M.; Arends, M.J.; Patel, K.J. Genotoxic consequences of endogenous aldehydes on mouse haematopoietic stem cell function. *Nature* **2012**, *489*, 571–575. [CrossRef]
57. Wang, Y.; Leung, J.W.; Jiang, Y.; Lowery, M.G.; Do, H.; Vasquez, K.M.; Chen, J.; Wang, W.; Li, L. FANCM and FAAP24 maintain genome stability via cooperative as well as unique functions. *Mol. Cell* **2013**, *49*, 997–1009. [CrossRef]
58. Clauson, C.; Schärer, O.D.; Niedernhofer, L. Advances in understanding the complex mechanisms of DNA interstrand cross-link repair. *Cold Spring Harb. Perspect. Biol.* **2013**, *5*, a012732. [CrossRef]
59. Kaina, B.; Christmann, M.; Naumann, S.; Roos, W.P. MGMT: Key node in the battle against genotoxicity, carcinogenicity and apoptosis induced by alkylating agents. *DNA Repair (Amst.)* **2007**, *6*, 1079–1099. [CrossRef]
60. Hiddinga, B.I.; Pauwels, P.; Janssens, A.; van Meerbeeck, J.P. O6-Methylguanine-DNA methyltransferase (MGMT): A drugable target in lung cancer? *Lung Cancer (Amst. Neth.)* **2017**, *107*, 91–99. [CrossRef]
61. Plummer, R. Perspective on the pipeline of drugs being developed with modulation of DNA damage as a target. *Clin. Cancer Res.* **2010**, *16*, 4527–4531. [CrossRef] [PubMed]
62. Fan, C.H.; Liu, W.L.; Cao, H.; Wen, C.; Chen, L.; Jiang, G. O(6)-methylguanine DNA methyltransferase as a promising target for the treatment of temozolomide-resistant gliomas. *Cell Death Dis.* **2013**, *4*, e876. [CrossRef] [PubMed]
63. Crow, Y.J.; Leitch, A.; Hayward, B.E.; Garner, A.; Parmar, R.; Griffith, E.; Ali, M.; Semple, C.; Aicardi, J.; Babul-Hirji, R.; et al. Mutations in genes encoding ribonuclease H2 subunits cause Aicardi-Goutières syndrome and mimic congenital viral brain infection. *Nat. Genet.* **2006**, *38*, 910–916. [CrossRef] [PubMed]
64. Pizzi, S.; Sertic, S.; Orcesi, S.; Cereda, C.; Bianchi, M.; Jackson, A.P.; Lazzaro, F.; Plevani, P.; Muzi-Falconi, M. Reduction of hRNase H2 activity in Aicardi-Goutières syndrome cells leads to replication stress and genome instability. *Hum. Mol. Genet.* **2015**, *24*, 649–658. [CrossRef] [PubMed]
65. Hiller, B.; Achleitner, M.; Glage, S.; Naumann, R.; Behrendt, R.; Roers, A. Mammalian RNase H2 removes ribonucleotides from DNA to maintain genome integrity. *J. Exp. Med.* **2012**, *209*, 1419–1426. [CrossRef] [PubMed]
66. Crow, Y.J.; Hayward, B.E.; Parmar, R.; Robins, P.; Leitch, A.; Ali, M.; Black, D.N.; van Bokhoven, H.; Brunner, H.G.; Hamel, B.C.; et al. Mutations in the gene encoding the 3′-5′ DNA exonuclease TREX1 cause Aicardi-Goutières syndrome at the AGS1 locus. *Nat. Genet.* **2006**, *38*, 917–920. [CrossRef]
67. Chowdhury, D.; Beresford, P.J.; Zhu, P.; Zhang, D.; Sung, J.S.; Demple, B.; Perrino, F.W.; Lieberman, J. The exonuclease TREX1 is in the SET complex and acts in concert with NM23–H1 to degrade DNA during granzyme A-mediated cell death. *Mol. Cell* **2006**, *23*, 133–142. [CrossRef]
68. Yang, Y.G.; Lindahl, T.; Barnes, D.E. Trex1 exonuclease degrades ssDNA to prevent chronic checkpoint activation and autoimmune disease. *Cell* **2007**, *131*, 873–886. [CrossRef]
69. Günther, C.; Kind, B.; Reijns, M.A.; Berndt, N.; Martinez-Bueno, M.; Wolf, C.; Tüngler, V.; Chara, O.; Lee, Y.A.; Hübner, N.; et al. Defective removal of ribonucleotides from DNA promotes systemic autoimmunity. *J. Clin. Invest.* **2015**, *125*, 413–424. [CrossRef]
70. Gekara, N.O. DNA damage-induced immune response: Micronuclei provide key platform. *J. Cell. Biol.* **2017**, *216*, 2999–3001. [CrossRef]
71. Gao, D.; Li, T.; Li, X.-D.; Chen, X.; Li, Q.-Z.; Wight-Carter, M.; Chen, Z. Activation of cyclic GMP-AMP synthase by self-DNA causes autoimmune diseases. *Proc. Natl. Acad. Sci. USA* **2015**, *112*, 5699–5705. [CrossRef] [PubMed]
72. Fucikova, J.; Moserova, I.; Urbanova, L.; Bezu, L.; Kepp, O.; Cremer, I.; Salek, C.; Strnad, P.; Kroemer, G.; Galluzzi, L.; et al. Prognostic and Predictive Value of DAMPs and DAMP-Associated Processes in Cancer. *Front. Immunol.* **2015**, *6*, 402. [CrossRef] [PubMed]
73. Zitvogel, L.; Galluzzi, L.; Kepp, O.; Smyth, M.J.; Kroemer, G. Type I interferons in anticancer immunity. *Nat. Rev. Immunol.* **2015**, *15*, 405–414. [CrossRef] [PubMed]
74. Galluzzi, L.; Buqué, A.; Kepp, O.; Zitvogel, L.; Kroemer, G. Immunogenic cell death in cancer and infectious disease. *Nat. Rev. Immunol.* **2016**, *17*, 97–111. [CrossRef]
75. Morita, M.; Stamp, G.; Robins, P.; Dulic, A.; Rosewell, I.; Hrivnak, G.; Daly, G.; Lindahl, T.; Barnes, D.E. Gene-targeted mice lacking the Trex1 (DNase III) 3′–>5′ DNA exonuclease develop inflammatory myocarditis. *Mol. Cell. Biol.* **2004**, *24*, 6719–6727. [CrossRef]

76. Stetson, D.B.; Ko, J.S.; Heidmann, T.; Medzhitov, R. Trex1 prevents cell-intrinsic initiation of autoimmunity. *Cell* **2008**, *134*, 587–598. [CrossRef]
77. Hasan, M.; Dobbs, N.; Khan, S.; White, M.A.; Wakeland, E.K.; Li, Q.Z.; Yan, N. Cutting edge: Inhibiting TBK1 by compound II ameliorates autoimmune disease in mice. *J. Immunol.* **2015**, *195*, 4573–4577. [CrossRef]
78. Zuo, L.; Prather, E.R.; Stetskiv, M.; Garrison, D.E.; Meade, J.R.; Peace, T.I.; Zhou, T. Inflammaging and Oxidative Stress in Human Diseases: From Molecular Mechanisms to Novel Treatments. *Int. J. Mol. Sci.* **2019**, *20*, 4472. [CrossRef]
79. Sedelnikova, O.A.; Redon, C.E.; Dickey, J.S.; Nakamura, A.J.; Georgakilas, A.G.; Bonner, W.M. Role of oxidatively induced DNA lesions in human pathogenesis. *Mutat. Res.* **2010**, *704*, 152–159. [CrossRef]
80. Efremov, Y.R.; Proskurina, A.S.; Potter, E.A.; Dolgova, E.V.; Efremova, O.V.; Taranov, O.S.; Ostanin, A.A.; Chernykh, E.R.; Kolchanov, N.A.; Bogachev, S.S. Cancer Stem Cells: Emergent Nature of Tumor Emergency. *Front. Genet.* **2018**, *9*, 544. [CrossRef]
81. Altieri, F.; Grillo, C.; Maceroni, M.; Chichiarelli, S. DNA damage and repair: From molecular mechanisms to health implications. *Antioxid. Redox Signal.* **2008**, *10*, 891–930. [CrossRef] [PubMed]
82. Kryston, T.B.; Georgiev, A.B.; Pissis, P.; Georgakilas, A.G. Role of oxidative stress and DNA damage in human carcinogenesis. *Mutat. Res.* **2011**, *711*, 193–201. [CrossRef] [PubMed]
83. Čolak, E.; Ignjatović, S.; Radosavljević, A.; Žorić, L. The association of enzymatic and non-enzymatic antioxidant defense parameters with inflammatory markers in patients with exudative form of age-related macular degeneration. *J. Clin. Biochem. Nutr.* **2017**, *60*, 100–107. [CrossRef]
84. Bokhari, B.; Sharma, S. Stress Marks on the Genome: Use or Lose? *Int. J. Mol. Sci.* **2019**, *20*, 364. [CrossRef]
85. Perl, A. Oxidative stress in the pathology and treatment of systemic lupus erythematosus. *Nat. Rev. Rheumatol.* **2013**, *9*, 674–686. [CrossRef]
86. Redza-Dutordoir, M.; Averill-Bates, D.A. Activation of apoptosis signalling pathways by reactive oxygen species. *Biochim. Biophys. Acta* **2016**, *1863*, 2977–2992. [CrossRef]
87. Doridot, L.; Jeljeli, M.; Chêne, C.; Batteux, F. Implication of oxidative stress in the pathogenesis of systemic sclerosis via inflammation, autoimmunity and fibrosis. *Redox Biol.* **2019**, 101122. [CrossRef]
88. Vona, R.; Giovannetti, A.; Gambardella, L.; Malorni, W.; Pietraforte, D.; Straface, E. Oxidative stress in the pathogenesis of systemic scleroderma: An overview. *J. Cell. Mol. Med.* **2018**, *22*, 3308–3314. [CrossRef]
89. Altindag, O.; Karakoc, M.; Kocyigit, A.; Celik, H.; Soran, N. Increased DNA damage and oxidative stress in patients with rheumatoid arthritis. *Clin. Biochem.* **2007**, *40*, 167–171. [CrossRef]
90. Gehrke, N.; Mertens, C.; Zillinger, T.; Wenzel, J.; Bald, T.; Zahn, S.; Tüting, T.; Hartmann, G.; Barchet, W. Oxidative Damage of DNA Confers Resistance to Cytosolic Nuclease TREX1 Degradation and Potentiates STING-Dependent Immune Sensing. *Immunity* **2013**, *39*, 482–495. [CrossRef]
91. Kemp, M.G.; Lindsey-Boltz, L.A.; Sancar, A. UV Light Potentiates STING (Stimulator of Interferon Genes)-dependent Innate Immune Signaling through Deregulation of ULK1 (Unc51-like Kinase 1). *J. Biol. Chem.* **2015**, *290*, 12184–12194. [CrossRef] [PubMed]
92. Gorgoulis, V.; Adams, P.D.; Alimonti, A.; Bennett, D.C.; Bischof, O.; Bishop, C.; Campisi, J.; Collado, M.; Evangelou, K.; Ferbeyre, G.; et al. Cellular Senescence: Defining a Path Forward. *Cell* **2019**, *179*, 813–827. [CrossRef] [PubMed]
93. Rai, P.; Onder, T.T.; Young, J.J.; McFaline, J.L.; Pang, B.; Dedon, P.C.; Weinberg, R.A. Continuous elimination of oxidized nucleotides is necessary to prevent rapid onset of cellular senescence. *Proc. Natl. Acad. Sci. USA* **2009**, *106*, 169–174. [CrossRef] [PubMed]
94. Lee, M.Y.; Wang, Y.; Vanhoutte, P.M. Senescence of cultured porcine coronary arterial endothelial cells is associated with accelerated oxidative stress and activation of NFkB. *J. Vasc. Res.* **2010**, *47*, 287–298. [CrossRef]
95. Ito, K.; Hirao, A.; Arai, F.; Takubo, K.; Matsuoka, S.; Miyamoto, K.; Ohmura, M.; Naka, K.; Hosokawa, K.; Ikeda, Y.; et al. Reactive oxygen species act through p38 MAPK to limit the lifespan of hematopoietic stem cells. *Nat. Med.* **2006**, *12*, 446–451. [CrossRef]
96. Li, G.; Luna, C.; Qiu, J.; Epstein, D.L.; Gonzalez, P. Alterations in microRNA expression in stress-induced cellular senescence. *Mech. Ageing Dev.* **2009**, *130*, 731–741. [CrossRef]
97. De Cecco, M.; Ito, T.; Petrashen, A.P.; Elias, A.E.; Skvir, N.J.; Criscione, S.W.; Caligiana, A.; Brocculi, G.; Adney, E.M.; Boeke, J.D.; et al. L1 drives IFN in senescent cells and promotes age-associated inflammation. *Nature* **2019**, *566*, 73–78. [CrossRef]

98. Watad, A.; Bragazzi, N.L.; Adawi, M.; Amital, H.; Toubi, E.; Porat, B.S.; Shoenfeld, Y. Autoimmunity in the Elderly: Insights from Basic Science and Clinics—A Mini-Review. *Gerontology* **2017**, *63*, 515–523. [CrossRef]
99. Agrawal, A.; Sridharan, A.; Prakash, S.; Agrawal, H. Dendritic cells and aging: Consequences for autoimmunity. *Expert Rev. Clin. Immunol.* **2012**, *8*, 73–80. [CrossRef]
100. Franceschi, C.; Garagnani, P.; Morsiani, C.; Conte, M.; Santoro, A.; Grignolio, A.; Monti, D.; Capri, M.; Salvioli, S. The Continuum of Aging and Age-Related Diseases: Common Mechanisms but Different Rates. *Front. Med. (Lausanne)* **2018**, *5*, 61. [CrossRef]
101. Corrales-Medina, V.F.; Alvarez, K.N.; Weissfeld, L.A.; Angus, D.C.; Chirinos, J.A.; Chang, C.C.; Newman, A.; Loehr, L.; Folsom, A.R.; Elkind, M.S.; et al. Association between hospitalization for pneumonia and subsequent risk of cardiovascular disease. *JAMA* **2015**, *313*, 264–274. [CrossRef] [PubMed]
102. Crooke, S.N.; Ovsyannikova, I.G.; Poland, G.A.; Kennedy, R.B. Immunosenescence and human vaccine immune responses. *Immun. Ageing* **2019**, *16*, 25. [CrossRef] [PubMed]
103. Kawane, K.; Ohtani, M.; Miwa, K.; Kizawa, T.; Kanbara, Y.; Yoshioka, Y.; Yoshikawa, H.; Nagata, S. Chronic polyarthritis caused by mammalian DNA that escapes from degradation in macrophages. *Nature* **2006**, *443*, 998–1002. [CrossRef] [PubMed]
104. Tsokos, G.C. Systemic lupus erythematosus. *N. Engl. J. Med.* **2011**, *365*, 2110–2121. [CrossRef]
105. Baccala, R.; Hoebe, K.; Kono, D.H.; Beutler, B.; Theofilopoulos, A.N. TLR-dependent and TLR-independent pathways of type I interferon induction in systemic autoimmunity. *Nat. Med.* **2007**, *13*, 543–551. [CrossRef]
106. Grieves, J.L.; Fye, J.M.; Harvey, S.; Grayson, J.M.; Hollis, T.; Perrino, F.W. Exonuclease TREX1 degrades double-stranded DNA to prevent spontaneous lupus-like inflammatory disease. *Proc. Natl. Acad. Sci. USA* **2015**, *112*, 5117–5122. [CrossRef]
107. Noble, P.W.; Bernatsky, S.; Clarke, A.E.; Isenberg, D.A.; Ramsey-Goldman, R.; Hansen, J.E. DNA-damaging autoantibodies and cancer: The lupus butterfly theory. *Nat. Rev. Rheumatol.* **2016**, *12*, 429–434. [CrossRef]
108. Senejani, A.G.; Liu, Y.; Kidane, D.; Maher, S.E.; Zeiss, C.J.; Park, H.J.; Kashgarian, M.; McNiff, J.M.; Zelterman, D.; Bothwell, A.L.; et al. Mutation of POLB causes lupus in mice. *Cell Rep.* **2014**, *6*, 1–8. [CrossRef]
109. Liu, L.; Tran, E.; Zhao, Y.; Huang, Y.; Flavell, R.; Lu, B. Gadd45 beta and Gadd45 gamma are critical for regulating autoimmunity. *J. Exp. Med.* **2005**, *202*, 1341–1347. [CrossRef]
110. Bassi, C.; Xavier, D.; Palomino, G.; Nicolucci, P.; Soares, C.; Sakamoto-Hojo, E.; Donadi, E. Efficiency of the DNA repair and polymorphisms of the XRCC1, XRCC3 and XRCC4 DNA repair genes in systemic lupus erythematosus. *Lupus* **2008**, *17*, 988–995. [CrossRef]
111. Lin, Y.J.; Wan, L.; Huang, C.; Chen, S.; Huang, Y.; Lai, C.; Lin, W.; Liu, H.P.; Wu, Y.S.; Chen, C.M.; et al. Polymorphisms in the DNA repair gene XRCC1 and associations with systemic lupus erythematosus risk in the Taiwanese Han Chinese population. *Lupus* **2009**, *18*, 1246–1251. [CrossRef] [PubMed]
112. Sheng, Y.J.; Gao, J.P.; Li, J.; Han, J.W.; Xu, Q.; Hu, W.; Pan, T.; Cheng, Y.; Yu, Z.Y.; Ni, C.; et al. Follow-up study identifies two novel susceptibility loci PRKCB and 8p11.21 for systemic lupus erythematosus. *Rheumatology (Oxford)* **2011**, *50*, 682–688. [CrossRef] [PubMed]
113. Han, J.W.; Zheng, H.F.; Cui, Y.; Sun, L.; Ye, D.Q.; Hu, Z.; Xu, J.H.; Cai, Z.M.; Huang, W.; Zhao, G.P.; et al. Genome-wide association study in a Chinese Han population identifies nine new susceptibility loci for systemic lupus erythematosus. *Nat. Genet.* **2009**, *41*, 1234–1237. [CrossRef] [PubMed]
114. Tsao, B.P.; Cantor, R.M.; Grossman, J.M.; Shen, N.; Teophilov, N.T.; Wallace, D.J.; Arnett, F.C.; Hartung, K.; Goldstein, R.; Kalunian, K.C. PARP alleles within the linked chromosomal region are associated with systemic lupus erythematosus. *J. Clin. Invest.* **1999**, *103*, 1135–1140. [CrossRef] [PubMed]
115. Mireles-Canales, M.P.; González-Chávez, S.A.; Quiñonez-Flores, C.M.; León-López, E.A.; Pacheco-Tena, C. DNA Damage and Deficiencies in the Mechanisms of Its Repair: Implications in the Pathogenesis of Systemic Lupus Erythematosus. *J. Immunol. Res.* **2018**, *2018*, 8214379. [CrossRef]
116. Ballestar, E.; Esteller, M.; Richardson, B.C. The epigenetic face of systemic lupus erythematosus. *J. Immunol.* **2006**, *176*, 7143–7147. [CrossRef]
117. Davies, R.C.; Pettijohn, K.; Fike, F.; Wang, J.; Nahas, S.A.; Tunuguntla, R.; Hu, H.; Gatti, R.A.; McCurdy, D. Defective DNA double-strand break repair in pediatric systemic lupus erythematosus. *Arthritis Rheum.* **2012**, *64*, 568–578. [CrossRef]
118. McConnell, J.R.; Crockard, A.D.; Cairns, A.P.; Bell, A.L. Neutrophils from systemic lupus erythematosus patients demonstrate increased nuclear DNA damage. *Clin. Exp. Rheumatol.* **2002**, *20*, 653–660.

119. Kaplan, M. Neutrophils in the pathogenesis and manifestations of SLE. *Nat. Rev. Rheumatol.* **2011**, *7*, 691–699. [CrossRef]
120. Lood, C.; Blanco, L.P.; Purmalek, M.M.; Carmona-Rivera, C.; De Ravin, S.S.; Smith, C.K.; Malech, H.L.; Ledbetter, J.A.; Elkon, K.B.; Kaplan, M.J. Neutrophil extracellular traps enriched in oxidized mitochondrial DNA are interferogenic and contribute to lupus-like disease. *Nat. Med.* **2016**, *22*, 146–153. [CrossRef]
121. Lee, H.M.; Sugino, H.; Aoki, C.; Nishimoto, N. Underexpression of mitochondrial-DNA encoded ATP synthesis-related genes and DNA repair genes in systemic lupus erythematosus. *Arthritis Res. Ther.* **2011**, *13*, R63. [CrossRef] [PubMed]
122. Delgado-Vega, A.M.; Martínez-Bueno, M.; Oparina, N.Y.; López Herráez, D.; Kristjansdottir, H.; Steinsson, K.; Kozyrev, S.V.; Alarcón-Riquelme, M.E. Whole Exome Sequencing of Patients from Multicase Families with Systemic Lupus Erythematosus Identifies Multiple Rare Variants. *Sci. Rep.* **2018**, *8*, 8775. [CrossRef] [PubMed]
123. Cerboni, B.; Morozzi, G.; Galeazzi, M.; Bellisai, F.; Micheli, V.; Pompucci, G.; Sestini, S. Poly(ADP-ribose) polymerase activity in systemic lupus erythematosus and systemic sclerosis. *Hum. Immunol.* **2009**, *70*, 487–491. [CrossRef]
124. Lee, K.J.; Dong, X.; Wang, J.; Takeda, Y.; Dynan, W.S. Identification of human autoantibodies to the DNA ligase IV/XRCC4 complex and mapping of an autoimmune epitope to a potential regulatory region. *J. Immunol.* **2002**, *169*, 3413–3421. [CrossRef]
125. Fell, V.L.; Schild-Poulter, C. The Ku heterodimer: Function in DNA repair and beyond. *Mutat. Res. Rev. Mutat. Res.* **2015**, *763*, 15–29. [CrossRef]
126. Luo, H.; Wang, L.; Bao, D.; Wang, L.; Zhao, H.; Lian, Y.; Yan, M.; Mohan, C.; Li, Q.Z. Novel Autoantibodies Related to Cell Death and DNA Repair Pathways in Systemic Lupus Erythematosus. *Genom. Proteom. Bioinform.* **2019**, *17*, 248–259. [CrossRef]
127. Mahajan, A.; Herrmann, M.; Muñoz, L.E. Clearance Deficiency and Cell Death Pathways: A Model for the Pathogenesis of SLE. *Front. Immunol.* **2016**, *7*, 35. [CrossRef]
128. Tsokos, G.C.; Lo, M.S.; Costa Reis, P.; Sullivan, K.E. New insights into the immunopathogenesis of systemic lupus erythematosus. *Nat. Rev. Rheumatol.* **2016**, *12*, 716–730. [CrossRef]
129. Park, J.W.; Moon, S.Y.; Lee, J.H.; Park, J.K.; Lee, D.S.; Jung, K.C.; Song, Y.W.; Lee, E.B. Bone marrow analysis of immune cells and apoptosis in patients with systemic lupus erythematosus. *Lupus* **2014**, *23*, 975–985. [CrossRef]
130. Cohen, P.L.; Caricchio, R.; Abraham, V.; Camenisch, T.D.; Jennette, J.C.; Roubey, R.A.; Earp, H.S.; Matsushima, G.; Reap, E.A. Delayed apoptotic cell clearance and lupus-like autoimmunity in mice lacking the c-mer membrane tyrosine kinase. *J. Exp. Med.* **2002**, *196*, 135–140. [CrossRef]
131. Gabrielli, A.; Avvedimento, E.V.; Krieg, T. Scleroderma. *N. Engl. J. Med.* **2009**, *360*, 1989–2003. [CrossRef] [PubMed]
132. Sambo, P.; Baroni, S.S.; Luchetti, M.; Paroncini, P.; Dusi, S.; Orlandini, G.; Gabrielli, A. Oxidative stress in scleroderma: Maintenance of scleroderma fibroblast phenotype by the constitutive up-regulation of reactive oxygen species generation through the NADPH oxidase complex pathway. *Arthritis Rheum.* **2001**, *44*, 2653–2664. [CrossRef]
133. Tsou, P.S.; Talia, N.N.; Pinney, A.J.; Kendzicky, A.; Piera-Velazquez, S.; Jimenez, S.A.; Seibold, J.R.; Phillips, K.; Koch, A.E. Effect of oxidative stress on protein tyrosine phosphatase 1B in scleroderma dermal fibroblasts. *Arthritis Rheum.* **2012**, *64*, 1978–1989. [CrossRef] [PubMed]
134. Servettaz, A.; Goulvestre, C.; Kavian, N.; Nicco, C.; Guilpain, P.; Chéreau, C.; Vuiblet, V.; Guillevin, L.; Mouthon, L.; Weill, B.; et al. Selective oxidation of DNA topoisomerase 1 induces systemic sclerosis in the mouse. *J. Immunol.* **2009**, *182*, 5855–5864. [CrossRef]
135. Svegliati, S.; Marrone, G.; Pezone, A.; Spadoni, T.; Grieco, A.; Moroncini, G.; Grieco, D.; Vinciguerra, M.; Agnese, S.; Jüngel, A.; et al. Oxidative DNA damage induces the ATM-mediated transcriptional suppression of the Wnt inhibitor WIF-1 in systemic sclerosis and fibrosis. *Sci. Signal.* **2014**, *7*, ra84. [CrossRef]
136. Mak, A.C.; Tang, P.L.; Cleveland, C.; Smith, M.H.; Kari Connolly, M.; Katsumoto, T.R.; Wolters, P.J.; Kwok, P.Y.; Criswell, L.A. Brief Report: Whole-Exome Sequencing for Identification of Potential Causal Variants for Diffuse Cutaneous Systemic Sclerosis. *Arthritis Rheumatol.* **2016**, *68*, 2257–2262. [CrossRef]
137. Shao, L. DNA Damage Response Signals Transduce Stress From Rheumatoid Arthritis Risk Factors Into T Cell Dysfunction. *Front. Immunol.* **2018**, *9*, 3055. [CrossRef]

138. Firestein, G.S.; Nguyen, K.; Aupperle, K.R.; Yeo, M.; Boyle, D.L.; Zvaifler, N.J. Apoptosis in rheumatoid arthritis: p53 overexpression in rheumatoid arthritis synovium. *Am. J. Pathol.* **1996**, *149*, 2143.
139. Firestein, G.S.; Echeverri, F.; Yeo, M.; Zvaifler, N.J.; Green, D.R. Somatic mutations in the p53 tumor suppressor gene in rheumatoid arthritis synovium. *Proc. Natl. Acad. Sci. USA* **1997**, *94*, 10895–10900. [CrossRef]
140. Yamanishi, Y.; Boyle, D.L.; Rosengren, S.; Green, D.R.; Zvaifler, N.J.; Firestein, G.S. Regional analysis of p53 mutations in rheumatoid arthritis synovium. *Proc. Natl. Acad. Sci. USA* **2002**, *99*, 10025–10030. [CrossRef]
141. Bottini, N.; Firestein, G.S. Duality of fibroblast-like synoviocytes in RA: Passive responders and imprinted aggressors. *Nat. Rev. Rheumatol.* **2013**, *9*, 24–33. [CrossRef] [PubMed]
142. Simelyte, E. DNA mismatch repair enzyme expression in synovial tissue. *Ann. Rheum. Dis.* **2004**, *63*, 1695–1699. [CrossRef] [PubMed]
143. Martelli-Palomino, G.; Paoliello-Paschoalato, A.B.; Crispim, J.C.; Rassi, D.M.; Oliveira, R.D.; Louzada, P.; Lucisano-Valim, Y.M.; Donadi, E.A. DNA damage increase in peripheral neutrophils from patients with rheumatoid arthritis is associated with the disease activity and the presence of shared epitope. *Clin. Exp. Rheumatol.* **2017**, *35*, 247–254. [PubMed]
144. Hajizadeh, S.; DeGroot, J.; TeKoppele, J.M.; Tarkowski, A.; Collins, L.V. Extracellular mitochondrial DNA and oxidatively damaged DNA in synovial fluid of patients with rheumatoid arthritis. *Arthritis Res. Ther.* **2003**, *5*, R234. [CrossRef] [PubMed]
145. Shao, L.; Fujii, H.; Colmegna, I.; Oishi, H.; Goronzy, J.J.; Weyand, C.M. Deficiency of the DNA repair enzyme ATM in rheumatoid arthritis. *J. Exp. Med.* **2009**, *206*, 1435–1449. [CrossRef]
146. Bashir, S.; Harris, G.; Denman, M.A.; Blake, D.R.; Winyard, P.G. Oxidative DNA damage and cellular sensitivity to oxidative stress in human autoimmune diseases. *Ann. Rheum. Dis.* **1993**, *52*, 659–666. [CrossRef]
147. Li, Y.; Shen, Y.; Jin, K.; Wen, Z.; Cao, W.; Wu, B.; Wen, R.; Tian, L.; Berry, G.J.; Goronzy, J.J.; et al. The DNA Repair Nuclease MRE11A Functions as a Mitochondrial Protector and Prevents T Cell Pyroptosis and Tissue Inflammation. *Cell Metab.* **2019**, *30*, 477–492. [CrossRef]
148. Mouw, K.W.; Goldberg, M.S.; Konstantinopoulos, P.A.; D'Andrea, A.D. DNA Damage and Repair Biomarkers of Immunotherapy Response. *Cancer Discov.* **2017**, *7*, 675–693. [CrossRef]
149. Li, A.; Yi, M.; Qin, S.; Chu, Q.; Luo, S.; Wu, K. Prospects for combining immune checkpoint blockade with PARP inhibition. *J. Hematol. Oncol.* **2019**, *12*, 98. [CrossRef]
150. McNally, J.P.; Millen, S.H.; Chaturvedi, V.; Lakes, N.; Terrell, C.E.; Elfers, E.E.; Carroll, K.R.; Hogan, S.P.; Andreassen, P.R.; Kanter, J.; et al. Manipulating DNA damage-response signaling for the treatment of immune-mediated diseases. *Proc. Natl. Acad. Sci. USA* **2017**, *114*, E4782–E4791. [CrossRef]
151. Mishra, N.; Reilly, C.M.; Brown, D.R.; Ruiz, P.; Gilkeson, G.S. Histone deacetylase inhibitors modulate renal disease in the MRL-lpr/lpr mouse. *J. Clin. Invest.* **2003**, *111*, 539–552. [CrossRef] [PubMed]
152. Regna, N.L.; Vieson, M.D.; Luo, X.M.; Chafin, C.B.; Puthiyaveetil, A.G.; Hammond, S.E.; Caudell, D.L.; Jarpe, M.B.; Reilly, C.M. Specific HDAC6 inhibition by ACY-738 reduces SLE pathogenesis in NZB/W mice. *Clin. Immunol.* **2016**, *162*, 58–73. [CrossRef] [PubMed]
153. Ren, J.; Liao, X.; Vieson, M.D.; Chen, M.; Scott, R.; Kazmierczak, J.; Luo, X.M.; Reilly, C.M. Selective HDAC6 inhibition decreases early stage of lupus nephritis by down-regulating both innate and adaptive immune responses. *Clin. Exp. Immunol.* **2018**, *191*, 19–31. [CrossRef] [PubMed]
154. Waibel, M.; Christiansen, A.J.; Hibbs, M.L.; Shortt, J.; Jones, S.A.; Simpson, I.; Light, A.; O'Donnell, K.; Morand, E.F.; Tarlinton, D.M.; et al. Manipulation of B-cell responses with histone deacetylase inhibitors. *Nat. Commun.* **2015**, *6*, 6838. [CrossRef] [PubMed]
155. Vojinovic, J.; Damjanov, N.; D'Urzo, C.; Furlan, A.; Susic, G.; Pasic, S.; Iagaru, N.; Stefan, M.; Dinarello, C.A. Safety and efficacy of an oral histone deacetylase inhibitor in systemic-onset juvenile idiopathic arthritis. *Arthritis Rheum.* **2011**, *63*, 1452–1458. [CrossRef]
156. Li, Y.; Shen, Y.; Hohensinner, P.; Ju, J.; Wen, Z.; Goodman, S.B.; Zhang, H.; Goronzy, J.J.; Weyand, C.M. Deficient Activity of the Nuclease MRE11A Induces T Cell Aging and Promotes Arthritogenic Effector Functions in Patients with Rheumatoid Arthritis. *Immunity* **2016**, *45*, 903–916. [CrossRef]

International Journal of
Molecular Sciences

Review

DNA Oxidation and Excision Repair Pathways

Tae-Hee Lee and Tae-Hong Kang *

Department of Biological Science, Dong-A University, Busan 49315, Korea; thlee@donga.ac.kr
* Correspondence: thkang@dau.ac.kr; Tel.: +82-51-200-7261; Fax: +82-51-200-7269

Received: 16 November 2019; Accepted: 2 December 2019; Published: 3 December 2019

Abstract: The physiological impact of the aberrant oxidation products on genomic DNA were demonstrated by embryonic lethality or the cancer susceptibility and/or neurological symptoms of animal impaired in the base excision repair (BER); the major pathway to maintain genomic integrity against non-bulky DNA oxidation. However, growing evidence suggests that other DNA repair pathways or factors that are not primarily associated with the classical BER pathway are also actively involved in the mitigation of oxidative assaults on the genomic DNA, according to the corresponding types of DNA oxidation. Among others, factors dedicated to lesion recognition in the nucleotide excision repair (NER) pathway have been shown to play eminent roles in the process of lesion recognition and stimulation of the enzyme activity of some sets of BER factors. Besides, substantial bulky DNA oxidation can be preferentially removed by a canonical NER mechanism; therefore, loss of function in the NER pathway shares common features arising from BER defects, including cancer predisposition and neurological disorders, although NER defects generally are nonlethal. Here we discuss recent achievements for delineating newly arising roles of NER lesion recognition factors to facilitate the BER process, and cooperative works of BER and NER pathways in response to the genotoxic oxidative stress.

Keywords: reactive oxygen species (ROS); nucleotide excision repair (NER); base excision repair (BER); oxidative DNA damage

---

1. Introduction

The integrity of the genome is endlessly threatened by reactive oxygen species (ROS) formed in living cells as metabolic byproducts. Therefore, maintenance of genomic integrity against ROS is a prerequisite for proper cell function and, hence, maintaining homeostasis. Essential cellular functions, such as oxidative phosphorylation and lipid peroxidation, produce ROS that can induce oxidative DNA damage. It has been estimated that as many as ten thousand oxidation reactions harm DNA per each human cell per day [1], which eventually produces a plethora of non-bulky (purine/pyrimidine base oxidations) and bulky (crosslinks, strand breaks, and cyclic bases) DNA lesions. These lesions, if not timely repaired, can interfere with essential DNA metabolisms, including transcription, recombination, and replication, which ultimately can give rise to unfavorable outcomes like cellular senescence and mutagenesis.

Unlike other cellular macromolecules, damaged DNA cannot be replaced and solely depends on repair to remain intact. In order to counteract oxidative DNA lesions, base excision repair (BER) is thought to be the primary pathway to remove non-bulky modifications formed largely on the bases of the DNA, including 8-oxo-7,8-dihydroguanine (8-oxoG), the most prevailing purine base oxidation with highly mutagenic potential, which is sometimes compared with thymine glycol, the most frequent pyrimidine one with relatively limited mutation frequency [2]. Meanwhile, nucleotide excision repair (NER), the most versatile DNA repair system in human cells, has been demonstrated to actively operate to neutralize DNA oxidation, especially with bulky oxidative lesions such as cyclopurines [3].

When damage is extensive, these repair processes are accompanied by cell-cycle checkpoint activation, which provides cells with sufficient time to either complete the repair or initiate apoptosis [4–7].

2. DNA Excision Repairs and Implication on Human Health

DNA lesions arising from the ROS attack can generate both non-bulky (non-helix distorting) and bulky (helix distorting) lesions. In human cells, BER and NER are the two DNA excision repair pathways responsible for the removal of non-bulky and bulky DNA lesions, respectively. Both repair pathways share three common steps, which include 1) lesion recognition, 2) excision of damaged nucleotide, and 3) resynthesis using error-free DNA polymerases (Figure 1).

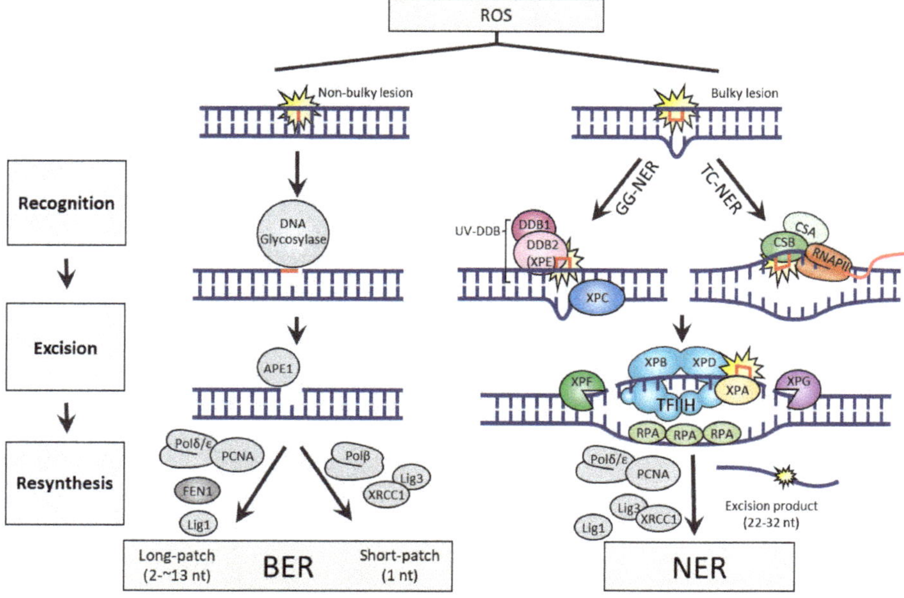

Figure 1. DNA excision repair mechanisms for bulky or non-bulky DNA lesions caused by reactive oxygen species (ROS).

2.1. Base Excision Repair

Many of the genes involved in BER are highly conserved from bacteria to humans [8], indicating that BER is a fundamental repair pathway in most living organisms. The BER pathway is specialized to fix non-bulky single-base lesions in the form of small chemical modifications, including oxidation, alkylation, and deamination damage. Base modifications are pro-mutagenic and/or cytotoxic, depending on how they interfere with the template function of the DNA during replication and transcription. To initiate the procedure, BER employs a specific DNA glycosylase for lesion recognition and elimination of the damaged base. Although every DNA glycosylase has a distinct structure and substrate specificity, all glycosylases share a common mode of action for damage recognition; 1) flipping the affected base out of the DNA helix, which facilitates a sensitive detection of even minor base modifications, 2) catalyzing the cleavage of an N-glycosidic bond, releasing a free base and creating an abasic site (apurinic/apyrimidinic site or AP site). DNA glycosylases can be either monofunctional or bifunctional. Monofunctional DNA glycosylases possess only the glycosylase activity, which includes UNG (uracil-N glycosylase), SMUG1 (single-strand-specific monofunctional uracil DNA glycosylase), MBD4 (methyl-binding domain glycosylase 4), TDG (thymine DNA glycosylase), MYH (MutY homolog DNA glycosylase), MPG (methylpurine glycosylase). In contrast, bifunctional DNA glycosylases have an intrinsic 3′ AP lyase activity accompanying with the glycosylase activity, which includes OGG1 (8-oxoguanine DNA glycosylase), NTH1 (endonuclease III-like), and NEIL1 (endonuclease VIII-like glycosylase). The processes following DNA glycosylase are common to the BER mechanism irrespective of the identity of the glycosylase. Base removal by a DNA glycosylase generates an AP site in DNA, which is then further processed by AP endonuclease 1 (APE1), which cleaves the DNA backbone 5′ to the abasic site, generating a 3′-hydroxyl and a 5′-2-deoxyribose-5′-phosphate (5′-dRP). DNA polymerase β (Polβ) utilizes the 3′-hydroxyl to fill the gap through template-directed synthesis. Depending on the number of nucleotides added, either short-patch (where a single nucleotide is replaced) or long patch BER (where 2-13 nucleotides are synthesized) pathways operate to complete the repair process. In short-patch BER, the intrinsic dRP-lyase activity of Polβ removes the 5′-dRP. The addition of more than one nucleotide (up to 13) constitutes long-patch BER and requires the assistance of flap endonuclease 1 (FEN1) to remove the displaced 5′-flap structure (Figure 1).

2.2. General Features of BER Defect

Genetic loss or mutation in key genes of the BER process, such as *APE1* [9], *Polβ* [10], *FEN1* [11], or the *DNA ligase 3* [12], has been shown to have embryonic lethality in mice, while the phenotype of DNA glycosylase disruptions in mice is usually rather moderate, the only known exception being the TDG, which was essential for embryonic development in mice [13,14].

Neurons encounter particularly high levels of oxidative stress because of the high metabolic rate required to support their electrical and synaptic functions. Thus, the integrity and capacity of systems that repair oxidative DNA damage would be expected to be critical for the survival and proper function of neurons, particularly under conditions of increased oxidative stress that occur during catastrophic pathological conditions, including ischemic stroke. In a study with a mouse model of focal cerebral ischemic brain injury performed on normal and $OGG1^{-/-}$ mice, the potential role of OGG1 in ameliorating the detrimental effect of oxidative DNA damage to neurons was evaluated. The results indicate that after cerebral ischemia, the accumulation of brain oxidative DNA base lesions was significantly greater in *OGG1*-deficient mice, and was associated with greater brain damage and poorer behavioral outcomes, revealing an important role for OGG1 in brain BER capacity, which contributes to neuronal survival after experimental stroke [15].

Neurons from *OGG1*-deficient mice are sensitive to oxidative stress and reduced OGG1 levels have also been associated with Alzheimer's disease (AD) [16]. Indeed, levels of expression of UNG, OGG1, and Polβ are lower in brain tissue from patients with AD than in brain tissue from age-matched controls without AD [17]. $Pol\beta^{+/-}$ heterozygote displayed impaired synaptic and cognitive functions, linking the loss of heterozygous BER function in the progression of AD [18].

While a mouse lacking *UNG* develops B-cell lymphomas, there is often not a clear phenotype in a mouse for a single DNA glycosylase mutation, presumably due to the substrate redundancy. For instance, the *OGG1*-deficient mice are viable and fertile without any visible phenotype [19,20], which suggests overlapping activities for the repair of 8-oxoG lesions. However, interestingly, additional deletion of *MYH* in *OGG1*-deficient mice predisposes 65.7% of mice to tumors, predominantly lung and ovarian tumors, and lymphomas [21]. 8-oxoG indeed is a common substrate of MYH and OGG1.

2.3. Nucleotide Excision Repair

NER is famous for the unique repair pathway in humans to remove photolesions produced by UV radiation (sun exposure) that mainly forms cyclobutane pyrimidine dimer (CPD), a non-bulky lesion and pyrimidine-(6,4)-pyrimidone product (6-4PP), a bulky lesion. Besides, it also efficiently eliminates an extremely broad range of structurally unrelated DNA lesions, including bulky chemical adducts and intrastrand crosslinks [22]. The basis of the versatility of NER originates that it circumvents recognition of the lesion itself, instead, the lesion recognizing NER factors detects the presence of unpaired single-stranded DNA opposite the damaged strand [23].

Owing to the distinct damage recognition events, NER mechanisms can be further specified into two subpathways, global genome NER (GG-NER) and transcription-coupled NER (TC-NER). The former is responsible for eliminating lesions throughout the whole genome, while the latter is for those in the transcribing strand of active genes. During GG-NER, XPC (xeroderma pigmentosum C) or UV-DDB (UV-damaged DNA binding protein; a heterodimeric complex with DDB1 and DDB2) initiate the recognition of the damage. Structural analysis showed certain lesions, such as CPD, which do not significantly distort the DNA helix, are first recognized by DDB2 (also known as XPE) to extrude the lesions into its binding pocket, and thereby create a kink that is now recognized by XPC [24]. If the lesion is in a transcribed gene, it is sensed as a blockage to RNA polymerase II (RNAPII) and requires Cockayne syndrome B (CSB) and CSA to initiate the TC-NER process [25]. Regardless of the damage recognition mechanisms, the downstream events are conserved in both NER mechanisms. Damage verification is executed by XPA and helix unwinding is carried out by TFIIH (complexed with the XPB and XPD helicases). Lesion excision is catalyzed by the structure-specific endonucleases XPF and XPG, which incise the damaged strand at 5′ and 3′ from the lesion, respectively, which promotes releasing out of the lesion containing 22–32 nt-long oligomers. Final DNA gap-filling synthesis and ligation are executed by the replication proteins proliferating cell nuclear antigen (PCNA), Polδ, Polε, and DNA ligase 1 or XRCC1 (X-ray repair cross-complementing protein 1)–DNA ligase 3 complex (Figure 1).

2.4. General Features of NER Defect

Hereditary mutations in NER-associated genes are nonlethal and associated with disorders that are characterized by UV sensitivity and cancer predisposition, such as XP, CS, and trichothiodystrophy (TTD) [26]. XP comprises seven complementation groups (*XPA-XPG*) with defective GG-NER. Five of these groups also exhibit defective TC-NER, whereas *XP-C* and *XP-E* patients are TC-NER-proficient. The XP patient shows hypersensitivity on minimal sun exposure, pigmentary accumulations at exposed skin regions, and multiple early age skin cancers. Progressive neuronal degeneration is also observed in approximately one-third of XP cases, generally after the appearance of cutaneous signs [27]. However, unlike GG-NER-deficient XP patients, CS patients belong to one of two complementation groups (*CS-A* or *CS-B*), and those that are completely defective in TC-NER are not cancer-prone but exhibit a drastic reduction in life span [28]. In addition, CS patients display a number of neurological and developmental abnormalities as well as hypersensitivity to sun exposure. Although the molecular basis that leads to the diverse features of CS remains largely unknown, a reduced ability of cells to relieve oxidative stress has been proposed to be a leading cause [29–31]. Since cells from CS patients were found to be hypersensitive to oxidative DNA damage, a role for the CS proteins in the response to oxidized bases has been proposed [32]. Mutations in the *CSB* account for the majority of CS cases [33].

Clinical heterogeneity in disorders with NER mutations opens the question of whether defects in this pathway are solely due to impaired repair of helix-distorting DNA lesions. XP patients along with TCR defects (caused by some specific alterations in XPB, XPD, and XPG) present, besides increased skin cancer risk, accelerated neurodegeneration, and CS symptoms (XP/CS). The causative relationship between mutations and the CS clinical features in XP/CS cases is complex and must not only involve the NER defects but also the other functions of the NER proteins. Several studies have demonstrated that transcription impairment, oxidative repair, and energy metabolism alteration, as well as genotoxic stress, may explain the combined XP/CS phenotype [34,35]. Neuronal death might be due to accumulated endogenous damage, and indeed a growing body of evidence indicates that NER proteins participate in the processing of oxidative DNA lesions that are produced by the normal cell metabolism. The role of NER proteins in different pathways might explain the heterogeneity in disorders with NER mutations [36].

3. Facilitated BER Kinetics by the NER Lesion Recognition Factors

The role of NER factors in the control of oxidative DNA damage is reviewed elsewhere [36]. Intriguingly, the key lesion recognition factors CSB (for TC-NER) and UV-DDB and XPC (for GG-NER) are implicated in facilitating the enzyme activity of key BER factors, such as DNA glycosylases and APE1, thereby they contribute to the protection of cells against non-bulky oxidative DNA lesion, such as 8-oxoG. Surprisingly; however, this initial recruitment of NER factors does not trigger the downstream NER process, which includes the lesion verification by the XPA and TFIIH complex [37].

3.1. XPC

XPC, the main lesion sensor in GG-NER presents as a heterotrimeric complex with HR23B (human RAD23 homolog B) and centrin 2 proteins. HR23B stabilizes the XPC protein [38], and the XPC-HR23B heterodimer is sufficient to reconstitute the cell-free NER reaction [39], whereas centrin 2 appears to facilitate the damage-specific DNA binding activity of the XPC complex [40]. This complex binds to various types of bulky lesions, thus triggering GG-NER and it also participates in the repair of non-bulky base lesions. Consequently, *XPC* deficiency not only results in decreased GG-NER but has also been linked to disturbed redox homeostasis due to the accumulating DNA oxidations. The XPC complex functionally interacts with OGG1, MPG, and TDG that initiate BER of oxidation, alkylation, and deamination products, respectively. D'Errico et al. show that the XPC–HR23B complex acts as a cofactor in the BER pathway via mediating the OGG1 loading and turnover kinetics, thereby freeing OGG1 to react with remaining lesions [41]. While analyzing the biochemical properties behind mutations found in *XP-C* patients, a critical single amino acid substitution at position 334 (P334H) weakening the interaction with OGG1 is defined. Cells from this patient exhibit low efficiency of UV-induced unscheduled DNA synthesis and a decreased OGG1 cleavage activity, indicating that the OGG1 activity is stimulated by XPC through direct interaction with its N-terminal part that encompasses the P334 surrounding region. This patient is also one of the rare XP-C patients who exhibit neurological symptoms [42].

The interaction of the MPG protein with XPC-HR23B proteins stimulates the DNA glycosylase activity and is correlated to the increased binding affinity of the MPG-HR23B protein complex for the substrate [43]. Biochemical studies demonstrate that XPC-HR23B also participates in BER of guanine/thymine or guanine/uridine mismatches, which are mainly derived from hydrolytic deamination of 5-methylcytosines or cytosines, respectively. The BER of these mismatches is initiated by TDG. The XPC complex is capable of stimulating TDG activity by promoting the release of TDG following the excision of the mismatched T base [44]. In addition, XPC stimulates the glycosylase activities of TDG and SMUG1, both of which interact physically with XPC [45].

3.2. UV-DDB

DNA in eukaryotes is packaged in tandemly arrayed nucleosomes that, together with numerous DNA- and nucleosome-associated enzymes and regulatory factors, make up chromatin [46]. Because DNA lesions that result from ROS can occur both within and outside of nucleosomes, the repair efficiency is necessarily dependent on the accessibility and structural requirements for enzyme catalysis to overcome the hindrance presented by the location of a DNA lesion. Therefore, it is reasonable that chromatin remodeling in the vicinity of damaged DNA is critical for enabling efficient repair and the subsequent repackaging of DNA into nucleosomes. Indeed several studies have shown that DNA lesions in the nucleosome are limiting access for glycosylases and APE1.

While XPC is required for GG-NER, it has little or no affinity for CPD lesions and does not recognize 6-4PP in the context of chromatin [47]. XPC recruitment to chromatin is facilitated by the UV-DDB (DDB1 and DDB2) complex [48]. In the absence of DDB2, XPC remains localized to 6-4PP and to a lesser extent to CPDs with substantially delayed kinetics [47]. UV-DDB, as part of CUL4A-RBX E3 ubiquitin ligase, has been shown to modify core histones around the sites of UV lesions [49].

The glycosylases remain bound to their AP site product until displaced by APE1, a step which is rate-limiting for most mammalian glycosylases [50]. Recently, Jang et al. defined the specific roles of UV-DDB in the early steps of BER mechanisms and proposed that UV-DDB is a general sensor of DNA damage in both NER and BER pathways, facilitating damage recognition in the context of chromatin. Specifically, they find that UV-DDB facilitates both OGG1 and APE1 strand cleavage and promotes Polβ-mediated gap-filling activity by 30-fold. The single-molecule real-time imaging technique reveals the dynamic interaction between UV-DDB and OGG1 or APE1, which facilitates turnover rates of OGG1 and APE1 from DNA, hence increasing BER capacity. Furthermore, in light of a novel chemoptogenetic approach, the dynamic recruitment of UV-DDB to locally-induced 8-oxoG sites in telomeric regions of DNA is detected in vivo [51].

3.3. CSB

A transcription elongation factor CSB (also known as ERCC6), in a complex with RNAPII, strongly binds to RNAPII when it is blocked by a bulky lesion and alters nucleosome structure near its occupancy sites by wrapping the DNA around the protein itself to trigger TC-NER [52]. However, 8-oxoG lesions, which only cause minor helix-distortions, do not block RNAPII elongation unless processed by its specific glycosylase OGG1, implying that transcription-coupled BER, if it exists, may not be directly triggered by stalled RNAPII on the oxidative lesions themselves [53].

Evidence for the role of CSB in the BER process has been provided by several groups, which report that cellular extracts from CSB null cells demonstrate reduced incision activity of oxidative DNA lesions in vitro [54–57]. CSB (but not downstream core NER factors) accumulates at sites of locally-induced oxidative damage in vivo, in a transcription-dependent manner, with similar kinetics as the OGG1 [37]. An interesting finding pinpoints that lysine (K) 991 in CSB is subject to being ubiquitinated and this ubiquitination selectively occurs in response to oxidative damage, shedding new light on the critical role of CSB on the discrimination of the different repair choices [58].

PARP1 is believed to stimulate the BER process by recruiting the DNA repair apparatus to the single-strand breaks and is found in complex with the BER protein XRCC1, DNA ligase 3, and Polβ. Thorslund et al. demonstrate that CSB is a novel substrate for PARP1 poly-ADP-ribosylation and that this modification inhibits the catalytic ATPase activity of CSB, while the meaning of poly-ADP-ribosylation of CSB remains to be answered [59]. In the meantime, it was reported that poly-ADP-ribosylated PARP1 is required for retention of CSB at sites of oxidative DNA damage, so that CSB promotes PARP1 displacement from damaged DNA to facilitate BER [60,61].

4. Oxidative DNA Damages that Can Be Readily Repaired by a Canonical NER

4.1. Bulky Lesions

ROS-induced covalent modifications to DNA encompass tandem base modification (a form of intrastrand crosslinks), purine 5′,8-cyclonucleosides, interstrand cross-links, and DNA–protein crosslinks [62,63]. Biochemical studies demonstrated that these lesions could markedly block DNA replication and transcription and that these lesions are repaired by the NER pathway [63,64]. A tandem base lesion G [8–5m]T is structurally similar to the CPD [65]. The XPA-deficient human brain and mouse liver contain higher levels of G[8–5m]T but not ROS-induced simple base lesions [66]. 5′,8-cyclo-2′-deoxyadenosine (cdA) and 5′,8-cyclo-2′-deoxyguanosine (cdG) are tandem lesions produced by the attack of hydroxyl radicals to the purine bases of DNA [67]. These lesions are also repaired primarily by the NER pathway [68].

Covalent DNA–protein crosslinks (DPCs, also known as protein adducts) represent an important class of DNA damage that may be produced according to the different mechanisms by certain enzymes that form covalent reaction intermediates with DNA, chemotherapeutics, and various endogenous and exogenous sources [69]. DPCs are highly toxic as they interfere with nearly all chromatin-based processes. Model studies have shown that 2-deoxyribonolactone (dL), an oxidized abasic site, is able to undergo crosslink formation with enzymes of the BER pathway, including Polβ [70]. Bifunctional DNA glycosylases possessing AP lyase activity, such as OGG1 and NTH1, have also been shown to form DPC in vitro with both dL and with its β–elimination product, butenolide [54]. It has been postulated that proteolytic digestion of the covalently-bound enzyme to DNA could be implicated in the initial repair process prior to being completed by NER enzymes [71].

The abasic site aldehyde is reactive and may progress to an interstrand crosslink (ICL) to a purine on the opposing strand [72,73]. Since the ICL lesion affects both strands of the DNA it is considered as a highly toxic DNA lesion that prevents transcription and replication by inhibiting DNA strand separation. So far, NER, translesion DNA synthesis, homologous recombination, and the Fanconi anemia pathway are identified to be involved in ICL repair in a coordinated fashion [74].

4.2. Non-Bulky Lesions

The 8-oxoG lesions tend to be easily further oxidized to form stereoisomeric spiroiminodihydantoin (Sp) and guanidinohydantoin (Gh) lesions [75,76]. These are still small DNA lesions that are generally efficiently recognized and excised by DNA glycosylases. In vitro analysis demonstrates the hydantoin lesions are not only excellent BER substrates but are also excised by the NER complex prepared as cell-free extract [77]. Recently, Shafirovich et al., using a cell-based assay, demonstrated that the BER and NER pathways compete with one another in intact human cells and can catalyze both Gh and Sp lesions [78]. The relative contribution of either process in intact cells depends on the local availability of the primary NER and BER factors that recognize and bind to the same lesions in a competitive fashion.

In addition, abundant 8-oxoG and thymine glycol lesion, both are canonical BER substrates, can be removed by NER in vitro as fast as CPD lesion [79]. To investigate 8-oxoG repair in intact living cells, a laser-assisted procedure to locally inflict oxidative DNA lesions was developed by Menoni et al. [37]. In light of this in vivo study, strong and very rapid recruitment of CSB and XPC to 8-oxoG lesions was observed. Interestingly, CSB exhibited a direct transcription-dependent repair of oxidative lesions associated with different RNA polymerases (RNAPI and RNAPII), but not involving other NER proteins.

The repair of AP lesions takes place predominantly by the APE1-mediated BER pathway. However, among chemically heterogeneous AP lesions formed in DNA, some are resistant to APE1 and thus refractory to BER [80]. Using reporter constructs accommodating stable APE1-resistant AP lesions, Kitsera et al. demonstrates that NER efficiently removes BER-resistant AP lesions and significantly enhances the repair of APE1-sensitive ones as well [81].

5. Concluding Remarks

The impact of ROS-induced oxidative DNA damage on human health and counteracting DNA excision repair pathways is discussed. Inevitable DNA oxidation reactions can interrupt essential DNA metabolisms and, thereafter, can evoke various stressful cellular responses including replication- and transcription stress response. Therefore, timely relief of the stresses by error-free DNA excision repair systems (BER and NER) is essential to maintain genomic integrity and proper cell function. While the causal relationship between human disorder and the loss of function of a specific NER gene is relatively clear, fewer connections have been made between impaired BER and human diseases. This is likely due to the multitude of backup systems like NER in the removal of small non-bulky lesions.

In light of the state-of-the-art technologies, including CRISPR/CAS9, optogenetics, and next-generation sequencing, great progress has been made towards understanding the precise mechanisms underlying the enhanced BER activity by NER-initiating factors. Besides, through the studies aimed at elucidating the complex mechanisms that underlie the NER- or BER-related phenotypes, now we are getting a better understanding of the fundamentals of diseases such as cancer, aging, and neuropathology. One of the major future challenges is to translate these valuable findings into human health benefits in the clinic for individualized cancer therapies based on precision medicine, and for the development of a well-aging strategy as well. Synthetic lethal strategies targeting DNA repair defects, using small molecule inhibitors such as PARP inhibitors, are shedding light on cancer treatment with low adverse-effects and more tumor-selective killing [82]. Small molecule modulators of NER/BER activity or specific protein–protein interactions could be new tools for chemical biology studies and might lead to new therapeutic approaches [83].

Author Contributions: Conceptualization, T.-H.K. and T.-H.L.; writing—original draft preparation, T.-H.L.; writing—review and editing, T.-H.K.; funding acquisition, T.-H.K.

Funding: This research was supported by the Basic Science Research Program through the National Research Foundation of Korea (NRF) funded by the Ministry of Education (NRF-2018R1D1A3B07043817 and NRF-2015R1D1A1A01056994).

Conflicts of Interest: The authors declare no conflicts of interest.

Abbreviations

5′-dRP	5′-2-deoxyribose-5′-phosphate
6-4PP	Pyrimidine-(6,4)-pyrimidone product
8-oxoG	8-oxo-7,8-dihydroguanine
Abasic site	Apurinic/apyrimidinic site or AP site
AD	Alzheimer′s disease
APE1	AP endonuclease 1
BER	Base excision repair
cdA	5′,8-cyclo-2′-deoxyadenosine
cdG	5′,8-cyclo-2′-deoxyguanosine
CPD	Cyclobutane pyrimidine dimer
CSB	Cockayne syndrome B
dL	2-deoxyribonolactone
DPCs	DNA–protein crosslinks
FEN1	Flap endonuclease 1
GG-NER	Global genome NER
Gh	Guanidinohydantoin
HR23B	Human RAD23 homolog B
ICL	Interstrand crosslink
MBD4	Methyl-binding domain glycosylase 4
MPG	Methylpurine glycosylase
MYH	MutY homolog DNA glycosylase
NEIL1	Endonuclease VIII-like glycosylase
NER	Nucleotide excision repair
NTH1	Endonuclease III-like glycosylase
OGG1	8-oxoguanine DNA glycosylase
PCNA	Proliferating cell nuclear antigen
Polβ	DNA polymerase β
RNAPII	RNA polymerase II
ROS	Reactive oxygen species
SMUG1	Single-strand-specific monofunctional uracil DNA glycosylase
Sp	Spiroiminodihydantoin
TC-NER	Transcription-coupled NER
TDG	Thymine DNA glycosylase
TFIIH	Transcription Factor II H; complexed with the XPB and XPD helicases
TTD	Trichothiodystrophy
UNG	Uracil-N glycosylase
UV-DDB	UV-damaged DNA binding protein; a heterodimeric complex with DDB1 and DDB2
XPC	Xeroderma pigmentosum C
XRCC1	X-ray repair cross-complementing protein 1

References

1. Ciccia, A.; Elledge, S.J. The DNA damage response: Making it safe to play with knives. *Mol. Cell* **2010**, *40*, 179–204. [CrossRef] [PubMed]
2. Svilar, D.; Goellner, E.M.; Almeida, K.H.; Sobol, R.W. Base excision repair and lesion-dependent subpathways for repair of oxidative DNA damage. *Antioxid. Redox Signal.* **2011**, *14*, 2491–2507. [CrossRef] [PubMed]
3. Kuraoka, I.; Bender, C.; Romieu, A.; Cadet, J.; Wood, R.D.; Lindahl, T. Removal of oxygen free-radical-induced 5′,8-purine cyclodeoxynucleosides from DNA by the nucleotide excision-repair pathway in human cells. *Proc. Natl. Acad. Sci. USA* **2000**, *97*, 3832–3837. [CrossRef] [PubMed]
4. Choi, J.Y.; Joh, H.M.; Park, J.M.; Kim, M.J.; Chung, T.H.; Kang, T.H. Non-thermal plasma-induced apoptosis is modulated by ATR- and PARP1-mediated DNA damage responses and circadian clock. *Oncotarget* **2016**, *7*, 32980–32989. [CrossRef] [PubMed]
5. Park, J.M.; Choi, J.Y.; Yi, J.M.; Chung, J.W.; Leem, S.H.; Koh, S.S.; Kang, T.H. NDR1 modulates the UV-induced DNA-damage checkpoint and nucleotide excision repair. *Biochem. Biophys. Res. Commun.* **2015**, *461*, 543–548. [CrossRef] [PubMed]
6. Kang, T.H.; Leem, S.H. Modulation of ATR-mediated DNA damage checkpoint response by cryptochrome 1. *Nucleic Acids Res.* **2014**, *42*, 4427–4434. [CrossRef]
7. Choi, J.Y.; Park, J.M.; Yi, J.M.; Leem, S.H.; Kang, T.H. Enhanced nucleotide excision repair capacity in lung cancer cells by preconditioning with DNA-damaging agents. *Oncotarget* **2015**, *6*, 22575–22586. [CrossRef]
8. Robertson, A.B.; Klungland, A.; Rognes, T.; Leiros, I. DNA repair in mammalian cells: Base excision repair: The long and short of it. *Cell. Mol. Life Sci. CMLS* **2009**, *66*, 981–993. [CrossRef]
9. Xanthoudakis, S.; Smeyne, R.J.; Wallace, J.D.; Curran, T. The redox/DNA repair protein, Ref-1, is essential for early embryonic development in mice. *Proc. Natl. Acad. Sci. USA* **1996**, *93*, 8919–8923. [CrossRef]
10. Gu, H.; Marth, J.D.; Orban, P.C.; Mossmann, H.; Rajewsky, K. Deletion of a DNA polymerase beta gene segment in T cells using cell type-specific gene targeting. *Science* **1994**, *265*, 103–106. [CrossRef]
11. Kucherlapati, M.; Yang, K.; Kuraguchi, M.; Zhao, J.; Lia, M.; Heyer, J.; Kane, M.F.; Fan, K.; Russell, R.; Brown, A.M.; et al. Haploinsufficiency of Flap endonuclease (Fen1) leads to rapid tumor progression. *Proc. Natl. Acad. Sci. USA* **2002**, *99*, 9924–9929. [CrossRef] [PubMed]
12. Puebla-Osorio, N.; Lacey, D.B.; Alt, F.W.; Zhu, C. Early embryonic lethality due to targeted inactivation of DNA ligase III. *Mol. Cell. Biol.* **2006**, *26*, 3935–3941. [CrossRef] [PubMed]
13. Cortazar, D.; Kunz, C.; Selfridge, J.; Lettieri, T.; Saito, Y.; MacDougall, E.; Wirz, A.; Schuermann, D.; Jacobs, A.L.; Siegrist, F.; et al. Embryonic lethal phenotype reveals a function of TDG in maintaining epigenetic stability. *Nature* **2011**, *470*, 419–423. [CrossRef] [PubMed]
14. Cortellino, S.; Xu, J.; Sannai, M.; Moore, R.; Caretti, E.; Cigliano, A.; Le Coz, M.; Devarajan, K.; Wessels, A.; Soprano, D.; et al. Thymine DNA glycosylase is essential for active DNA demethylation by linked deamination-base excision repair. *Cell* **2011**, *146*, 67–79. [CrossRef] [PubMed]
15. Liu, D.; Croteau, D.L.; Souza-Pinto, N.; Pitta, M.; Tian, J.; Wu, C.; Jiang, H.; Mustafa, K.; Keijzers, G.; Bohr, V.A.; et al. Evidence that OGG1 glycosylase protects neurons against oxidative DNA damage and cell death under ischemic conditions. *J. Cereb. Blood Flow Metab. Off. J. Int. Soc. Cereb. Blood Flow Metab.* **2011**, *31*, 680–692. [CrossRef]
16. Lillenes, M.S.; Rabano, A.; Stoen, M.; Riaz, T.; Misaghian, D.; Mollersen, L.; Esbensen, Y.; Gunther, C.C.; Selnes, P.; Stenset, V.T.; et al. Altered DNA base excision repair profile in brain tissue and blood in Alzheimer's disease. *Mol. Brain* **2016**, *9*, 61. [CrossRef]
17. Weissman, L.; Jo, D.G.; Sorensen, M.M.; de Souza-Pinto, N.C.; Markesbery, W.R.; Mattson, M.P.; Bohr, V.A. Defective DNA base excision repair in brain from individuals with Alzheimer's disease and amnestic mild cognitive impairment. *Nucleic Acids Res.* **2007**, *35*, 5545–5555. [CrossRef]
18. Sykora, P.; Misiak, M.; Wang, Y.; Ghosh, S.; Leandro, G.S.; Liu, D.; Tian, J.; Baptiste, B.A.; Cong, W.N.; Brenerman, B.M.; et al. DNA polymerase beta deficiency leads to neurodegeneration and exacerbates Alzheimer disease phenotypes. *Nucleic Acids Res.* **2015**, *43*, 943–959. [CrossRef]
19. Klungland, A.; Rosewell, I.; Hollenbach, S.; Larsen, E.; Daly, G.; Epe, B.; Seeberg, E.; Lindahl, T.; Barnes, D.E. Accumulation of premutagenic DNA lesions in mice defective in removal of oxidative base damage. *Proc. Natl. Acad. Sci. USA* **1999**, *96*, 13300–13305. [CrossRef]

20. Minowa, O.; Arai, T.; Hirano, M.; Monden, Y.; Nakai, S.; Fukuda, M.; Itoh, M.; Takano, H.; Hippou, Y.; Aburatani, H.; et al. Mmh/Ogg1 gene inactivation results in accumulation of 8-hydroxyguanine in mice. *Proc. Natl. Acad. Sci. USA* **2000**, *97*, 4156–4161. [CrossRef]
21. Xie, Y.; Yang, H.; Cunanan, C.; Okamoto, K.; Shibata, D.; Pan, J.; Barnes, D.E.; Lindahl, T.; McIlhatton, M.; Fishel, R.; et al. Deficiencies in mouse Myh and Ogg1 result in tumor predisposition and G to T mutations in codon 12 of the K-ras oncogene in lung tumors. *Cancer Res.* **2004**, *64*, 3096–3102. [CrossRef] [PubMed]
22. Marteijn, J.A.; Lans, H.; Vermeulen, W.; Hoeijmakers, J.H. Understanding nucleotide excision repair and its roles in cancer and ageing. *Nat. Rev. Mol. cell Biol.* **2014**, *15*, 465–481. [CrossRef] [PubMed]
23. Kusakabe, M.; Onishi, Y.; Tada, H.; Kurihara, F.; Kusao, K.; Furukawa, M.; Iwai, S.; Yokoi, M.; Sakai, W.; Sugasawa, K. Mechanism and regulation of DNA damage recognition in nucleotide excision repair. *Genes Environ. Off. J. Jpn. Environ. Mutagen Soc.* **2019**, *41*, 2. [CrossRef] [PubMed]
24. Scrima, A.; Konickova, R.; Czyzewski, B.K.; Kawasaki, Y.; Jeffrey, P.D.; Groisman, R.; Nakatani, Y.; Iwai, S.; Pavletich, N.P.; Thoma, N.H. Structural basis of UV DNA-damage recognition by the DDB1-DDB2 complex. *Cell* **2008**, *135*, 1213–1223. [CrossRef] [PubMed]
25. Lans, H.; Hoeijmakers, J.H.J.; Vermeulen, W.; Marteijn, J.A. The DNA damage response to transcription stress. *Nat. Rev. Mol. Cell Biol.* **2019**, *20*, 766–784. [CrossRef] [PubMed]
26. Park, J.M.; Kang, T.H. Transcriptional and Posttranslational Regulation of Nucleotide Excision Repair: The Guardian of the Genome against Ultraviolet Radiation. *Int. J. Mol. Sci.* **2016**, *17*, 1840. [CrossRef]
27. Lehmann, A.R.; McGibbon, D.; Stefanini, M. Xeroderma pigmentosum. *Orphanet J. Dis.* **2011**, *6*, 70. [CrossRef]
28. Wang, Y.; Chakravarty, P.; Ranes, M.; Kelly, G.; Brooks, P.J.; Neilan, E.; Stewart, A.; Schiavo, G.; Svejstrup, J.Q. Dysregulation of gene expression as a cause of Cockayne syndrome neurological disease. *Proc. Natl. Acad. Sci. USA* **2014**, *111*, 14454–14459. [CrossRef]
29. Pascucci, B.; Lemma, T.; Iorio, E.; Giovannini, S.; Vaz, B.; Iavarone, I.; Calcagnile, A.; Narciso, L.; Degan, P.; Podo, F.; et al. An altered redox balance mediates the hypersensitivity of Cockayne syndrome primary fibroblasts to oxidative stress. *Aging Cell* **2012**, *11*, 520–529. [CrossRef]
30. Andrade, L.N.; Nathanson, J.L.; Yeo, G.W.; Menck, C.F.; Muotri, A.R. Evidence for premature aging due to oxidative stress in iPSCs from Cockayne syndrome. *Hum. Mol. Genet.* **2012**, *21*, 3825–3834. [CrossRef]
31. Cleaver, J.E.; Bezrookove, V.; Revet, I.; Huang, E.J. Conceptual developments in the causes of Cockayne syndrome. *Mech. Ageing Dev.* **2013**, *134*, 284–290. [CrossRef] [PubMed]
32. Stevnsner, T.; Muftuoglu, M.; Aamann, M.D.; Bohr, V.A. The role of Cockayne Syndrome group B (CSB) protein in base excision repair and aging. *Mech. Ageing Dev.* **2008**, *129*, 441–448. [CrossRef] [PubMed]
33. Troelstra, C.; van Gool, A.; de Wit, J.; Vermeulen, W.; Bootsma, D.; Hoeijmakers, J.H. ERCC6, a member of a subfamily of putative helicases, is involved in Cockayne's syndrome and preferential repair of active genes. *Cell* **1992**, *71*, 939–953. [CrossRef]
34. van Hoffen, A.; Kalle, W.H.; de Jong-Versteeg, A.; Lehmann, A.R.; van Zeeland, A.A.; Mullenders, L.H. Cells from XP-D and XP-D-CS patients exhibit equally inefficient repair of UV-induced damage in transcribed genes but different capacity to recover UV-inhibited transcription. *Nucleic Acids Res.* **1999**, *27*, 2898–2904. [CrossRef] [PubMed]
35. Soltys, D.T.; Rocha, C.R.; Lerner, L.K.; de Souza, T.A.; Munford, V.; Cabral, F.; Nardo, T.; Stefanini, M.; Sarasin, A.; Cabral-Neto, J.B.; et al. Novel XPG (ERCC5) mutations affect DNA repair and cell survival after ultraviolet but not oxidative stress. *Hum. Mutat.* **2013**, *34*, 481–489. [CrossRef]
36. Pascucci, B.; D'Errico, M.; Parlanti, E.; Giovannini, S.; Dogliotti, E. Role of nucleotide excision repair proteins in oxidative DNA damage repair: An updating. *Biochem. Biokhimiia* **2011**, *76*, 4–15. [CrossRef]
37. Menoni, H.; Hoeijmakers, J.H.; Vermeulen, W. Nucleotide excision repair-initiating proteins bind to oxidative DNA lesions in vivo. *J. Cell Biol.* **2012**, *199*, 1037–1046. [CrossRef]
38. Ng, J.M.; Vermeulen, W.; van der Horst, G.T.; Bergink, S.; Sugasawa, K.; Vrieling, H.; Hoeijmakers, J.H. A novel regulation mechanism of DNA repair by damage-induced and RAD23-dependent stabilization of xeroderma pigmentosum group C protein. *Genes Dev.* **2003**, *17*, 1630–1645. [CrossRef]
39. Araujo, S.J.; Tirode, F.; Coin, F.; Pospiech, H.; Syvaoja, J.E.; Stucki, M.; Hubscher, U.; Egly, J.M.; Wood, R.D. Nucleotide excision repair of DNA with recombinant human proteins: Definition of the minimal set of factors, active forms of TFIIH, and modulation by CAK. *Genes Dev.* **2000**, *14*, 349–359.

40. Nishi, R.; Okuda, Y.; Watanabe, E.; Mori, T.; Iwai, S.; Masutani, C.; Sugasawa, K.; Hanaoka, F. Centrin 2 stimulates nucleotide excision repair by interacting with xeroderma pigmentosum group C protein. *Mol. Cell. Biol.* **2005**, *25*, 5664–5674. [CrossRef]
41. D'Errico, M.; Parlanti, E.; Teson, M.; de Jesus, B.M.; Degan, P.; Calcagnile, A.; Jaruga, P.; Bjoras, M.; Crescenzi, M.; Pedrini, A.M.; et al. New functions of XPC in the protection of human skin cells from oxidative damage. *EMBO J.* **2006**, *25*, 4305–4315. [CrossRef] [PubMed]
42. Bernardes de Jesus, B.M.; Bjoras, M.; Coin, F.; Egly, J.M. Dissection of the molecular defects caused by pathogenic mutations in the DNA repair factor XPC. *Mol. Cell. Biol.* **2008**, *28*, 7225–7235. [CrossRef] [PubMed]
43. Miao, F.; Bouziane, M.; Dammann, R.; Masutani, C.; Hanaoka, F.; Pfeifer, G.; O'Connor, T.R. 3-Methyladenine-DNA glycosylase (MPG protein) interacts with human RAD23 proteins. *J. Biol. Chem.* **2000**, *275*, 28433–28438. [CrossRef] [PubMed]
44. Shimizu, Y.; Iwai, S.; Hanaoka, F.; Sugasawa, K. Xeroderma pigmentosum group C protein interacts physically and functionally with thymine DNA glycosylase. *EMBO J.* **2003**, *22*, 164–173. [CrossRef]
45. Shimizu, Y.; Uchimura, Y.; Dohmae, N.; Saitoh, H.; Hanaoka, F.; Sugasawa, K. Stimulation of DNA Glycosylase Activities by XPC Protein Complex: Roles of Protein-Protein Interactions. *J. Nucleic Acid* **2010**, *2010*. [CrossRef]
46. Odell, I.D.; Wallace, S.S.; Pederson, D.S. Rules of engagement for base excision repair in chromatin. *J. Cell. Physiol.* **2013**, *228*, 258–266. [CrossRef]
47. Sugasawa, K.; Okamoto, T.; Shimizu, Y.; Masutani, C.; Iwai, S.; Hanaoka, F. A multistep damage recognition mechanism for global genomic nucleotide excision repair. *Genes Dev.* **2001**, *15*, 507–521. [CrossRef]
48. Nishi, R.; Alekseev, S.; Dinant, C.; Hoogstraten, D.; Houtsmuller, A.B.; Hoeijmakers, J.H.; Vermeulen, W.; Hanaoka, F.; Sugasawa, K. UV-DDB-dependent regulation of nucleotide excision repair kinetics in living cells. *DNA Repair* **2009**, *8*, 767–776. [CrossRef]
49. Fischer, E.S.; Scrima, A.; Bohm, K.; Matsumoto, S.; Lingaraju, G.M.; Faty, M.; Yasuda, T.; Cavadini, S.; Wakasugi, M.; Hanaoka, F.; et al. The molecular basis of CRL4DDB2/CSA ubiquitin ligase architecture, targeting, and activation. *Cell* **2011**, *147*, 1024–1039. [CrossRef]
50. Esadze, A.; Rodriguez, G.; Cravens, S.L.; Stivers, J.T. AP-Endonuclease 1 Accelerates Turnover of Human 8-Oxoguanine DNA Glycosylase by Preventing Retrograde Binding to the Abasic-Site Product. *Biochemistry* **2017**, *56*, 1974–1986. [CrossRef]
51. Jang, S.; Kumar, N.; Beckwitt, E.C.; Kong, M.; Fouquerel, E.; Rapic-Otrin, V.; Prasad, R.; Watkins, S.C.; Khuu, C.; Majumdar, C.; et al. Damage sensor role of UV-DDB during base excision repair. *Nat. Struct. Mol. Biol.* **2019**, *26*, 695–703. [CrossRef] [PubMed]
52. Beerens, N.; Hoeijmakers, J.H.; Kanaar, R.; Vermeulen, W.; Wyman, C. The CSB protein actively wraps DNA. *J. Biol. Chem.* **2005**, *280*, 4722–4729. [CrossRef] [PubMed]
53. Menoni, H.; Wienholz, F.; Theil, A.F.; Janssens, R.C.; Lans, H.; Campalans, A.; Radicella, J.P.; Marteijn, J.A.; Vermeulen, W. The transcription-coupled DNA repair-initiating protein CSB promotes XRCC1 recruitment to oxidative DNA damage. *Nucleic Acids Res.* **2018**, *46*, 7747–7756. [CrossRef] [PubMed]
54. Khobta, A.; Epe, B. Repair of oxidatively generated DNA damage in Cockayne syndrome. *Mech. Ageing Dev.* **2013**, *134*, 253–260. [CrossRef] [PubMed]
55. Tuo, J.; Muftuoglu, M.; Chen, C.; Jaruga, P.; Selzer, R.R.; Brosh, R.M., Jr.; Rodriguez, H.; Dizdaroglu, M.; Bohr, V.A. The Cockayne Syndrome group B gene product is involved in general genome base excision repair of 8-hydroxyguanine in DNA. *J. Biol. Chem.* **2001**, *276*, 45772–45779. [CrossRef]
56. Dianov, G.; Bischoff, C.; Sunesen, M.; Bohr, V.A. Repair of 8-oxoguanine in DNA is deficient in Cockayne syndrome group B cells. *Nucleic Acids Res.* **1999**, *27*, 1365–1368. [CrossRef]
57. Tuo, J.; Jaruga, P.; Rodriguez, H.; Bohr, V.A.; Dizdaroglu, M. Primary fibroblasts of Cockayne syndrome patients are defective in cellular repair of 8-hydroxyguanine and 8-hydroxyadenine resulting from oxidative stress. *FASEB J. Off. Publ. Fed. Am. Soc. Exp. Biol.* **2003**, *17*, 668–674. [CrossRef]
58. Ranes, M.; Boeing, S.; Wang, Y.; Wienholz, F.; Menoni, H.; Walker, J.; Encheva, V.; Chakravarty, P.; Mari, P.O.; Stewart, A.; et al. A ubiquitylation site in Cockayne syndrome B required for repair of oxidative DNA damage, but not for transcription-coupled nucleotide excision repair. *Nucleic Acids Res.* **2016**, *44*, 5246–5255. [CrossRef]

59. Thorslund, T.; von Kobbe, C.; Harrigan, J.A.; Indig, F.E.; Christiansen, M.; Stevnsner, T.; Bohr, V.A. Cooperation of the Cockayne syndrome group B protein and poly(ADP-ribose) polymerase 1 in the response to oxidative stress. *Mol. Cell. Biol.* **2005**, *25*, 7625–7636. [CrossRef]
60. Scheibye-Knudsen, M.; Mitchell, S.J.; Fang, E.F.; Iyama, T.; Ward, T.; Wang, J.; Dunn, C.A.; Singh, N.; Veith, S.; Hasan-Olive, M.M.; et al. A high-fat diet and NAD (+) activate Sirt1 to rescue premature aging in cockayne syndrome. *Cell Metab.* **2014**, *20*, 840–855. [CrossRef]
61. Boetefuer, E.L.; Lake, R.J.; Dreval, K.; Fan, H.Y. Poly(ADP-ribose) polymerase 1 (PARP1) promotes oxidative stress-induced association of Cockayne syndrome group B protein with chromatin. *J. Biol. Chem.* **2018**, *293*, 17863–17874. [CrossRef] [PubMed]
62. Wang, Y. Bulky DNA lesions induced by reactive oxygen species. *Chem. Res. Toxicol.* **2008**, *21*, 276–281. [CrossRef] [PubMed]
63. Cadet, J.; Ravanat, J.L.; TavernaPorro, M.; Menoni, H.; Angelov, D. Oxidatively generated complex DNA damage: Tandem and clustered lesions. *Cancer Lett.* **2012**, *327*, 5–15. [CrossRef] [PubMed]
64. Melis, J.P.; van Steeg, H.; Luijten, M. Oxidative DNA damage and nucleotide excision repair. *Antioxid. Redox Signal.* **2013**, *18*, 2409–2419. [CrossRef]
65. Gu, C.; Zhang, Q.; Yang, Z.; Wang, Y.; Zou, Y.; Wang, Y. Recognition and incision of oxidative intrastrand cross-link lesions by UvrABC nuclease. *Biochemistry* **2006**, *45*, 10739–10746. [CrossRef]
66. Wang, J.; Cao, H.; You, C.; Yuan, B.; Bahde, R.; Gupta, S.; Nishigori, C.; Niedernhofer, L.J.; Brooks, P.J.; Wang, Y. Endogenous formation and repair of oxidatively induced G[8–5 m]T intrastrand cross-link lesion. *Nucleic Acids Res.* **2012**, *40*, 7368–7374. [CrossRef]
67. Krokidis, M.G.; Terzidis, M.A.; Efthimiadou, E.; Zervou, S.K.; Kordas, G.; Papadopoulos, K.; Hiskia, A.; Kletsas, D.; Chatgilialoglu, C. Purine 5′,8-cyclo-2′-deoxynucleoside lesions: Formation by radical stress and repair in human breast epithelial cancer cells. *Free Radic. Res.* **2017**, *51*, 470–482. [CrossRef]
68. Kropachev, K.; Ding, S.; Terzidis, M.A.; Masi, A.; Liu, Z.; Cai, Y.; Kolbanovskiy, M.; Chatgilialoglu, C.; Broyde, S.; Geacintov, N.E.; et al. Structural basis for the recognition of diastereomeric 5′,8-cyclo-2′-deoxypurine lesions by the human nucleotide excision repair system. *Nucleic Acids Res.* **2014**, *42*, 5020–5032. [CrossRef]
69. Stingele, J.; Bellelli, R.; Boulton, S.J. Mechanisms of DNA-protein crosslink repair. *Nat. Rev. Mol. Cell Biol.* **2017**, *18*, 563–573. [CrossRef]
70. DeMott, M.S.; Beyret, E.; Wong, D.; Bales, B.C.; Hwang, J.T.; Greenberg, M.M.; Demple, B. Covalent trapping of human DNA polymerase beta by the oxidative DNA lesion 2-deoxyribonolactone. *J. Biol. Chem.* **2002**, *277*, 7637–7640. [CrossRef]
71. Reardon, J.T.; Cheng, Y.; Sancar, A. Repair of DNA-protein cross-links in mammalian cells. *Cell Cycle* **2006**, *5*, 1366–1370. [CrossRef] [PubMed]
72. Catalano, M.J.; Liu, S.; Andersen, N.; Yang, Z.; Johnson, K.M.; Price, N.E.; Wang, Y.; Gates, K.S. Chemical structure and properties of interstrand cross-links formed by reaction of guanine residues with abasic sites in duplex DNA. *J. Am. Chem. Soc.* **2015**, *137*, 3933–3945. [CrossRef] [PubMed]
73. Price, N.E.; Johnson, K.M.; Wang, J.; Fekry, M.I.; Wang, Y.; Gates, K.S. Interstrand DNA-DNA cross-link formation between adenine residues and abasic sites in duplex DNA. *J. Am. Chem. Soc.* **2014**, *136*, 3483–3490. [CrossRef] [PubMed]
74. Deans, A.J.; West, S.C. DNA interstrand crosslink repair and cancer. *Nat. Rev. Cancer* **2011**, *11*, 467–480. [CrossRef] [PubMed]
75. Luo, W.; Muller, J.G.; Rachlin, E.M.; Burrows, C.J. Characterization of spiroiminodihydantoin as a product of one-electron oxidation of 8-Oxo-7,8-dihydroguanosine. *Org. Lett.* **2000**, *2*, 613–616. [CrossRef] [PubMed]
76. Niles, J.C.; Wishnok, J.S.; Tannenbaum, S.R. Spiroiminodihydantoin is the major product of the 8-oxo-7,8-dihydroguanosine reaction with peroxynitrite in the presence of thiols and guanosine photooxidation by methylene blue. *Org. Lett.* **2001**, *3*, 963–966. [CrossRef]
77. Shafirovich, V.; Kropachev, K.; Anderson, T.; Liu, Z.; Kolbanovskiy, M.; Martin, B.D.; Sugden, K.; Shim, Y.; Chen, X.; Min, J.H.; et al. Base and Nucleotide Excision Repair of Oxidatively Generated Guanine Lesions in DNA. *J. Biol. Chem.* **2016**, *291*, 5309–5319. [CrossRef]
78. Shafirovich, V.; Kropachev, K.; Kolbanovskiy, M.; Geacintov, N.E. Excision of Oxidatively Generated Guanine Lesions by Competing Base and Nucleotide Excision Repair Mechanisms in Human Cells. *Chem. Res. Toxicol.* **2019**, *32*, 753–761. [CrossRef]

79. Reardon, J.T.; Bessho, T.; Kung, H.C.; Bolton, P.H.; Sancar, A. In vitro repair of oxidative DNA damage by human nucleotide excision repair system: Possible explanation for neurodegeneration in xeroderma pigmentosum patients. *Proc. Natl. Acad. Sci. USA* **1997**, *94*, 9463–9468. [CrossRef]
80. Rosa, S.; Fortini, P.; Karran, P.; Bignami, M.; Dogliotti, E. Processing in vitro of an abasic site reacted with methoxyamine: A new assay for the detection of abasic sites formed in vivo. *Nucleic Acids Res.* **1991**, *19*, 5569–5574. [CrossRef]
81. Kitsera, N.; Rodriguez-Alvarez, M.; Emmert, S.; Carell, T.; Khobta, A. Nucleotide excision repair of abasic DNA lesions. *Nucleic Acids Res.* **2019**, *47*, 8537–8547. [CrossRef] [PubMed]
82. Pilie, P.G.; Tang, C.; Mills, G.B.; Yap, T.A. State-of-the-art strategies for targeting the DNA damage response in cancer. *Nat. Rev. Clin. Oncol.* **2019**, *16*, 81–104. [CrossRef] [PubMed]
83. Gavande, N.S.; VanderVere-Carozza, P.S.; Hinshaw, H.D.; Jalal, S.I.; Sears, C.R.; Pawelczak, K.S.; Turchi, J.J. DNA repair targeted therapy: The past or future of cancer treatment? *Pharmacol. Ther.* **2016**, *160*, 65–83. [CrossRef] [PubMed]

MDPI
St. Alban-Anlage 66
4052 Basel
Switzerland
Tel. +41 61 683 77 34
Fax +41 61 302 89 18
www.mdpi.com

International Journal of Molecular Sciences Editorial Office
E-mail: ijms@mdpi.com
www.mdpi.com/journal/ijms

www.ingramcontent.com/pod-product-compliance
Lightning Source LLC
LaVergne TN
LVHW070923170726
843515LV00005B/849